Forestry in the Midst of Global Changes

Forestry in the Midst of Global Changes

Edited by
Christine Farcy
Inazio Martinez de Arano
Eduardo Rojas-Briales

CRC Press
Taylor & Francis Group
Boca Raton London New York

CRC Press is an imprint of the
Taylor & Francis Group, an **informa** business

CRC Press
Taylor & Francis Group
6000 Broken Sound Parkway NW, Suite 300
Boca Raton, FL 33487-2742

First issued in paperback 2020

ISBN 13: 978-0-367-57073-6 (pbk)
ISBN 13: 978-1-138-19708-4 (hbk)

Library of Congress Cataloging-in-Publication Data

Names: Farcy, Christine, editor. | Martinez de Arano, Inazio, editor. | Rojas-Briales, Eduardo, editor.
Title: Forestry in the midst of global changes / [edited by] Christine Farcy, Inazio Martinez de Arano & Eduardo Rojas-Briales.
Description: Boca Raton : Taylor & Francis, 2018. | Includes bibliographical references and index.
Identifiers: LCCN 2018020245 | ISBN 9781138197084 (hardback : alk. paper)
Subjects: LCSH: Forests and forestry--Social aspects. | Forests and forestry--Political aspects. | Forest management--Social aspects.
Classification: LCC SD387.S55 F674 2018 | DDC 634.9--dc23
LC record available at https://lccn.loc.gov/2018020245

**Visit the Taylor & Francis Web site at
http://www.taylorandfrancis.com**

**and the CRC Press Web site at
http://www.crcpress.com**

We dedicate this book to our colleague and friend Franz Schmithüsen, who tirelessly explored new frontiers and inspired useful transgressions.

Contents

SECTION I   Setting the Scene

SECTION II   Urbanization of the Society

SECTION III   Tertiarization of the Economy

SECTION IV   Globalization

SECTION V Lessons Learned

Preface

It is hard for many to remember a slower time. For those raised in the age of techno-logical innovation, this is especially true. A combination of today's overwhelming global challenges, vast amounts of information, and a rapidly changing pace creates an effect that can almost be dizzying.

Under this pressure, many of us seek an escape: an escape from the noise of urban life or new demands, a refuge that is often found in nature. Perhaps, it is the stillness and peace found in a forest that creates the illusion of isolation. An assumption that this entity is free from external influences, but as we gear up to tackle global chal-lenges, it is an assumption that must be let go.

Forests are intertwined with all changes across time. Forests carry legacy—the history of previous choices and management decisions which find expression in structure and ecosystem functionality. Forests carry the present—a reflection of humanity's present-day valuation and a showing of our chosen priorities. Forests, however, do not carry our future.

In today's complex world, it is our responsibility to be attuned to the unique mani-festations of global changes in the forestry context. It is our responsibility to ensure that our approach to today's interventions and tomorrow's new era is informed and inclusive. It is our responsibility to enable young people to contribute to and inform the dialogue that defines the actions of our future forests and their inherited world.

It is why *Forestry in the Midst of Global Changes* comes at a critical time. By identifying and analyzing three key drivers of change challenging forestry—the urbanization of society, tertiarization of the economy, and globalization—it provides needed insight. This not only informs and instructs but also exemplifies the spirit of inclusivity needed to cocreate our future. It is for this reason that I am proud to share that the introductory chapters of each key driver are framed and written by a young voice.

Young people are not just future leaders but also today's—researching, manag-ing, policy making—ever present in all spectrums of our field. It is vital that our forward-facing discussions include these voices and solutions. I state with great joy that the youth today are rising up to grasp momentum and drive meaningful change. Despite current pressures and a seemingly dwindling hourglass, I hope that this book inspires all to be courageous in applying new insights and an intergenerational effort to shaping our future forests.

Salina Abraham
International Forestry Students' Association President

Acknowledgment

This book was made possible thanks to the support of the Minister of Forest of the Walloon Region (Belgium).

Editors

Christine Farcy is a holder of an MSc in agricultural engineering (forest and water) at University of Louvain (UCL) (1994) and a PhD in agricultural sciences and biological engineering at UCL (2005). She is currently a senior researcher, invited lecturer, and international forest policy advisor. Her expertise are in forestry, national and international forest policy, forest planning, remote sensing, information systems, bridge between social and natural sciences, and science and policy cross-fertilization. She is currently, among others, the chair of the European Forestry Commission of the Food and Agriculture Organization of the United Nations (FAO) and the vice chair of the board of the European Forest Institute (EFI), a chair or member of international evaluation panels, and a reviewer of scientific disciplinary and interdisciplinary journals and projects.

Inazio Martinez de Arano is the head of the Office at European Forest Institute Mediterranean Facility since June 2013. He holds a degree in ecology (Basque Country University), a postdegree in landscape management (Polytechnic University of Valencia), and an MSc in forestry (Universidad Austral de Chile). He has worked in forest research (sustainable forest operations) and has served as president of the union of forest owners of southern Europe, following policy and market developments in Europe, and is active in multilateral policy forums (Forest Europe, United Nations Economic Commission for Europe Committee on Forests and the Forest Industry, Silva Mediterranea, the European Union's civil dialogue). His current focus is on understanding the challenges and opportunities for innovating in the Mediterranean bioeconomy.

Eduardo Rojas-Briales is a holder of an MSc in forestry at the University of Freiburg (1985) and a PhD in forestry at the Polytechnical University of Valencia (UPV). He was involved in forest inventory and management at Deutsche Forstservice GmbH (1988–1990). He was the director of the Catalan Forest Owners' Association and the Spanish representative at the Confederation of European Forest Owners (CEPF) (1992–1998), an associate professor at the University of Lleida (1994–2000), a leader of the forest policy group at CTFC (1996–1999), a member of the Scientific Advisory Board of EFI (1998–2002), a forest consultant of Silvamed SL (1999–2003), a professor at UPV (2003–2010, 2016–), the vice dean of the Life Sciences Faculty (UPV) (2004–2010), the assistant director-general and head of the Forestry Department at FAO (2010–2015), the chair of the Collaborative Partnership on Forests (2010–2015), the cochair of the UN Programme on Reducing Emissions from Deforestation and Forest Degradation (2014), the UN commissioner general for Expo Milano 2015 (2013–2015), the chair of the Spanish MSc Forestry Board (2016–), and a member of the Programme for the Endorsement of Forest Certification (PEFC) International Board (2016–).

Contributors

Louise Adam
Earth and Life Institute, Environmental
  Sciences (ELI-e)
University of Louvain
Louvain-la-Neuve, Belgium

Yemi Adeyeye
Faculty of Forestry and Liu Institute
  for Global Issues
University of British Columbia
Vancouver, Canada

Paul Arnould
University of Lyron
École normale supérieure de Lyon
Laboratoire Environnement Ville
  Société
Lyon, France

Adrina C. Bardekjian
Department of Forest Resources
  Management
University of British Columbia
Vancouver, Canada

Olivier Baudry
Earth and Life Institute, Environmental
  Sciences (ELI-e)
University of Louvain
Louvain-la-Neuve, Belgium

Martí Boada
Institute of Environmental Science
  and Technology
Autonomous University of Barcelona
Cerdanyola del Vallès, Spain

Miguel Cabrera-Bonet
Higher Technical School of Agronomic
  Engineering and the Natural
  Environment
Polytechnic University of Valencia
Valencia, Spain

Jennifer DeBoer
Faculty of Forestry
University of British Columbia
Vancouver, Canada

Rafael Delgado-Artés
Higher Polytechnic School of Gandia
Polytechnic University of Valencia
Grau de Gandia, Spain

Tahia Devisscher
Department of Forest Resources
  Management
University of British Columbia
Vancouver, Canada

Jennie Dey de Pryck
Independent
Rome, Italy

Marlène Elias
Bioversity International
Rome, Italy

Christine Farcy
Earth and Life Institute, Environmental
  Sciences (ELI-e)
University of Louvain
Louvain-la-Neuve, Belgium

Pierre Fastrez
Institute Language and Communication
University of Louvain
Louvain-la-Neuve, Belgium

Anne-Marie Granet
Department of Sustainable Forest
    Management
Office National des Forêts
Paris, France

Patrice Harou
Pinchot Institute
Washington, DC

and

Université de Lorraine à Strasbourg
AgroParisTech/CNRS/INRA/BETA
Nancy, France

Naomie Herpin-Saunier
Department of Forest and Conservation
    Sciences
University of British Columbia
Vancouver, Canada

Lauri Hetemäki
European Forest Institute
Joensuu, Finland

Nicole Huybens
Fundamental Sciences Department
University of Quebec in Chicoutimi
Chicoutimi, Canada

Ingrid Jarvis
Department of Forest and Conservation
    Sciences
University of British Columbia
Vancouver, Canada

Cecil Konijnendijk van den Bosch
Department of Forest Resources
    Management
University of British Columbia
Vancouver, Canada

Maribel Lozano
Institute of Environmental Science
    and Technology
Autonomous University of Barcelona
Cerdanyola del Vallès, Spain

Roser Maneja
Institute of Environmental Science
    and Technology
Autonomous University of Barcelona
Cerdanyola del Vallès, Spain

Inazio Martinez de Arano
Mediterranean Facility
European Forest Institute
Barcelona, Spain

Mauro Masiero
Department of Land, Environment,
    Agriculture and Forestry
University of Padova
Legnaro, Italy

Julie Matagne
Institute Language and Communication
University of Louvain
Louvain-la-Neuve, Belgium

Patrick Meyfroidt
Fonds de la Recherche Scientifique–
Fonds National de la Recherche
    Scientifique
Brussels, Belgium

and

Earth and Life Institute
Earth and Climate Research Centre
Université catholique de Louvain
Louvain-la-Neuve, Belgium

Maria J. Murcia
Faculty of Forestry
University of British Columbia
Vancouver, Canada

and

IAE Business School
Universidad Austral
Pilar, Argentina

Sylvie Nail
Faculty of Foreign Languages
 and Cultures
University of Nantes
Nantes, France

Annukka Näyhä
Jyväskylä University School of Business
 and Economics
Jyväskylä, Finland

Lorien Nesbitt
Department of Forest Resources
 Management
University of British Columbia
Vancouver, Canada

Päivi Pelli
School of Forest Sciences
University of Eastern Finland
Joensuu, Finland

Davide Pettenella
Department of Land, Environment,
 Agriculture and Forestry
University of Padova
Legnaro, Italy

Pauline Pirlot
Institute of Political Sciences
 Louvain-Europe
University of Louvain
Louvain-la-Neuve, Belgium

Elena Pisani
Department of Land, Environment,
 Agriculture and Forestry
University of Padova
Legnaro, Italy

Philippe Polomé
University of Lyon, Centre National
 de la Recherche Scientifique,
 Groupe d'Analyse et de Théorie
 Economique–Lyon St-Etienne
Lyon, France

Ewald Rametsteiner
Food and Agriculture Organization
 of the United Nations
Rome, Italy

Eduardo Rojas-Briales
Higher Technical School of Agronomic
 Engineering and the Natural
 Environment
Polytechnic University of Valencia
Valencia, Spain

Laura Secco
Department of Land, Environment,
 Agriculture and Forestry
University of Padova
Legnaro, Italy

Meike Siegner
Faculty of Forestry
University of British Columbia
Vancouver, Canada

Bimbika Sijapati Basnett
Center for International Forestry
 Research
Bogor, Indonesia

Sarah Sra
Faculty of Forestry
University of British Columbia
Vancouver, Canada

Matilda van den Bosch
Department of Forest and Conservation
   Sciences
University of British Columbia
Vancouver, Canada

Jiadong Ye
Faculty of Forestry
University of British Columbia
Vancouver, Canada

Section I

Setting the Scene

1 Introduction

Christine Farcy, Inazio Martinez de Arano,
and Eduardo Rojas-Briales

This book is born from the shared will of further developing some very preliminary ideas drafted for a particular editorial project (Farcy et al. 2016) while remaining in a format willing to inspire further thinking, dialogues, and initiatives rather than to share or state "definitive" considerations.

Like other sectors traditionally relying on material goods, forestry is indeed facing significant societal changes, which are inducing such deep challenges that they need to be firstly addressed taking a distance, in an open and inclusive way, off the beaten track, even inciting the emergence of contradictions. This should progressively allow other steps of later deepening and further conceptual or methodological innovations. This book constitutes an attempt in such a perspective.

Forestry is understood in this book as generically embracing both theory and practice of all that constitutes the management and use of forests, including their preservation and restoration. Made by sets of rules, implicit or explicit knowledge, and skills, forestry constitutes a broad toolbox dedicated to orienting and supporting human intentional interventions involving forests. It is often associated with a specific economic sector dedicated to valorizing forest products and to framing their exchanges. We will refer to this sector as *forestry sector* in order to also take on board immaterial goods, which are often not explicitly included in the more familiar terms *forest sector, forest-based sector,* or, more recently, *forest bioeconomy sector.*

Forests, in their polysemic meaning, have contributed to the development of human societies and civilizations throughout history (i.e., Fernow 1911; Holmes 1975; Meiggs 1982; Westoby 1989; List 2000; Ramakrishnan 2007; Chalvet 2011; Grebner et al. 2013; Schmithüsen 2013; Roberts et al. 2017). So obviously, forestry was never monolithic and was always, in a more or less explicit way, under continuous processes of change and adaptation.

In today's world, besides environmental concerns that are often under the spotlight of scientific, political, and news media, forestry is challenged by some major global societal drivers of change, less studied and addressed, while having undoubtedly deep and long-lasting impacts on the relationship not only between humans and forests but also between humans themselves when being related to forests. We identify in particular three major interconnected global societal changes, which are the pillars structuring the book.

The first one is the urbanization of the society, a process that has been developing since the first step of industrialization and which is particularly significant today when the world's urban population is exceeding 50%. The ongoing urbanization is changing lifestyles and practices with a tendency to dematerialization and is not only offering new opportunities and perspectives on and for forests and trees but also

3

inducing or revealing deep changes in individual perceptions and social representations of forests. Urbanity is also the mirror of rurality where forestry took place and developed for centuries and which is somehow orphan and worried today.

The second one is the tertiarization of the economy. Today, the sector of services largely dominates the economy and involves a major part of the active population in the world. This ongoing process modifies professional modalities and ways of life and not only opens new doors to forests and trees through the immaterial goods they provide but is also deeply changing the framework, the rules, the processes, and the modalities of production and exchanges between economic actors, and the ways of conceiving innovation.

The third one is undoubtedly globalization in its economic, social, and political components. By shortening distances, overcoming borders, accelerating exchanges, standardizing practices, flattening hierarchical structures, pushing for interdependence, globalization impacts everyone everywhere in multiple ways. Forestry does not escape this multifaceted bottom blade.

Thus, the book focuses on global drivers of changes from the perspective of their relationships with the functioning of the society. Environmental drivers are not addressed as such but are implicitly or indirectly considered, thanks to the systemic approach framing all the reasoning.

The purpose of the book is, by digging deeper into those changes through multidisciplinary, interdisciplinary, or even transdisciplinary and crosscutting contributions by authors from diverse backgrounds and horizons, to collegially contribute to the design of the forestry of tomorrow. The intention is to show how usual or familiar borders and traditional comfort zones are porous, shaky, and moving, obliging to review or update practices, skills, knowledge, and even principles and to open eyes and arms. Each contribution has its own coherence and autonomy and relies on its own disciplinary pillars, while being interconnected with the others through a shared hill structuring the overall questioning. Contradictions between schools of thought could thus appear.

In Section I, after very briefly and succinctly recalling the hidden complexity of forests in Chapter 2, the main successive milestones that forged forestry are presented in Chapter 3, serving as the starting reference. This section closes with Chapter 4 with an original and innovative contribution expounding the different ethical models coexisting today on the forest scene. While this chapter could have been part of the closing section of the book, placing it in the introductory chapter allows readers to have these models firmly in mind when progressing through the various contributions prepared by colleagues, each of them inspired by his or her respective ethics. Indeed, no ethical model dominates the others in this editorial project.

Then, the following three sections, Sections II through IV, are each dedicated to one of the drivers of societal change that are at the heart of the book. Links between these sections will, however, often appear when reading, given the interdependence between the three processes. Each section begins with a chapter presenting the process and some related findings and trends in a very general way. These chapters were prepared by three teams of young forestry colleagues who have agreed to share a fresh look at the domain they have just started to navigate, and we thank them for that.

Section II focuses on the process of the urbanization of the society as understood not only literally but also figuratively (Chapter 5). The first focus is on forests and the urban lifestyle (Chapter 6). Chapter 7 offers an immersion in the shared, while paradoxical, vision of forest mainly owned by today's urban society. The reverse of urbanization is rural desertification. Through the Spanish case, the vast and complex consequences of this process are analyzed in Chapter 8. Afterward, an original contribution, Chapter 9, argues in favor of recreating or renewing the role of culture (dendroculture) and art in today's society. Finally, an innovative understanding of forest communication is presented and detailed in Chapter 10.

Section III develops the issue of the tertiarization of the economy in Chapter 11 and continues in Chapter 12 with a useful analysis of and debate on the increasing role of services in forestry, both upstream and downstream of forest management. Three chapters then focus on services directly related to forests. Exemplifying the issue, a very extensive contribution, Chapter 13, focuses on the case of the tertiary role of forests in human health. Then, the section ends with two chapters on economics: the first, Chapter 14, gives the floor to forest economists, allowing them to explain how they conceive the framework of investments in forest services. In a mirror game, the second, Chapter 15, relies on behavioral economics and offers a nuanced debate on various levers of motivation of forest service producers.

Section IV focuses on the globalization process. As this section deals with a broad and complex process, in Chapter 16, priority is given to illustrating some of its various components. Chapter 17 focuses on the recent and controversial issue of financialization of the forestry sector. Then, in Chapter 18, technological versus social innovations are debated on. Afterward, a very extensive chapter (Chapter 19) analyzes the impact of globalization on gender employment, in particular in agroforestry systems in the global south. Finally, a very innovative and challenging vision of the fragmented international forest policy landscape is developed in Chapter 20.

Section V brings together elements of discussion and concludes with a view to open further dialogues and debates.

REFERENCES

Chalvet, M. 2011. *Une Histoire de la Forêt*. Paris: Editions du Seuil.

Farcy, C., de Camino, R., Martinez de Arano, I., and E. Rojas-Briales. 2016. External drivers of changes challenging forestry: Political and social issues at stake. In *Ecological Forest Management Handbook*, ed. G. Larocque, 87–105. Boca Raton, FL: Taylor & Francis Group/CRC Press.

Fernow, B. E. 1911. *A Brief History of Forestry: Europe, the United States and Other Countries*. Toronto: University Press.

Grebner, D. L., Bettinger, P. and J. C. Siry. 2013. *Introduction to Forestry and Natural Resources*. Cambridge, MA: Elsevier Academic Press.

Holmes, G. D. 1975. History of forestry and forest management. *Philosophical Transactions of the Royal Society of London, Series B, Biological Sciences* 271(911):69–80.

List, P. C. 2000. *Environmental Ethics and Forestry*. Philadelphia, PA: Temple University Press.

Meiggs, R. 1982. *Trees and Timber in the Ancient Mediterranean World*. Oxford: Oxford University Press.

Ramakrishnan, P. S. 2007. Traditional forest knowledge and sustainable forestry: A northeast India perspective. *Forest Ecology and Management* 249:91–99.

Roberts, P., Hunt, C., Arroyo-Kalin, M., Evans, D. and N. Boivin. 2017. The deep human prehistory of global tropical forests and its relevance for modern conservation. *Nature Plants* 3:17093.

Schmithüsen, F. 2013. Three hundred years of applied sustainability in forestry. *Unasylva* 240(64):3–11.

Westoby, J. 1989. *Introduction to World Forestry*. Oxford, UK: Basil Blackwell.

2 What Is a Forest?

Paul Arnould

CONTENTS

2.1 INTRODUCTION

A forest is for many people something obvious: the domain of trees, as opposed to grass landscapes such as lawns, meadows, savannahs, and steppes, identified on every continent (Figure 2.1).

In France, for example, not only nonspecialist dictionaries such as Larousse or Littré but also specialized ones on geography, ecology, or agronomy refer to a forest as "a wide area covered by trees." Starting from this apparent simplicity will help us recall the polysemic nature of forests (Lund 2002, 2012), in this short introductory chapter whose extension scope has no ambition to exhaustivity.

2.2 WHAT IS A TREE?

A tree is undoubtedly the backbone of a forest. Since the time of the Greek Theophrastus (Amigues 1988) nearly 2500 years ago, it has been known that a tree differs from a bush and a shrub by having a well-individualized trunk and by reaching at least 10 m in height under normal conditions of development. In contrast to grass growth, tree growth is cumulative (Le Play 1996). A tree makes wood, which allows growth in height, diameter, and volume, like a gigantic nesting doll with successive annual envelopes from the heart to the bark, and the outer shell is usually smooth in the young age and then more and more cracked.

The notion of trees is problematic as to size, age, shape, and biology. Hallé (2005) clearly posed the tricky problems of characterizing a tree. Bamboos and palms by not being woody are not considered trees in the strict sense. They give, however, landscapes similar to those of forests (Figure 2.2). Moreover, some trees, such as redwoods (*Sequoia gigantea* and *Sequoia sempervirens*) and eucalyptus (*Eucalyptus* sp.),

FIGURE 2.1   Polish landscape of the tree (the forest), the grass (the meadow), and the built (road and farm). (Courtesy of Creative Commons Zero CCO.)

FIGURE 2.2   Mountain "forest" of tree fern, Reunion Island. (Photo by Paul Arnould.)

are unquestionably some of the highest (over 100 m), the biggest (over 1500 t), and the oldest (5000 years) living beings in the current world.

In contrast, the heights of natural *bonsais*, with their tormented shapes, being exposed to the wind, cold, and drought, do not exceed 2 or 3 m. This is the case of the Canadian spruce (*Picea glauca*), which grows on the edge of the polar tundra, of beeches (*Fagus sylvestris*) and cembro pines (*Pinus cembra*), which are scattered on lawns of high-altitude mountains, such as the Alps, near the limit of the persistent snows, the limit called "timberline" by Anglo-Saxon authors, or of some acacias growing in arid zones.

2.3   HOW MANY TREE SPECIES ARE THERE ON THE GLOBE?

The number of plant species present on the surface of the planet is still questionable and a fortiori that of trees. "Major challenges to filling the gap in knowledge on plant species include frequent synonymy, the difficulty of discriminating certain species by morphology alone, and the fact that many undiscovered species are small in size, difficult to find, or have a small geographic range (Scheffers et al. 2012)" cited by the Food and Agriculture Organization of the United Nations (FAO) (2014a).

Examples from Canada illustrate the issue of synonymies in tree species. On the eastern side, spruce is also called fir. And white spruce has at least two Latin names, namely, *Picea glauca* and *Picea alba* or *Picea canadensis*, while black spruce is called *Picea mariana* or *Picea nigra*. Another well-known example on the western side of North America is the Douglas fir, first called *Pseudotsuga douglassi* before being requalified as *Pseudotsuga menziessi* from the name of the botanist accompanying the Vancouver expedition during which the tree was discovered (Claire 2010).

> The answer to the question [. . .] remains very rough, varying from 50,000 (National Academy of Sciences 1991) to between 80,000 and 100,000 species (Oldfield, Lusty and MacKinven, 1998; Turok and Geburek, 2000). These estimates are even more confusing in light of the different definitions of a tree (FAO 2014a).

For example, Beech et al. (2017), using International Union for Conservation of Nature (IUCN) definition,[1] referred in 2017 to the number of 65,065 species, representing 20% of all plant species. Following this study, Brazil is, with 8,715 species, the country with the largest diversity of trees followed by Colombia (5,776) and Indonesia (5,142).

FAO's *State of the World's Forest Genetic Resources 2013* refers to a range of 80,000–100,000 as the most widely used estimate for the number of tree species and indicates "the need for further efforts in botanic assessment to obtain more accurate figure" (FAO 2014a).

2.4   HOW MANY TREES ARE THERE ON THE PLANET?

The total number of trees on the planet is another subject of assessments resulting in very different but impressive numbers. Numbering trees is, however, misleading because it depends on the parameters and categories of size, height, or diameter used.

The latest study (Crowther et al. 2015) refers to an assessment based on images from the American *Modis* satellite and more than 400,000 field surveys and suggests the number of more than 3,041 billion trees, which means 422 trees per inhabitant of the planet. Previous study solely based on the exploitation of satellite images, proposed the number of 400 billion trees, representing 61 trees per inhabitant. The still questionable "prize list" of this type of study places Russia at the top with 641.6 billion trees, ahead of Canada (318.2) and Brazil (301.8).

2.5   WHERE ARE THE FORESTS?

There are forests all around the globe, from the northern polar circle to the equator (Figure 2.3). Oversimplifying, three large zonal forest complexes can be distinguished

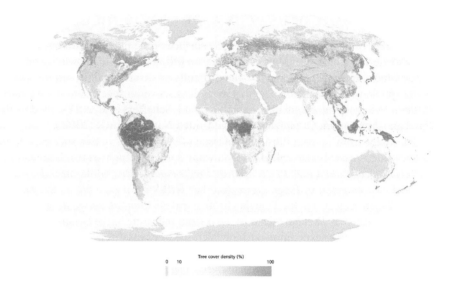

FIGURE 2.3   World forest map 2010. (From FAO, *Forest Resource Assessment*, http://www
.fao.org/forestry/fra/80298/en/, 2010.)

(Boulier and Simon 2009). A boreal forest, called taiga in Russia, where it covers
hundreds of millions of hectares, is also mainly present in Canada and in the Nordic
countries. Four temperate forests extend on the eastern and western sides of the
North American and Eurasian continents. Three dense tropical forests are respectively in the Amazon, the Congo basin, and South East Asia (Indonesia, Malaysia,
Papua-New Guinea), each of these sets covering about 1 billion ha.

There are also other less extensive and more dispersed categories. First of all,
the five Mediterranean forests, the eponymous forests on the shores of the sea of the
same name; the Californian beside Los Angeles and San Francisco; the Chilean in
the hinterland of Valparaiso and Santiago; the South African in the Cape region; and
the Australian in the vicinity of Perth. It is useful to note that in Spain and Middle
East and North Africa regions, forests include extensive grasslands, as expressed
through the Spanish word *monte*, synonymous to forest.

Next are the many mountain forests in the Alps, the Andes, the Rocky Mountains,
Himalaya, Urals, the Appalachians, the Pyrenees, and the Balkans. Coastal forests,
of which mangroves are a tropical archetype of strong originality, also largely escape
the zonal logic of the previous large sets.

Various forest types develop in the subtropical crown, and finally, forests are also
diffusely present in drylands. Despite their importance to climate and social issues,
they have only recently been recognized as specific entities. In the framework of a
global collaborative analysis using remote sensing images of very high spatial and
temporal resolution, Bastin et al. reported in FAO (2016a) and further published in
the journal *Science* (Bastin et al. 2017) an estimate of the global forest extent in
dryland biome, which 40–47% higher than the existing one, with 467 million ha of
forest never previously documented.

2.6   WHICH FOREST COVER?

Between 1960 and 1980, Allen and Barnes (1985) identified more than 20 publications offering assessments of global forest areas. They ranged from 2.393 billion ha for the lowest estimate to 6.050 billion ha for the highest. It is obvious that researchers did not measure the same reality, using different definitions and tools and implementing other methodologies.

Public policies nevertheless require reference to a shared and agreed definition, and the FAO definition is the reference in this field: "land spanning more than 0.5 hectares with trees higher than 5 meters and a canopy cover of more than 10%, or trees able to reach these thresholds *in situ*. It does not include land that is predominantly under agricultural or urban land use" (FAO 2015).[2] Based on the FAO assessment, forest plant formations constitute a major spatial system and occupy nearly 4 billion ha on the surface of the continents, representing 30.6% of global land area (FAO 2015). The world's largest forest country remains to be Russia with more than 815 million ha (20% of global forest area), followed by Brazil with 493.5 million ha and Canada with 347 million ha.

2.7   FORESTS AT THE CROSSROAD OF SYSTEMS

Far away from the initial definition, a forest shows its complexity, even more when opening the scope to human and social levels and referring to various interrelated forest systems (Arnould 2004; Farcy and Devillez 2005; Mazoyer and Roudart 1997).

A forest is indeed an ecosystem made not only of trees but also of undergrowth plants and of animals, a biocenosis, installed on soils more or less deeply exploited by the root systems of the largest of its constituents, forming a biotope (Figure 2.4).

FIGURE 2.4   Old forest of Wiener Wald (Austria). (Photo by Paul Arnould.)

Forests are biodiversity hotspots worldwide with 70% of terrestrial biodiversity being included in forested landscapes.

In reaction against a vision of the ecosystem without place and history, geographers and historians defined the forest as a geo-system or better as a silvo-system, which means a coproduction by nature and society, a state of the landscape and the environment at a given moment (Houzard 1985; Amat et al. 2008). The state forest of Verdun, which developed where the battle of the same name took place in 1916, is a particularly telling illustration.[3]

A forest constitutes a land system with diverse property regimes. Following FAO (2015), 76% of global forest area is publicly owned, 20% is private, and 4% is of unknown ownership (FAO 2015). Forests are state properties in many countries, often accompanied by complex schemes of customary rights supposed to benefit local people (Le Roy et al. 1996; FAO 2016b). Forests are sometimes owned, through a diversity of modalities, by inhabitants or municipalities. They are also appropriated by private companies such as, for example, in South Africa, New Zealand, Chile, or, as in Europe, by millions of private owners (Nichiforel et al. 2018).

A forest is a political system where usages were formerly codified in "customaries," where practices were governed by edicts, codes, or decrees, whose inheritances interfere with the new national and especially international legislative arsenals. Evolutions are slow and reflect the great inertia of legal arrangements.

Sociologists describe forests as social systems, managed by individuals whose life stories are significant and structured by rivalries between social groups seeking to control forest areas (FAO 2014b). They perceive forests in terms of usages in the Ancient Regime, of practices in the current period, and of issues and conflicts at all times.

Forests are well known as economic systems framed by various needs and expectations of the agrarian, industrial, and postindustrial societies that have developed in the course of history and are still coexisting today (Nair 2004; Mermet and Farcy 2011; FAO 2014a) and on which depend the livelihoods of the 1.2 billion to 1.5 billion forest inhabitants and forest-dependent people (Chao 2012).

Finally, a forest is a psychosystem. It is viewed, imagined, dreamed or hated, idealized, represented, and symbolized in whole or in parts, the tree, or even the leaf, the fruit, or the root (Corvol et al. 1997; Dereix et al. 2016). Forests are set to music and painted, are objects in myths and legends, advertising, fashions, and speeches (Schama 1995; Mottet 2017). More often associated with traditional societies (Descola 2005), these complex systems are becoming more important in today's urban world where some magical or mythical thought about forests is emerging (Harrison 1992).

2.8 CONCLUSION

What is a forest? Apparently, this is a simple question whose complexity has just been evoked in those lines. Alternately seen as an obstacle or a catalyst, forests accompanied humanity from its beginning and, for that reason, are invested with deeply rooted symbolic values, which are decisive when it comes to debating its future and destiny.

REFERENCES

Allen, J. C. and D. F. Barnes. 1985. The causes of deforestation in developing countries. *Annals of the Association of American Geographers* 75:163–184.

Amat, J.-P., Dorize, L., and C. Le Coeur. 2008. *Eléments de Géographie Physique*. Rosny-sous-Bois: Bréal.

Amigues, S. 1988. Théophraste, Recherches sur les plantes, Tome I: Livres I–II. Texte établi et traduit (Collection des Universités de France, Budé.). Paris: Les Belles Lettres. forêt, Comité des Travaux Scientifiques et Historiques:13–30.

Arnould, P. 2004. Nouvelles forêts, vieilles forêts, forêts de l'entre-deux, au XIXe et XXe siècle: entre rationalité économique et fertilité symbolique. In *Les forêts d'Occident du Moyen Age à nos jours*, ed. A. Corvol-Dessert, 253–277. Toulouse: Presses Universitaires du Mirail.

Bastin, J.-F., Berrahmouni, N., Grainger, A. et al. 2017. The extent of forest in dryland biomes. *Science* 356:635–638.

Beech, E., Rivers, M., Oldfield, S., and P. P. Smith. 2017. GlobalTreeSearch: The first complete global database of tree species and country distributions. *Journal of Sustainable Forestry* 36:454–489.

Boulier, J. and L. Simon. 2009. *Atlas des forêts dans le monde. Protéger, développer, gérer une ressource vitale*. Paris: Editions Autrement.

Chao, S. 2012. *Forest peoples: Numbers across the world. Moreton-in-Marsh: Forest peoples programme*. Moreton-in-Marsh: Forest Peoples. http://www.forestpeoples.org/sites/fpp/files/publication/2012/05/forest-peoples-numbers-across-world-final_0.pdf (Accessed March 7, 2018).

Claire, R. 2010. Le Douglas, un Arbre Exceptionnel. Drulingen: Scheuer.

Corvol, A., Hotyat, M., and P. Arnould. 1997. *La Forêt: Perceptions et Représentations*. Paris: L'Harmattan.

Crowther, T. W., Glick, H. B., Covey, K. R. et al. 2015. Mapping tree density at a global scale. *Nature* 525:201–207.

Dereix, C., Farcy, C., and F. Lormant. 2016. *Forêt et Communication: Héritages, Représentations et Défis*. Paris: L'Harmattan.

Descola, P. 2005. *Par-delà Nature et Culture*. Paris: Gallimard.

FAO (Food and Agriculture Organization of the United Nations). 2010. World's forest 2010. In *Forest Resource Assessment*. Rome: FAO http://www.fao.org/forestry/fra/80298/en/ (Downloaded April 16, 2018).

FAO. 2014a. *The State of the World's Forest Genetic Resources*. Rome: FAO. http://www.fao.org/3/a-i3825e.pdf (Accessed April 10, 2018).

FAO. 2014b. *State of the World Forest: Enhancing the Socioeconomic Benefits from Forests*. Rome: FAO.

FAO. 2015. *Global Forest Resource Assessment 2015*. Rome: FAO.

FAO. 2016a. *Trees, Forests and Land Use in Drylands: The First Global Assessment*. Rome: FAO. http://www.fao.org/3/a-i5905e.pdf (Accessed April 10, 2018). Rome: FAO.

FAO. 2016b. *Forty Years of Community-Based Forestry: A Review of Its Extent and Effectiveness*. Rome: FAO.

Farcy, C. and F. Devillez. 2005. New orientations of forest management planning from an historical perspective of the relations between man and nature. *Forest Policy and Economics* 7:85–95.

Hallé, F. 2005. *Plaidoyer Pour L'arbre*. Arles: Actes Sud.

Harrison, R. P. 1992. *Forests: The Shadow of Civilization*. Chicago, IL: Chicago University Press.

Houzard, G. 1985. Sylvosystème et sylvofaciès: Essai d'étude globale du milieu forestier. Colloques phytosociologiques. *Phytosociologie et Foresterie.* 14:231–236.

Le Play, F. 1996. Des forêts, considérées dans leurs rapports avec la constitution physique du globe et l'économie des sociétés. IDF, Collection Sociétés, Espaces, Temps. Lyon: ENS Editions.

Le Roy, E., Karsenty, A., and A. Bertrand. 1996. *La Sécurisation Foncière en Afrique, Pour une Gestion Viable des Ressources Renouvelables.* Paris: Karthala.

Lund, G. H. 2002. Comings to terms with politicians and definitions. In *IUFRO Occasional Paper no. 14 Forest Terminology: Living Expert Knowledge. How to Get Society to Understand Forest Terminology*, ed. International Union of Forest Research Organizations (IUFRO), 23–44. Vienne: IUFRO.

Lund, H. G. 2012. *National Definitions of Forest/Forestland Listed by Country.* Forest Information Service. http://home.comcast.net/~gyde/lundpub.htm (Accessed March 8, 2018).

Mazoyer, M. and L. Roudart. 1997. *Histoire des Agricultures du Monde. Du Néolithique à la Crise Contemporaine.* Paris: Seuil.

Mermet, L. and C. Farcy. 2011. Contexts and concepts of forest planning in a diverse and contradictory world. *Forest Policy and Economics* 13:361–365.

Mottet, J. 2017. *La Forêt Sonore. De L'esthétique à L'écologie.* Champ Vallon: Ceyzérieu.

Nair, C. T. S. 2004. Que réserve l'avenir pour l'enseignement forestier? *Unasylva* 216(55):3–9.

Nichiforel, L., Keary, K., Deuffic, P. et al. 2018. How private are Europe's private forests? A comparative property rights analysis. *Land Use Policy.* In Press.

Schama, S. 1995. *Landscape and Memory.* New York: AA Knopf.

ENDNOTES

1. Tree definition agreed on by IUCN's Global Tree Specialist Group: "a woody plant with usually a single stem growing to a height of at least two meters, or if multi-stemmed, with at least one vertical stem having five centimeters in diameter at breast height."

2. Parallelly, "the United Nations Framework Convention on Climate Change (UNFCCC) threshold values for forest are a minimum area of 0.01–1.0 ha, a minimum tree height of 2–5 m and a minimum crown cover 10–30%, while the United Nations Environment Program (UNEP) defines forest based on a minimum crown cover of 40%" (Chao 2012).

3. http://www.onf.fr/enforet/verdun/@@index.html.

3 Main Milestones in Forestry Evolution

Christine Farcy, Inazio Martinez de Arano, and Eduardo Rojas-Briales

CONTENTS

3.1 INTRODUCTION

Forests, in their polysemic meaning, have contributed to the development of human's societies and civilizations throughout history. So obviously, forestry, which is at the crossroad between humans and natural systems, was not univocal and was framed by the various roles successively played by forests in and for human societies and by the related waves of expansion or degradation of forest covers and composition depending on human demands and technological developments.

This chapter, far for being a strict historical one, rather offers an overview of some of the main milestones in forestry mutations induced by or in response to paradigm shifts understood as changes in thoughts, perceptions, or values granted by human societies to forests and trees (Capra 1982). Those milestones were often not taking place everywhere or maybe not at the same time, but we considered them to be significant enough due to their respective contributions to the development of forestry principles and pillars.

3.2 MILLENNIA OF *IMPLICIT* COMMUNITY-BASED FORESTRY

Since the deep past, worldwide forests were used and modified by groups of hunter-gatherers [i.e., the studies by Fritzbøger and Søndergaard (1995), Bahuchet and Betsch (2012), and Roberts et al. (2017)]. In a forest-dependent society, few of which remain

15

today, characterized by low population density and levels of income, "forests are used to meet basic needs such as for woodfuel, medicines, food, and construction materials. These societies have limited capacity to alter the forest environment drastically. Forests permeate cultural, social and religious beliefs and perceptions" (Nair 2004).

Neolithic agriculture spread through slash-and-burn farming systems that have persisted for millennia. With the increase in population, or where deforestation was a consequence of overcultivation, postforest agrarian systems progressively developed following the evolution of knowledge and techniques [i.e., the book by Mazoyer and Roudard (1997) and the study by Bahuchet and Betsch (2012)]. Pasturing cattle, horses, or pigs; harvesting timber, wood bark, deadwood, litter, and leaves; and shifting cultivation were effective economic components of rural farming [i.e., the books by Letrange (1909), Le Roy et al. (1996), and Chalvet (2011) and the study by Agnoletti et al. (2009)].

In agrarian societies, "forests are viewed as space to expand agriculture, including livestock; as a source of low-cost inputs for agriculture and of woodfuel, fodder, medicines and other non-wood forest products such as bush meat. Sale of products also supplements income to farming communities. With settled cultivation, the service functions of forests (e.g., watershed protection, arresting land degradation) become important" (Nair 2004). Forests and trees are integral part of daily life and are closely framing activities which are guided by local inherited or shared knowledge and often governed by local land tenure framework relying on uses, even temporary ones [i.e., the book by Le Roy et al. (1996) and studies by Robaye (1997), Nair (2004), and Parrotta and Trosper (2012); Figure 3.1].

FIGURE 3.1   Detail of *Forest Landscape (The Rest on the Flight into Egypt)*, Jan Brueghel, 1607. The State Hermitage Museum, St. Petersburg. (Courtesy of The State Hermitage Museum/Konstantin Sinyavsky.)

Tribal or postfeudal societies developed diverse forest management regimes in adaptive and dynamic ways [i.e., the books by Letrange (1909), Le Roy et al. (1996), Chalvet (2011), and Watkins (2014) and the studies by Goblet d'Alviella (1925), Odera (2004), Wyatt (2004), and FAO (2016a)]. Such agrarian uses of forest, also qualified as preindustrial by Mather (2001) or agricultural forest of sustenance by Ernst (1998), based on labor and land as main production factors, prevailed until around the end of the eighteenth century in many European regions (Mather 2001). It is still the case today in many rural regions in the global south where rural families or communities are getting their livelihoods from an extensive work of the land.

Associated practices might often be embedded into the contemporary framework of agroforestry [i.e., the books by Mazoyer and Roudart (1997) and Ramachandran Nair and Garrity (2012) and the studies by Arnold (2001), Garrity et al. (2010), Pye-Smith (2010), Bahuchet and Betsch (2012), and FAO (2016a)]. Moreover, the logic of the agrarian society could still affect forestry today due to legislative inertia in Western countries by way of legislation related to access or heritage ownership rights.

3.3 LONG HEGEMONY OF INDUSTRIAL OR MONOFUNCTIONAL FORESTRY

The key role played by wood in the industrial revolution in Europe was a turning point in the emergence of formalized forestry. In turn, it was the main source of domestic and industrial energy, support for mining galleries, basic material for building, and even shipbuilding in coastal countries (Larrère et Nougarède 1990), wood became a strategically important resource in Europe since the seventeenth century. In fact, the Spanish and Portuguese term for wood (*madera*) is derived from the Latin *materia*. Forests, being the place of production, suffered rising pressures and received increasing attention. Existing usable forests were intensively exploited, and potential ones were actively prospected while recurrent attempts were made by the holders of authority to regulate and restrict the customary rights enjoyed by local population or to make their exercises more and more difficult (Farcy 2012). At the dawn of the eighteenth century, "the demand for wood could no longer be met by expansion into previously unused forests" (Schmithüsen 2013).

Those tensions gave way in the second half of the eighteenth century to the process of formalization and systematization of knowledge in an effort to develop a scientific disciplinary body dedicated to ensuring steady wood supplies through time (Watkins 2014). The main object of this so called modern forestry was the forest, considered as a set or collection of trees, while arboriculture dealt with individual trees (Pinchot 1900). Modern forestry mainly focused on forests in rural areas, often located in remote and marginalized places.

Scientifically forged as a response to the needs of the emerging industrial society (Schmithüsen 2013), modern forestry emerged under the umbrella of enlightenment (Agnoletti et al. 2009; von Carlowitz 1713). *D'Alembert and Diderot's Encyclopedia*, mirror of the conquests made by humans in a world that was in itself unknown, represented the list of human possessions and the inventory of appropriated objects and justified the right to possess, to own, and to trade physical objects while considering irrelevant the world regarded outside the realm of human activity

(Farcy and Devillez 2005). The contemporary Napoleon's civil code further formalized the conception of nature being an external object at the disposal of humankind (Feltz 2001). "Human's role is therefore to become the master and proprietor of nature which he strives to understand, perfect and transform by scientific means for the general wellbeing of all mankind" (Farcy and Devillez 2005).

Eminently anthropocentric and utilitarian, and relying on the "gospel of efficiency" (Nelson 2013), modern forestry (also called by some authors as scientific or industrial forestry) suggested that forests are natural resources that can be rationally managed over long periods according to explicit objectives (Mather 1991, 2001; Agnoletti et al. 2009; Chalvet 2011). Mainly in the German speaking area, by outstanding people such as Carlowitz, Cotta, Hartig, Hundesagen, or Faustmann and in France by not less renowned Buffon, Dralet, Duhamel de Monceau, Reamur or Lorentz, modern forestry achieved its ideal model by the way of the theory of the "normal forest" (Agnoletti et al. 2009). This ideal forest should be structured in a balanced way in terms of distribution of ages of its forest stands or trees size that allows to harvest periodically the average growth of the given forest.

Although numerous examples demonstrate the interest that foresters of that period have had for other uses, goods, or services, in particular in south and eastern Europe and in mountain areas, where the focus was mostly given to restoration, modern forestry considered and modeled forests as tree factories or place for the cultivation of wood (Hartig 1805; EEA 2010). As further formalized under the thesis of the "wake effect," managing forests for sustained wood production was considered as implicitly implying the achievement of other purposes, such as soil protection, water quality, or public access (Dieterich 1953; Rupf 1960).

The core concept of modern forestry, sustained yield timber management, resulted from quantification and rationalization applied to both the description of nature and the regulation of economic practice (Lowood 1990). In the associated forest economics models further endorsed by neoclassical economics, the central role is given to the concept of maximization, "people's preferences are static, society is a mathematical aggregation of homogeneous rational agents, public inputs are through market signals, and there is no role for any institution other than the market" (Kant 2003). The underlying decision-making process was considered quite simple, linear, predictable, segmented, and exclusive.

As to the governance model, modern forestry relied on a "command and control" scheme seen as more efficient for securing wood supplies for strategic uses (Kennedy et al. 1998; Farcy and Devillez 2005; Harrington and Morgenstern 2007; Michel and Gil 2013). Further widespread adoption of military-inspired uniforms by state forest services illustrates this strategic orientation of which there are still reminiscences today.

A posteriori entitled monofunctional forestry (Mather 2001; Raum and Potter 2015), modern forestry, and the emerging forest sector contributed in designing a system characterized by institutional, social, and spatial partitioning where forest became "an island, i.e. a world apart, isolated from the rest of the rural landscape" (Plaisance 1979). The fact that most threats to forests such as fires, poaching, and grazing come from outside could also partly explain that fact.

Refined for decades by generations of botanists, economists, and experts on silviculture modern forestry spread worldwide since the nineteenth century as hegemonic set of ideas and practices through specific relays such as Brandis (India), Fernow or Pinchot (United States) or by the way of various structures of the European imperial and colonial powers (Corvol 2005a; Agnoletti et al. 2009; Boulier and Simon 2009; Schmithüsen 2013; FAO 2016a). This continuous and hegemonic dissemination led to the creation of state forestry departments and to the development of specific legal codes all over the world. It shaped specialized institutions and conditioned the disciplinary organization of knowledge, research, education, and training worldwide (Agnoletti et al. 2009; Ratnasingam et al. 2013; Raum and Potter 2015). In 1926, the first edition of the World Forestry Congress held in Rome focused on technical concerns related to the preservation of a timber resource, the further supply of commercial timber by way of plantation, and the establishment of the conditions for a sustained yield of timber production (Nail 2010).

3.4 SHIFTS AWAY FROM TIMBER PRIMACY

While the validity of modern forestry principles could not conceptually be questioned for industrial societies, their relevance for forest-dependent people (Chao 2012), in agrarian or postindustrial societies (Nair 2004) or as main components of a global paradigm for forestry, has clearly been challenged for several years (Ormerod 1997; Kennedy et al. 1998; Kant 2003, 2013; Kennedy and Koch 2004; Piketty 2013). The progressive shift away from the dominant model based on timber primacy followed several directions as reactions to changing contexts or to shortcomings induced by an unappropriated framework.

3.4.1 COMMUNITY-BASED FORESTRY

While the industrial scheme allowed in Europe, Russia, or North America, the recovery of forest area and related soils, stocks and increasing volumes of wood supply in some sustainable or affordable way, as well as the trustworthy provision of public goods in forests, the export of the model has seen resistance.

The expansion of modern forestry in overseas colonized countries from the middle of the nineteenth century was relying on the imposition of "scientific forestry." Selecting the best forests, demarcating them as state forests, extinguishing or limiting customary rights, and guarding forests against incursion or unauthorized used were contributing to the main project of strategically controlling the resource or managing forests to maximize timber production, for the benefit of the colonizing power and/or state (Arnold 2001; Odera 2004; Agnoletti et al. 2009; FAO 2016a; Figure 3.2).

Many local communities have been displaced by colonial forest administration, even through the use of force, from forests that they had traditionally occupied and depended on for their livelihood. Communities were relocated and denied access to newly stated "protected areas." Exclusion went hand in hand with specialization and compartmentalization of land uses, in particular after the introduction of cash crops such as cocoa or sugarcane (Odera 2004).

Despite opposition and even resistance by local people, the same scheme, where central governments assumed all rights over forest access and management, was adopted after the independence from the middle of the twentieth century, by many

FIGURE 3.2   Boundary markers of forest estate in Algeria. (Photo by Christine Farcy.)

postcolonial authorities. However, with increased populations, declining food production in farmlands, and forest land degradation, centralized management systems
became unable to sustain the system simply through enforcement. "Command and
control" and "set apart" schemes showed particular inefficiency for patrolling the
extensive porous forest boundaries and when imposing unbalanced weight to poor
surrounding communities (Odera 2004; Agnoletti et al. 2009).

An explicit version of community-based forestry emerged during the 1970s and
1980s in an effort "to response partly to a perceived failure of the forest industry development model to lead the socioeconomic development, and partly to the
increasing rate of deforestation and forest land degradation in developing countries"
[Gilmour et al. (1989) cited by the Food and Agriculture Organization of the United
Nations (FAO) (2016a)].

Inspired by works such as that by Ostrom (1990), who showed the positive links between property rights granted to local communities and autonomous regulation of the use of natural resources, community-based forestry became an evolving branch of forestry mostly dedicated to rural areas and concerns, and relying on inclusiveness. It is considered today as including "initiatives, sciences, policies, institutions and processes that are intended to increase the role of local people in governing and managing forest resources" [RECOFTC (2013) cited by FAO (2016a)]. The heir of less formalized framework, such as social forestry and participatory forestry, its scope and driving forces can be illustrated by keywords such as livelihoods, local/traditional knowledge, decentralization, participation, and secure tenure (Arnold 2001; Odera 2004; Charnley and Poe 2007).

Being formalized, community-based forestry firstly focused mainly on collaborative management of forest commons, in rural regions of the global south but more recently also embraced forest management by smallholder family forestry (Wiersum et al. 2005) that is considered more aligned with collaborative forestry (FAO 2016a). Eminently linked to and rooted in rural areas, community-based forestry often relies on and benefits from agroforestry techniques (Lundgren 1982; Mazoyer and Roudart 1997; Montambault and Alavalapati 2005; Rigueiro-Rodriguez et al. 2009; Ramachandran Nair and Garrity 2012) while not encompassing the broader land use system with which they are associated (Arnold 2001; FAO 2016a).

3.4.2 TOWARD POSTINDUSTRIAL FORESTRY

In the 1970s, the myth of technical progress as a factor of well-being and quality of life is shattered by the awareness of the harmful or even irreversible influence of human on nature, which has been underestimated and even ignored for a long time. The Cartesian view of humans being external to nature proved to be irrelevant (Farcy and Devillez 2005). Concomitantly, postproductivist transition in "advanced" societies induced the related increased interest for other values than materialist ones when basic needs are satisfied (Inglehart 1971) while increasing urbanization of the society strongly modified lifestyle and values, in particular by losing rural roots (Fritzbøger and Søndergaard 1995; Kennedy et al. 1998; Schmithüsen 1999; Mather 2001; Kennedy and Koch 2004; Kant 2013).

Following a process also impacting other primary sectors such as agriculture, postindustrial forestry results from a progressive and multifactor shift away from timber preeminence toward increasing focus on services and rising primacy on cultural, environmental, and aesthetic values (Mather 2001; Nair 2004). Where it was considered for a long time as an exclusive place for the production of a material good, forest of postindustrial societies rather became the scene for consumption of services by a largely urban population (Mather 2001). In such a scheme, there is a shift from mass production to meeting needs of smaller markets with increased emphasis on customer and customization. Dematerialized production is mainly relying on critical resources such as information and knowledge (Kennedy et al. 1998; Nair 2004). Possessing or private ownership may persist but increasingly circumscribed by public and social regulations (pluralism mode) written or not.

3.4.2.1   Multipurpose Forestry

3.4.2.1.1   Multiple Use and Multifunctional Forestry

The evolution toward postindustrial forestry emerged in the second half of the twentieth century with the "multiple use" concept. Confirming emergent interest and needs on the issue, the fifth World Forestry Congress held in the United States of America under FAO auspices in 1960, focused on "multiple use of forest and associated lands." This use relies on "the management of the forest in a manner that, while conserving the basic land resource, will yield a high level of production in the five major uses—wood, water, forage, recreation, and wildlife—for the benefit of the greatest number of people in the long run" (FAO 1960). The model considered that a use should be dominant and that the secondary ones should not be chosen at its detriment. The underlying logic was still mainly economic, the purpose being "to select a set of uses that maximized the long run human value to society of the uses of the forest" (Nelson 2013). Participants to the most important gathering of the world forestry sector expressed great interest for the concept and acknowledged its great potential while recognizing it as challenging, especially if the single income was wood.

The nearby concept of forest function was formalized in the same period. Launched in 1953 by Dieterich, it is considered as a societal demand posted to forest. Today, the meaning is rather converging toward the idea of a task assigned to forests and forestry by the society (Bader and Riegert 2011). Implicitly questioning Dietrich's wake effect theory that was framing industrial forestry, Hasel (1968) opened the way to integrative multifunctional forestry where resource function, protective function, and recreation function should be implemented together while even differing in importance. Further distinction was made between the strict multifunctional model according to which all functions are to be fulfilled everywhere and the segregation scheme based on specialized zoning and further implementation of multifunctionality at largest scale (Borchers 2010). In the frame of debate also taking place in agriculture, trade-offs became progressively a keyword for fulfilling several functions in a context where wood production remained the main financial sources (Raum and Potter 2015).

Such a paradigm shift induced progressive changes in the means of forestry, in its ends and in the ethics of its professionals. It was also questioning the previous linearity of the decision-making process and the main role and responsibility of the forestry administration (Kennedy et al. 1998; Mather 2001).

3.4.2.1.2   Sustainable Forestry

A deepening of those trends took place in the 1990s under the combined pressure not only of domestic demands from the society but also mostly of commitments taken and obligations adopted under the umbrella of international processes (Raum and Potter 2015). The United Nations conferences held in Rio in 1992 constituted significant milestones when mainstreaming the concept of sustainable development. The forest community considered appropriate to take initiatives on the issue and launched sustainable forestry. The sustainable forest management (SFM) concept was then formalized under the umbrella of forest-related international forum or regional processes. For example, SFM was defined in Europe "as the stewardship and use of forests, maintaining their biodiversity, productivity, regeneration capacity and vitality, to fulfil, now and in the future, ecological, economic and social functions" (MCPFE 1993).

Sustainable forestry explicitly endorsed new strategic objectives such as soil protection and water quality, the preservation of biodiversity and of the quality of landscapes, carbon stocking, the recreational, and cultural values of forests while acknowledging that space and time in their ecological essence are fundamentally antimechanist (Prigogine and Stengers 1984; Andersson et al. 2000; Buttoud 2000; Figure 3.3).

Sustainability also deepened multifunctionality by opening the time horizons to social responsibility drivers in particular toward new generations. Whereas Schmithüsen (2013) described how sustainability was understood by former modern forestry in particular in reference to sustainable wood production, sustainability of the 1990s elevates this pursuit of equity between generations to the rank of a founding principle and strategic objective. Uncertainty associated to concerns ahead called for adaptive approaches in management and planning (Lessard 1998).

Sustainability strongly questioned mainstream forest economics, opening new answers as, for example, on nonlinearities, multiple equilibrium, rationality of choices, behavioral drivers or intertemporal ethics, and experts are, until now, arguing for new frontiers for forest economics (Kant 2003, 2013; Zang and Pearse 2011; Hyde 2013).

In parallel, sustainability enriched the decision-making framework in line with the emergence of governance concept [i.e., the books by Graz (2004) and Rayner et al. (2010) and the study by Tuomasjukka (2010)]. The frequent conflicts between the forest as a "shared heritage of humanity" and the forest as an "appropriated space," where the owner reigns supreme, often illustrate the rupture and the diversification of the owner's prerogatives to the benefit of new communities of users (Comby 1997; Brédif and Boudinot 2001). Far from the "command and control" model, foresters are to become public servants able to manage the relationship between people and forest in a framework where the majority of public forest

FIGURE 3.3   Megaphones to hear forest—Estonia. (Courtesy of Tõnu Tunnel.)

stakeholders to be served is yet to be born (Magill 1988; Kennedy et al. 1998; Kennedy and Koch 2004).

Multiple use, multifunctional forestry and sustainable forestry were all forged in and by actors of the forest community. Today, in practice, while still focusing on rural areas and on forest rather than trees, they are embedded in each other (EASAC 2017) or often considered similar even synonymous, as illustrated by Cubbage et al. (2007) or Sarre and Sabogal (2013).

3.4.2.2 Urban Forestry

As confirmed by its selection by the Collaborative Partnership on Forest as the main topic for celebrating the 2018 International Day of Forest, the role of forests and trees in urban context is increasingly acknowledged. In a world of accelerating urbanization (Kennedy et al. 1998), with 70% of the population expected to be urban by 2050 (United Nations 2015), forests and trees gain importance for their potential contributions to more balanced, safe, pleasant, and healthy urban development [i.e., the studies by Ulrich (1990), Dwyer et al. (1991), Hodge (1995), Konijnendijk et al. (2005), Randrup et al. (2005); and FAO (2016b) and the books by Kowarik and Körner (2005) and Nail (2006)].

Not only urban woodlands but also parks, gardens, and trees have long been associated with cities, including since ancient civilizations [i.e., the studies by Haque (1987), Corvol (2005b), and Forrest and Konijnendijk (2005) and the books by Hobhouse (1992) and Chalvet (2011)]. Rather than for utilitarian purpose, trees and forests were mostly accompanying cities for their aesthetic, cultural, and spiritual values [i.e., the studies by Kuchelmeister and Braatz (1993) and Brosseau (2011) and the book by Forrest and Konijnendijk (2005); Figure 3.4].

Considered desirable but not necessarily essential for a long time, urban forestry is being formalized, originating in North America during the 1960s and 1970s

FIGURE 3.4 Flowered undergrowth of Azaleas in the Botanical Garden of New York. (Photo by Christine Farcy.)

(Jorgensen 1970; Konijnendijk 1997; Konijnendijk et al. 2005) while the social side of forestry was emerging and giving increasing emphasis to societal needs and demands (Westoby 1989). "Many different interpretations have existed and initially considerable opposition from different sides was encountered. Arboriculturists and other green area professionals were hesitant about introducing the term as a way for foresters to extend their domain to urban areas. Foresters themselves, however, were often not convinced of having a mission in managing small-scale green areas or even single trees in urban areas" [e.g., the studies by Ball (1997) and Miller (2001) cited by Randrup et al. (2005)].

Defined as a specialized branch of forestry, merging arboriculture and ornamental horticulture, whose objective is to contribute to the physiological, sociological, and economic well-being of urban society (Kuchelmeister and Braatz 1993; FAO 2016b), urban (and periurban) forestry is more clearly outlined and gave rise to scientific journals or specific academic training.

3.4.2.3 Recent External and Broader Approaches

Some developments still in progress were recently emerging from international fora dealing with environmental matters. They were conceptually influencing the forest scene, and authors such as Mather (2001) or Raum and Potter (2015) are already considering them as part of postindustrial forestry.

3.4.2.3.1 *Ecosystem Approach and Model Forests*

Specific frameworks relying on the concept of ecosystem developed since the 1970s, with the view to both better take on board the complexity of the relationship between nature and the society and increase public interest and awareness in biodiversity conservation (Slocombe 1993; Johnson 1999; Bengtsson et al. 2000; Mather 2001; Goméz-Baggethun et al. 2010, Waylen et al. 2014).

The ecosystem approach is rooted in the UN Convention on Biological Diversity (CBD) and was adopted by the second meeting of the Conference of the Parties in Jakarta in 1995 "as the primary framework for action under the convention" (Waylen et al. 2014). Developed as a strategy, the ecosystem approach, instead of being precisely defined, is based on a set of principles, endorsed by the international community, "for the integrated management of land, water and living resources that promotes conservation and sustainable use in an equitable way" (CBD 2000).

Whereas its operationalization seemed to have been undermined by the confusion induced by its flexible nature (Waylen et al. 2014), the relevance of its principles was confirmed when applied to the *forest model* concept (Corrales et al. 2005). This "partnership-based process through which individuals and groups, representing a diversity of values, work together toward a common vision of sustainable development for a landscape in which forests are an important feature" firstly developed in Canada and spread worldwide by the way of a global network of model forests developed on the ground (IMFN 2008).

3.4.2.3.2 *Ecosystem Service Approach*

Some 10 years later after its adoption by the CBD, the ecosystem approach was further developed in the framework of the millennium ecosystem assessment (MEA) process dedicated to the evaluation of the state of global natural resources and the effects of ecosystem change on human well-being. In the framework developed, ecosystem structures and functioning are considered as directly leading to valued human welfare benefits (Campos et al. 2005). The illustration of such a close link, should contribute "to awake society to think more deeply about the importance of nature and its destruction" (Norgaard 2010).

While already existing before, the core element of the MEA, the ecosystem service concept, became successful due to its link with and interest in decision-making processes (Méral and Pesche 2016). It also leads to the development of innovative policy instruments allowing tackling, through economic incentives, the contribution of private actors and the production of public services such as water quality or biodiversity conservation. Forest-related schemes such as payment for environmental services develop in various regions of the world and are becoming important components of forest policy toolbox [i.e., the studies by Cubbage et al. (2007), Engel et al. (2008), and Redford and Adams (2009) and the book by Bennett et al. (2012)].

The ecosystem approach success was such that it leads to some dominant simplistic metaphors of nature being based on stock and flow while giving overemphasis on valuation and monetary incentivization. While considered as an interesting part of a larger solution, "its dominance in our characterization of our situation and the solution is blinding us to the ecological, economic, and political complexities of the challenges we actually face" (Norgaard 2010) and offers risks to pursuit single purposes (e.g., carbon sequestration) (Goméz-Baggethun et al. 2010; Waylen et al. 2014; Raum and Potter, 2015).

Tied with the concept of function for a long time, the forest community is working on conceptual gateways such as, for example, that of Bader and Riegert (2011), with wordings such as "task to forest," "natural effect produced by forest," and "effects satisfying tasks."

3.4.2.3.3   *Landscape Approach*

The landscape approach emerged early this decade considers the need to integrate all land uses as opposed to the classical sectorial approaches of agriculture, livestock, gracing, wildlife management, water, biodiversity preservation or environmental/ecosystem services. The emerging relevance of land use management and planning in disaster risk reduction promoted through the Hyogo Framework for Action (HFA) by the United Nations Office for Disaster Risk Reduction (UNISDR) and the progress in including the different land uses into the climate process under the Land Use, Land-Use Change and Forestry (LULUCF) regulation were undoubtedly contributing factors. The main venue advocating for the landscape approach has been the Global Landscape Forum organized since 2013 by the Center for International Forestry Research (CIFOR), which evolved from the successful six Forest Days arranged during the conferences of the parties (COPs) of the United Nations Framework Convention on Climate Change (UNFCCC) from 2007 until 2012 under the aegis of the Collaborative Partnership on Forests. As noted by Sayer et al. (2013) in their literature review, landscape approach implementation relies on a set of principles, often close to those embodied under ecosystem approach, emphasizing adaptive management, stakeholder involvement and multiple objectives in a shift from project-oriented actions to process-oriented activities. Landscape approach was also central to the last conference of the parties of the Convention on Biological Diversity (CBD) held in Mexico in 2016, centered in integration of biodiversity considerations into agriculture, forestry, fisheries, and tourism. Through its suggestive approach, there has been considerable reluctance in its acceptance (Reed et al. 2014), especially in the United Nations system, most likely due to the complexity in translating the landscape term from English into other official languages, misleading its two meanings (territory and scenic beauty). In fact, landscape approach might be in reality not new at all but quite close to the well-consolidated concept of watershed management.

3.5 CONCLUSION

The evolution of forestry was not linear, and this review has shown an overlay between old and new models, in particular during the last 25 years where changes were accelerating. Indeed, new approaches, including dominant ones, are not exclusive and often overlay rather than supersede previous models and so generating complexity (Surel 2000; Raum and Potter 2015) or even confusion as observed recently (Norgaard 2010; Petit et al. 2014; Waylen et al. 2014; FAO 2016a).

Different developments also occurred at the same time, but in different places, due to varying contexts, conjunctures, or drivers. In practice, a posteriori attempts to create gateways appeared as, for example, between community-based forestry and sustainable forestry or between sustainable forestry and ecosystem approach (Bader and Riegert 2011; Nelson 2013; Quine et al. 2013; FAO 2016a).

Last trends also relied on the need to correct or adapt wrong or inadequate framework heir from the past: community-based forestry to correct lack of inclusiveness and to link with agroforestry, postindustrial forestry to shift away from wood primacy and move towards forest as place for consumption of services, urban forestry to open the scope to urbanity and to encompass arboriculture; ecosystem and landscape approaches to boost and open up conservation concerns.

REFERENCES

Agnoletti, M., Dargavel, J., and E. Johann. 2009. History of forestry. In *The Role of Food, Agriculture, Forestry and Fisheries in Human Nutrition*, Online Encyclopedia of Life Support Systems (EOLSS). Oxford: Eolss Publishers Co. Ltd.

Andersson, F. O., Feger, K.-H., Hüttl, R. F. et al. 2000. Forest ecosystem research—Priorities for Europe. *Forest Ecology and Management* 132:111–119.

Arnold, J. E. M. 2001. *Forests and People: 25 Years of Community Forestry*. Rome: FAO.

Bader, A. and C. Riegert. 2011. Interdisciplinarity in 19th and early 20th century: Reflections on ecosystem series over forest. *Rupkatha Journal on Interdisciplinary Studies in Humanities* 3:87–98.

Bahuchet, S. and J.-M. Betsch. 2012. L'agriculture itinérante sur brûlis, une menace sur la forêt tropicale humide ? Savoirs et savoir-faire des Amérindiens en Guyane Française. In *Revue D'ethnoécologie* 1. http://ethnoecologie.revues.org/768. DOI: 10.4000/ethno ecologie.768.

Ball, J. 1997. On the urban edge: A new and enhanced role for foresters. *Journal of Forest* 95, 10:6–10.

Bengtsson, J., Nilsson, S. G., Franc, A., and P. Menozzi. 2000. Biodiversity, disturbances, ecosystem function and management of European forests. *Forest Ecology and Management* 132:39–50.

Bennett, G., Carroll, N., and K. Hamilton. 2012. *Charting New Waters: State of Watershed Payments 2012.* Washington, DC: Forest Trends. http://www.forest-trends.org/documents /files/doc_3308.pdf (Accessed January 20, 2018).

Borchers, J. 2010. Segregation versus mulifunktionalität in der Forstwirtschaft. *Forst und Holz* 65:44–49.

Boulier, J. and L. Simon. 2009. Atlas des forêts dans le monde. *Protéger, Développer et Gérer une Ressource Vitale.* Paris: Editions Autrement.

Brédif, H. and P. Boudinot. 2001. *Quelles Forêts Pour Demain? Eléments de Stratégie Pour une Approche Rénovée du Développement Durable.* Paris: L'Harmattan.

Brosseau, S. 2011. Du territoire rural aux parcs urbains: Métamorphoses du *satoyama* dans la métropole de Tokyo. Agriculture Métropolitaine/Métropole Agricole, 196–210.

Buttoud, G. 2000. Multipurpose management of mountain forests: Which approaches? *Forest Policy and Economics* 4:83–87.

Campos, J. J., Alpizar, F., Louman, B., and J. Parrotta. 2005. An integrated approach to forest ecosystem services. In *Forests in the Global Balance—Changing Paradigms*, eds. Mery, G., Alfaro, R., Kanninen, M. and M. Lovobikov, 97–116. Vienna: International Union of Forest Research Organizations (IUFRO), World Series Volume 17.

Capra, F. 1982. *The Turning Point: Science, Society and the Rising Culture.* New York: Simon and Schuster.

CBD (Convention on Biological Diversity). 2000. *Decision V/6 Ecosystem Approach. Decision adopted by the Conference of Parties to the Convention on Biological Diversity at Its Fifth Meeting, May 15–26, 2000, Nairobi, Kenya.* http://www.cbd.int /decision/cop/default.shtml?id=7148 (Accessed August 14, 2017).

Chalvet, M. 2011. *Une Histoire de la Forêt.* Paris: Editions du Seuil.

Chao, S. 2012. *Forest People: Numbers across the World.* Moreton-in-Marsh: Forest People Programme, http://www.forestpeoples.org/sites/default/files/publication/2012/05/forest -peoples-numbers-across-world-final_0.pdf (accessed August 14, 2017).

Charnley, S. and M. R. Poe. 2007. Community forestry in theory and practice: Where are we now? *Annual Review of Anthropology* 36:301–336.

Comby, J. 1997. La gestation de la propriété privée. In *Droits de Propriété et Environnement*, eds. Falque, M. and M. Massenet, 275–284. Paris: Dalloz.

Corrales, O. F., Carrera, F., and J. J. Campos. 2005. El bosque modelo: una plataforma territorial para la aplicación del enforque ecosistémico. *Recursos Naturales y Ambiente* 45:6–12.

Corvol, A. 2005a. *Les Arbres Voyageurs.* Paris: Robert Laffont.

Corvol, A. 2005b. Mutations et enjeux en forêt de Soignes: les années 1900. *Journal Forestier Suisse* 156(8):279–287.

Cubbage, F., Harou, P., and E. Sills. 2007. Policy instruments to enhance multi-functional forest management. *Forest Policy and Economics* 9:833–851.

Dieterich, V. 1953. *Forstwirtschaftspolitik. Eine Einführung.* Berlin: Paul Parey.

Dwyer, J. F., Schroeder, H. W., and P. H. Gobster. 1991. The significance of urban trees and forests: Toward a deeper understanding of values. *Journal of Arboriculture* 17:276–284.

EASAC (European Academies' Science Advisory Council). 2017. *Multi-Functionality and Sustainability in the European Union's Forests.* EASAC Policy Report 32. Halle: EASAC. http://www.easac.eu/home/reports-and-statements/detail-view/article/multi-fun .html (Accessed January 23, 2018).

EEA (European Environment Agency). 2010. *10 Messages for 2010: Forest Ecosystems.* Copenhagen: EEA.

Engel, S., Agiola, S. P., and S. Wunder. 2008. Designing Payments for Environmental Services in theory and practice: An overview of the issues. *Ecological Economics* 65:663–674.

Ernst, C. 1998. An ecological revolution? The Schlagwaldwirtschaft in western Germany in the eighteenth and nineteenth centuries. In *European Woods and Forests: Studies in Cultural History*, ed. C. Watkins, 83–92. Wallingford: CAB International.

FAO (Food and Agriculture Organization of the United Nations). 1960. Fifth World Forestry Congress. *Unasylva* 14.

FAO. 2016a. *Forty Years of Community-based Forestry: A Review of its Extent and Effectiveness*. Rome: FAO.

FAO. 2016b. *Guidelines on Urban and Peri-Urban Forestry*. FAO Forestry Paper 178. Rome: FAO.

Farcy, C. and F. Devillez. 2005. New orientations of forest management planning from an historical perspective of the relations between man and nature. *Forest Policy and Economics* 7:85–95.

Farcy, C. 2012. La forêt et l'agriculture: chronique d'une histoire inachevée. In *Agroécologie. Entre Pratiques et Sciences Sociales*, eds. Van Dam, D., Streith, M., Nizet J., and P. M. Stassart, 165–178. Dijon: Educagri Editions.

Feltz, B. 2001. Rapport homme/nature, développement durable et expertise scientifique. In *Population et Développement: Savoirs et Jeux D'acteurs Pour des Développements Durables*, 8(53):183–193. Louvain-la-Neuve: Academia-Bruylant.

Forrest, M. and C. Konijnendijk. 2005. History of urban forests and trees in Europe. In *Urban Forests and Trees*, eds. Konijnendijk, C. C., Nilsson, K., Randrup, Th. B. and J. Schipperijn, 23–48. Berlin: Springer-Verlag.

Fritzbøger, B. and P. Søndergaard. 1995. A short history of forest uses. In *Multiple-Use Forestry in the Nordic Countries*, ed. M. Hytönen, 11–41. Helsinki: METLA, Finnish Forest Research Institute, Helsinki Research Centre.

Garrity, D. P., Akinnifesi, F. K., Ajayi, O. C. et al. 2010. Evergreen agriculture: A robust approach to sustainable food security in Africa. *Food Security* 2:197–214.

Gilmour, D. A., King, G. C., and M. Hobley. 1989. Management of forests for local use in the hills of Nepal: Changing forest management paradigms. *Journal of World Forest Resource Management* 4:93–110.

Goblet d'Alviella, F. 1925. Notes sur l'histoire des forêts belges—Les droits d'usage en forêt. Bulletin de la Centrale Forestière de Belgique 28.

Gómez-Baggethun, E., de Groot, R., Lomas, P. L., and C. Montes. 2010. The history of ecosystem services in economic theory and practice: From early notions to markets and payment schemes. *Ecological Economics* 69:1209–1218.

Graz, J.-C. 2004. La *gouvernance de la Mondialisation*. Paris: La découverte.

Haque, F. 1987. Urban foresty: 13 City profiles. Unasylva 39:14–25.

Harrington, W. and R. D. Morgenstern. 2007. Economic incentives versus command and control: What's the best approach for solving environmental problems? In *Acid in the Environment. Lessons Learned and Future Prospects*, eds. Visgilio, G. R. and D. M. Whitelaw, 233–240. New York: Springer.

Hartig, G. L. 1805. Instructions pour la culture du bois. *Traduction de J. Baudrillart*. Paris: Levrault.

Hasel, K. 1968. Die Zukunft der Deutschen Forstwirtschaft. In *Jahrbuch des Deutschen Forstvereins*, 36–60. Hiltrup bei Münster (Westf.): Landwirtschaftsverlag.

Hobhouse, P. 1992. *Plants in Garden History: An Illustrated History of Plants and Their Influence on Garden Styles—From Ancient Egypt to the Present Day*. London: Pavilion.

Hodge, S. J. 1995. Creating and managing woodlands around towns. Forestry Commission handbook 11. London: HMSO.

Hyde, W. F. 2013. Twelve unresolved questions for forest economics and two crucial recommendations for forest policy analysis. In *Handbook of Forest Resource Economics*, eds. Kant, S. and S. Alavalapati, 538–552. Abingdon-on-Thames: Routledge.

International Model Forest Network (IMFN). 2008. *Model Forest Development Guide.* Ottawa: International Model Forest Network Secretariat. http://www.imfn.net/system /files/Model_Forest_Development_Guide_en.pdf (Accessed August 14, 2017).

Inglehart, R. 1971. The silent revolution in Europe: Intergenerational change in post-industrial societies. *American Political Science Review* 65(4):991–1017.

Johnson, B. L. 1999. Introduction to the special feature: Adaptive management—Scientifically sound, socially challenged? *Ecology and Society* 3:10.

Jorgensen, E. 1970. Urban forestry in Canada. Presented at the 46th International Shade Tree Conference.

Kant, S. 2003. Extending the boundaries of forest economics. *Forest Policy and Economics* 5:39–56.

Kant, S. 2013. New frontiers of forest economics. *Forest Policy and Economics* 35:1–8.

Kennedy, J. J., Dombeck, M. P., and N. E. Koch. 1998. Values, beliefs and management of public forests in the Western world at the close of the twentieth century. *Unasylva* 49:16–26.

Kennedy, J. J., and N. E. Koch. 2004. Viewing and managing natural resources as human-ecosystem relationships. *Forest Policy and Economics* 6:497–504.

Konijnendijk, C. C. 1997. *Urban Forestry: Overview and Analysis of European Forest Policies. Part 1: Conceptual Framework and European Urban Forestry History.* EFI Working Paper 12. Joensuu: European Forest Institute.

Konijnendijk, C. C., Nilsson, K., Randrup, Th. B., and J. Schipperijn. 2005. *Urban Forests and Trees.* Berlin: Springer-Verlag.

Kowarik, I. and S. Körner. 2005. *Wild Urban Woodlands: New Perspectives for Urban Forestry.* Berlin: Springer.

Kuchelmeister, G. and S. Braatz. 1993. Urban forestry revisited. *Unasylva* 173(44):3–12.

Larrère, R. and O. Nougarède. 1990. La forêt dans l'histoire des systèmes agraires: de la dissociation à la réinsertion? *Cahiers d'Economie et Sociologie Rurales* 17:15–16.

Le Roy, E., Karsenty, A., and A. Bertrand. 1996. *La Sécurisation Foncière en Afrique, Pour une Gestion Viable des Ressources Renouvelables.* Paris: Karthala.

Lessard, G. 1998. An adaptive approach to planning and decision-making. Landscape and *Urban Planning* 40:81–87.

Letrange, P. 1909. *Des Droits D'usage Dans les Forêts D'Ardenne.* Paris: A. Rousseau.

Lowood, H. E. 1990. The calculating forester: Quantification, cameral science and the emergence of scientific forestry in management in Germany. In *The Quantifying Spirit in the 18th century*, eds. Frängsmyr, T., Heilbron, J. L. and R. E. Rider, 315–342. Berkeley, CA: University of California Press.

Lundgren, B. 1982. Introduction. *Agroforestry Systems*, 1:7–12.

Magill, A. W. 1988. Natural resource professionals: The reluctant public servants. *Environmental Professional* 10:295–303.

Mather, A. S. 1991. Pressures on British Forest Policy: Prologue to the post-industrial forest? *Erea* 23:245–253.

Mather, A. S. 2001. Forests of consumption: Postproductivism, postmaterialism and the postindustrial forest. *Environment and Planning C: Government and Policy* 19:249–268.

Mazoyer, M. and L. Roudart. 1997. *Histoire des Agricultures du Monde: Du Néolithique à la Crise Contemporaine.* Paris: Seuil.

MCPFE (Ministerial Conference on the Protection of Forests in Europe). 1993. *Declaration and Resolutions—Helsinki.* Vienna: MCPFE.

Méral, P. and D. Pesche. 2016. *Les Services Ecosystémiques. Repenser les Relations Nature et Société.* Versailles: Editions Quae.

Michel, M. and L. Gil. 2013. La transformación histórica del paisaje forestal en la Comunidad Autónoma de Euskadi. Vitoria Gasteiz: Servicio de publicaciones del Gobierno Vasco. http://www.euskadi.eus/contenidos/libro/transforpaisaje/es_agripes/adjuntos /coleccion_lur18.pdf

Miller, R. W. 2001. Urban forestry in third level education—The US experience. In *Planting the Idea—The Role of Education in Urban Forestry*, eds. Collins K. D and C. Konijnendijk, 49–57. Dublin: Tree Council of Ireland.

Montambault, J. R. and J. R. R. Alavalapati. 2005. Socioeconomic research in agroforestry: A decade in review. *Agroforestry Systems* 65(5):151–161.

Nail, S. 2006. *Bosques Urbanos en América Latina*. Bogota: Universidad Externa de Colombia.

Nail, S. 2010. *Forest Policies and Social Changes in England*. Dordrecht, London: Springer.

Nair, C. T. S. 2004. Que réserve l'avenir pour l'enseignement forestier? *Unasylva* 216(55):3–9.

Nelson, H. N. 2013. Multiple-use forest management versus ecosystem forest management: A religious question? *Forest Policy and Economics* 35:9–20.

Norgaard, R. B. 2010. Ecosystem services: From eye-opening metaphor to complexity blinder. *Ecological Economics* 69:1219–1227.

Odera, J. 2004. *Lessons Learnt on Community Forest Management in Africa*. Nairobi: Royal Swedish Academy of Agriculture and Forestry (KSLA); Kumasi: African Forest Research Network (AFORNET); and Rome: FAO. www.afforum.org/sites/default/files/English/English_118.pdf (Accessed 14 August, 2017).

Ormerod, P. 1997. *The Death of Economics*. Hoboken, NJ: Wiley.

Ostrom, E. 1990. Governing the Commons: The Evolution of Institutions for Collective Action. Cambridge, UK: Cambridge University Press.

Parrotta, J. A. and R. L. Trosper. 2012. *Traditional Forest-related Knowledge: Sustaining Communities, Ecosystems and Biocultural Diversity*. Dordrecht: Springer Netherlands.

Petit, O., Hubert, B. and J. Theys. 2014. Science globale et interdisciplinarité: quand contagion des concepts rime avec confusion. *Natures Sciences Sociétés* 22:187–188.

Piketty, T. 2013. *Le Capital au XXIème Siècle*. Paris: Seuil.

Pinchot, G. 1900. A primer on forestry. Part 1—The forest. *Bulletin of the US Department of Agriculture, Bureau of Forestry* 24.

Plaisance, G. 1979. *La Forêt Française*. Paris: Denoël.

Prigogine, I. and I. Stengers. 1984. *Order Out of Chaos: Man's New Dialogues with Nature*. London: Heinemann.

Pye-Smith C. 2010. A rural revival in Tanzania: How agroforestry is helping farmers to restore the woodlands in Shinyanga Region. *Trees for Change 7*. Nairobi: World Agroforestry Centre.

Quine, C. P., Bailey, S. A., and K. Watts. 2013. Sustainable forest management in a time of ecosystem services frameworks: Common ground and consequences. *Journal of Applied Ecology* 50:863–867.

Ramachandran Nair, P. K. and D. Garrity. 2012. *Agroforestry—The Future of Global Land Use*. Dordrecht: Springer.

Randrup, T. B., Konijnendijk, C. C., Kaennel Dobbertin, M., and R. Prüller. 2005. The concept of urban forestry in Europe. In *Urban Forests and Trees*, eds. Konijnendijk, C. C., Nilsson, K., Randrup, T. B. and J. Schipperijn, 9–21. Berlin, Heidelberg: Springer-Verlag.

Ratnasingam, J., Ioras, F., Vacalie, C. C., and L. Wenming. 2013. The future of professional forestry education: Trends and challenges from the Malaysian perspective. *Notulae Botanicae Horti Agrobotanici Cluj-Napoca* 41(1):12–20.

Raum, S. and C. Potter. 2015. Forestry paradigms and policy change: The evolution of forestry policy in Britain in relation to the ecosystem approach. *Land Use Policy* 49:462–470.

Rayner, J., Buck, A., and P. Katila. 2010. Embracing Complexity: Meeting the Challenges of International Forest Governance. Vienna: IUFRO, World Series Volume 28.

RECOFTC (Center for People and Forests). 2013. *Community Forestry in Asia and the Pacific: Pathway to Inclusive Development*. Bangkok: RECOFTC.

Redford, K. H. and W. M. Adams. 2009. Payment for ecosystem services and the challenge of saving nature. *Conservation Biology* 23:785–787.

Reed, J., Deakin, L., and T. Sunderland. 2014. What are "integrated landscape approaches" and how effectively have they been implemented in the tropics: A systematic map protocol. *Environmental Evidence* 4:2.

Rigueiro-Rodriguez, A., McAdam, J., and M. R. Mosquera-Losada. 2009 *Agroforestry in Europe: Current status and Future Prospects.*. Dordrecht: Springer.

Robaye, R. 1997. Du dominium ex iure Quiritium à la propriété du Code civil des français. *Revue Internationale des Droits de l'Antiquité* 44:311–332.

Roberts, P., Hunt, C., Arroyo-Kalin, M., Evans D., and N. Boivin. 2017. The deep human prehistory of global tropical forests and its relevance for modern conservation. *Nature Plants* 3:17093.

Rupf, H. 1960. Wald und Mensch im Geschehen der Gegenwart. *Allgemeine Forstzeitschrift* 38:545–552.

Sarre, A. and C. Sabogal. 2013. Is SFM an impossible dream? *Unasylva* 240(64):2634.

Sayer, J., Sunderland, T., Ghazoul, J. et al. 2013. Ten principles for a landscape approach to reconciling agriculture, conservation, and other competing land uses. *Proceedings of the National Academy of Sciences of the United States of America* 110, 21:8349–8356.

Schmithüsen, F. 1999. Perceiving forests and their management. *Annales de Géographie* 108:479–508.

Schmithüsen, F. 2013. Three hundred years of applied sustainability in forestry. *Unasylva* 240(64):3–11.

Slocombe, D. S. 1993. Implementing ecosystem-based management. *BioScience* 43:612–622.

Surel, Y. 2000. The role of cognitive and normative frames in policy-making. *Journal of European Public Policy* 7:495–512.

Tuomasjukka, T. 2010. Forest policy and economics in support of good governance. *EFI Proceedings* 58.

Ulrich, R. S. 1990. The role of trees in well-being and health. Paper presented at Fourth Urban Forestry Conference.

United Nations. 2015. World Urbanization Prospects: The 2014 Revision. New York: United Nations, Department of Economic and Social Affairs, Population Division.

von Carlowitz, H. C. 1713. *Sylvicultura Oeconomica, oder Haußwirthliche Nachricht und Naturgemäße Anweisung zur Wilden Baum-Zucht.* Reprint of 2nd edition, 2009. Remagen-Oberwinter: Verlag Kessel.

Watkins, C. 2014. Trees, Woods and Forests: A Social and Cultural History. London: Reaktion.

Waylen, K. A., Hastings, E. J., Banks, E. A., Holstead, K. L., Irvine, R. J. and K. L. Blackstock. 2014. The need to disentangle key concepts from ecosystem-approach jargon. *Conservation Biology* 28:1215–1224.

Westoby, J. 1989. *Introduction to World Forestry.* Oxford: Basil Blackwell.

Wiersum, K. F., Elands, B. H. M., and A. H. Marjanke. 2005. Small-scale forest ownership across Europe: Characteristics and future potential. *Small-Scale Forest Economics, Management and Policy* 4(1):1–19.

Wyatt, S. 2004. Co-existence of Atikamekw and industrial forestry paradigms: Occupation and management of forestlands in the St-Maurice river basin, Québec. PhD Dissertation, Quebec City: Université Laval.

Zang, D. and P. H. Pearse. 2011. *Forest Economics.* Vancouver: UBC Press.

4 Main Current Ethical Models on the Scene

Nicole Huybens

CONTENTS

4.1 INTRODUCTION

Urbanization, globalization, and tertiarization of the world deeply permeate the present time. The increase in the global human population, in particular in cities, sometimes catastrophic consequences of past choices such as fossil energy impacts on climate or declining biodiversity, and rising social, environmental, and economic awareness influence both local socioenvironmental controversies and forest-related public policies and legislation.

Today, nature appears to be more often threatened than it threatens while at the same time human intervention on nature, particularly in forests, can also positively contribute and be part of the solution.

The thought of a world where economics has become a goal, where nature is called an environment for humans, and where humans are becoming rational and opportunistic tends to replace the thinking on what is inherently good, on sacred nature, and on yesterday's and today's various cultural symbolisms. However, the value recently placed on democratizing forest-related decision-making draws into public discourse notions such as economics serving communities, intrinsic value of nature, and feelings, spirituality, and multicultural values associated with trees and forests around the world.

This chapter offers from a philosophical perspective some thoughts on the various ethical rationales underlying laws, decisions, and views of the different actors involved in the management of natural areas, including forests, with the intention of providing landmarks for forest-related decision-making in the twenty-first century.

4.2   ON ETHICS

In order to understand the importance of ethics in decision-making, and how actors are identifying a particular forest management as being "good," there is a need for clearly distinguishing between ethical, scientific, economic, technical, aesthetic, and symbolic discourses. Differences between ethics, morals, and deontology should be also discussed.

4.2.1   Distinguishing the Various Discourses

Ethical reasoning can be distinguished from scientific, economic, technical, aesthetic, and symbolic ones by its goals.

Camerini (2003) makes a clear distinction between ethics and science: the former relates to good and bad, while the latter focuses on true and false.[1] In principle, an excellent description stemming from the best science contains no moral rule. Therefore, by their very nature, scientific descriptions are and should remain amoral. Science describes while ethics prescribes. For example, describing natural processes does not include the idea that it is "good" to imitate them. Science entails decision-making based on fact, whereas ethics requires decision-making based on value. Scientific knowledge not only certainly contributes to technical progress, but it also allowed the worst human inventions. "The possibility for Humanity to exterminate itself, non-existent until recently, is one of the (rare) positive effects of the power of the means of destruction. Even the most reckless technician is forced to think. Ethics has forced entry into the sanitized field of scientific research" (Reeves 1986).

Both ethical and scientific reasoning are socially and culturally rooted, and evolve with time.[2] However, as noted by Latour (2015), scientific discourse is used by many actors, and sometimes by scientists themselves, in a prescriptive way, contributing to befog the understanding of the respective scope of both discourses. Not only is discriminating the discourses needed, but also connecting them in an explicit way in order to avoid amalgam or "clandestine" link (Morin 2004).

The economic reasoning is also socially and ideologically rooted. Economic criteria seem so prevailing in the today globalizing world that "profitable" often tends to mean "good." Moreover, ethical reasoning is arduous and does not result in a "sovereign good," which means a "good" that would be universal and worth for all periods and cultures. Thus, often, private or public actors are using or referring to economic criteria for facilitating decision-making.

The symbolic discourse provides meaning, which in itself is neither good nor profitable nor true! For example, the "nature untouched" by human presence (even in appearance) has important symbolic value for modern environmentalists, as sacred forests and trees in some African cultures. Having these forests and trees protected against any human intervention for these reasons shows the link between symbolic and ethical discourses.

The aesthetic discourse informs on what is deemed beautiful or ugly, whereas technical discourse informs on what is achievable. Finally, legal discourse allows knowing what is permitted or prohibited, underlying norms being based on social consensus which are often based on scientific data.

The good, the true, the beautiful, the profitable, the achievable, the permitted, and the sacred all play a role in humans' decision-making in relation to today's nature, even if some are not explicit, being encrypted, or mixed within the most socially valorized discourses. The hegemony of science or of economy should be questioned when establishing an intercultural dialogue about forest management, for instance. But they should not however be replaced by a hegemony of ethics or symbolism: none is sufficient on its own to address today's responsibility of humans in today's nature. So whatever emphasis is granted to ethics in this text, it is important to keep in mind that human responsibility with regard to nature cannot rely on ethical reasoning only.

4.2.2 ETHICS, MORALS, AND DEONTOLOGY

"Practical philosophy, said Kant, is not based on "what is," but rather on "what needs to be" (Russ 1995)." Therein, ethics, morals, and deontology look alike, and they rely on the essential connections that tie us to others, whoever "the other" is. "The moral act is an action of reliance: reliance with the other, reliance with a community, reliance with a society, and even reliance with the human species" (Morin 2004). The "good" is not a personal preference.

It seems to be a consensus between authors consulted on the fact that morals and ethics are "time's children," which means not universal but rather linked to or rooted in a specific period and culture. The concept of deontology is also consensual: a set of duties related to a profession, often formalized in binding code of conduct whose transgression entails formalized penalties. In Quebec, for example, there is a code of deontology for the forest engineer (Box 4.1).

Spelling out the differences between ethics and morals seems to be more difficult. Ethics (*ethos*, Greek origin) and morals (*mores*, Latin origin) have similar general definitions: conduct or ways to act, determined by usage, mores, and custom. Ethics and morals therefore consist of questions and responses on the "good" way to behave with others in a group.

But more precise definitions are diverging. Morals would prescribe "what to do or not to do" in order to lead a "good" life as described by ethics. As a reflection on the foundations of moral prescriptions, ethics would be theoretical and morals rather pragmatic, by stating directly what is "good" and what is "bad" (Russ 1995). For Beauchamp (1993), ethics has a secular, rational, and suggestive character, whereas morals would rather have a constraining and prescriptive character often linked to a religion. Morals are a search for virtue often religious, while ethics would be a search for wisdom (Russ 1995).

BOX 4.1 PROFESSIONAL FORESTERS IN QUEBEC AND THEIR CODE OF DEONTOLOGY: FEW THOUGHTS (BOUTHILLIER 2017)

In Quebec, professional foresters are framed by a code of deontology whose last reformulation dates back to 1994. The purpose of this code was first to identify measures to protect the public. Of the 60 duties or obligations prescribed by the code, 42 relate to commitments to clients, 6 to relations between colleagues, and 4 to the advertising rules to which professionals must comply. One section reminds professional foresters that their priority duty is to take into account the consequences of their interventions on the environment and on the safety, health, and property of people. The code was inspired by an anthropocentric conception of environmental ethics where humans, through a rigorously scientific approach, must guide the forces of nature for the benefit of present and future human generations. Yet the context has changed, moving toward ecocentric ethics as stated by Sandström et al. (2016) who observed some shift toward the ecosystem approach to forest management for the past 20 years.

In the United States, the code of deontology of professional foresters was revised in 2000, linking the anthropocentric approach and the dialogical paradigm: professional foresters must commit themselves to encouraging dialogue and debates between them and with the interested public in order to ensure that forest policies guiding their actions are based on science as much as on social values and allow a plurality of voices to collectively define the desired forest.

Habermas (1992) is of the view that ethics is linked to daily practice, whereas the moral rules elaborated in a discussion process are more universal. Ethics would be an inner, free, and individual requirement for the realization of values, which is the ferment and the principle of progress of morals. Ethics would rather define a "you can" or "you cannot," while morals would state a "you must" or "you do not have to."

Ballet and De Bry (2001) propose to consider that it is media reasons and not conceptual ones that push back the word *morals* after the word *ethics*. Morals would be conformist, dogmatic, and imperative. Its religious connotation also makes it reject, for the benefit of ethics that seems more neutral and more able to solve the complex problems facing contemporary human.

The events and situations of life induce major contradictions between moral principles to which it is possible to give universal credit only if reflecting them in abstraction. We therefore retain with Blondel (1999) that morals is felt as a set of constraints and obligations that keeps in the austere path of duty and virtue, while ethics in its complexity would be a free reflection to lead to an ideal.

4.2.3 ETHICAL REASONING

The ethical reasoning is sometimes linked to values (deontological ethics or ethics of conviction) and/or to awareness of the positive and negative consequences of a choice (consequentialist or utilitarian ethics). The ethics of conviction is often that of environmental movements: the duty to protect nature allows them to make decisions with which they are consistent and responsible. On the other hand, to base one's ethical judgment on the consequences of the project would lead to considering job creation, economic growth, and the minimization of negative environmental consequences as "good." Ethics practiced (Legault 2006) articulates both convictions and consequences to develop the best judgment "in the circumstances."

4.3 ON ENVIRONMENTAL ETHICS

With the failure of the frames of reference leading to the idea of a "sovereign good" (Ballet and De Bry 2001), the needs for ethics became more localized and specified. Since the 1980s, ethics developed for the biology, finance, business, media, business, politics, advertising, international trade. Nature does not escape to this trend, and today, for Western societies, there are several ways often implicit and rarely put into dialogue to consider "doing well" with the forest, mountains, seals, or biodiversity.

Before addressing the ethical reasoning of the environment, it is useful to look at the terms *human* and *nature* because all humans carry cultural representations of nature and humanity that they have an unfortunate and natural tendency to confuse with reality.

4.3.1 NATURE AND CULTURES

On the basis of various conceptions of the relationships, identified in all cultures, between "physicality" and "interiority," or simply said, between the body (the external and the visible) and the spirit (the conscience, the soul, the memory, the inner world, and the invisible), Descola (2005) has proposed four ontologies: animism, totemism, analogism, and naturalism (Table 4.1).

In animistic cultures, plants and animals are believed to have the same attributes of interiority as humans but under a different physical appearance. Therefore, there

TABLE 4.1

Environmental Ethics under the Naturalist Ontology

		On the Physicality (the Body)			
		Discontinuity	Continuity		
On the Interiority (the Spirit)	Continuity	Animism	Totemism		
	Discontinuity	Analogism	Naturalism		
			Valuing humanity	Valuing any life	Valuing natural ecosystem

Source: Descola, P., *Diversité des Natures, Diversité des Cultures*, Bayard, Paris, 2005.

is continuity in the spirit and only the body distinguishes human from nonhuman. Social categories (kinship and friendship) are used to think about the relationship with nonhumans. The spirit of plants, animals, and abiotic elements, such as rivers and the wind, reveal itself in humans' dreams. Savard (2004) uses, for example, the narratives of Native peoples of northeastern Canada to illustrate this type of representation of the world. In animistic cultures, reciprocity between human and nonhuman is a moral rule, and all existing things are considered equivalent (there is no sense to superior and inferior): when we take something, we must give something back even if it is only symbolic, in the form of a ritual for instance.

Ethics and values in these cultures are relational. "They are based on restoring harmony in and between human, non human, and spiritual world. Being articulated as behavioural expectations, customs, taboos and rites, they are often explicitly exemplified in myth, story and legend" (Studley 2007). The animistic narrative of the world does not involve a fragmentation of the different facets of humanity: it links them to the contrary in a structured whole. These ways of knowing "keep the physical, environmental and spiritual aspects of life experience in harmony through ceremony, vision quests and dreams. They also accept multiple states of consciousness, namely, dreams, trances, visions, and natural or plant-induced meditative states" (Studley 2007).

Totemism refers to the conception in which sets of humans and nonhumans share the same physical and spiritual properties. For the aborigines of Australia, for example, the name of the totemic group ("the hopping" and "the watcher") refers to a feature attributed to a species and not to the species. The totem comes from a very ancient time when the "beings of the dream" disappeared leaving seeds, the "souls-children," in specific places. By frequenting these places, all the females (human and nonhuman) are fertilized by these seeds, and they breed totemic class tots from there, which all belong to the same "specie."

Discontinuity in spirits and bodies is the peculiarity of analogism. This conception is common in China, India, the Andes, pre-Columbian Mexico, West Africa, and Europe until the seventeenth century. Humans and nonhumans are diverse, and the world is divided into a multitude of forms separated by small gaps and ordered along a continuum. In Europe, before the Renaissance, there was a "chain of beings," from the most humble to the most perfect, the human being at the top of this hierarchy. In such a hierarchical world, what is "inferior" has a lower value and the sacrifice of inferior beings is not immoral.

Finally, naturalism refers to the conception according to which there is a continuity for the bodies (evolution of life) and a discontinuity between the spirits. This is the most familiar design in Western societies today. It is through their superior mind, their consciousness, and their ability to distinguish between the "good" and the "bad" that humans are different from nonhumans. The naturalistic design goes hand in hand with Western scientific knowledge, technical invention, and industrial exploitation. "Western science [. . .] evolved in a world that placed humanity apart from and above the natural world, and in command of apparently inexhaustible resources" (Studley 2007). The inequality of spirit between humans and nonhumans removes the moral obligation of reciprocity found in animist cultures. Humans owe nothing to nature and can use it to meet all their needs without restriction.

FIGURE 4.1   Screenshot from the video game "Never Alone" designed in collaboration with Inuit to discover their culture and their relationship with the world. (Adapted from Bonvoisin, D., Les jeux vidéo sont-ils plus verts que nature? Seminar presented at Point Culture, Louvain La Neuve, 2017.)

Dominant in this or that configuration, human nature identification schemes are not exclusive of each other and can coexist to varying degrees. For example, in Western societies, analogism and naturalism coexist.

The ethics of the environment in Western societies is mainly thought from the naturalist ontology. Interesting illustration can be found in a video game analysis by Bonvoisin (2017) from the point of view of the representation of nature. He notes that the very large majority of those games are built according to the rules of the dominant naturalism ontology while a very few are inspired by different visions as, for example, by animism (Figure 4.1).

The difficulty of thinking about the actualization of local or ancestral knowledge of other cultures in forest management, probably comes from a lack of knowledge of other ways of seeing the world that appear very foreign and "outdated." Likewise, from the point of view of animist societies, technical–scientific forest "management" is also a foreign process.

4.3.2   Naturalism and Environmental Ethics

Larrère (2006, 2009) and Huybens (2010) consider that there are three categories of reasoning for environmental ethics within the naturalist conception of the world: the anthropocentric, biocentric, and ecocentric approaches (Table 4.1). The first focuses on the human, humanity, and human communities. The second is rooted on the sacred character of life, regardless of what shape it takes. The last focuses on the functioning of ecosystems in their complexity when the human does not intervene to regulate them. These ways of perceiving the "good" and the "bad" must be considered within a naturalist ontology. Indeed, for the First Nations people of Canada or for the Pygmies, for example, it is impossible to imagine such distinction.

Anthropocentric reasoning separates the human (subject) and nature (object) and puts nature at humans' service. In the capitalist economic model, humans are absolute masters over nature: they can submit it to all their needs without compensation. The notion of sustainable development mitigates this position: humans become stewards. They must be able to meet their needs in nature and hand down to future generations, ecosystems that can satisfy their needs as well. Natural resources are seen as objects to be exploited and to make profit. They have no intrinsic value. This ethics has enabled the development of an efficient agriculture, science, and technology, while it has also left significant "footprint" on the environment due to the exclusion of nature from the scope of ethical consideration.

Ethics related to nature is therefore utilitarian: the rationale is to maximize positive consequences for humans at least cost and, more recently, with minimal impact on the environment. However, the principle of responsibility with respect to future generations derived from the idea of sustainable development has made ethical judgment more complex. Threatened species need to be protected so that the humans of tomorrow can benefit as well. Anthropocentric reasoning leads to anthropizing nature: in priority, to make it fit to meet human needs.

Biocentric reasoning promotes respect for all life, whoever it is. There is an intrinsic dignity to all living beings, regardless of their usefulness to humans. This is a radical questioning of the anthropocentric reasoning. It is based on a metaphysical ecology, and ethics is deontological: it is made of moral rules and prohibitions. The violence done by humans and their techniques to nature is denounced. The restitutionary justice (to give back for what one takes) is a categorical principle, for example. Kindness is also invoked as an ethical principle.

This reasoning leads to anthropomorphizing nature: treating it as if it had human faculties as, for example, with statements such as "Mother Earth is taking revenge." It is mostly used by people and groups who act in defense of animals and their right to live according to their nature. It is more the individuals, especially the suffering ones, than the ecosystems which are the subject of ethical concerns: hens locked up in cages and animals used in laboratories and for slaughter.

The ecocentric model allows considering the human being as only one among other elements of nature, placing emphasis on the common physicality between human and nature. Humans must therefore know and respect the laws of nature (by the superior "spirit") in order to maintain it in the state in which it is put without it. Humans become "Frankensteins" if they create something other than what nature would do ("Frankenfood" for genetically modified organisms, for example). Particular prominence is given to experts who know the laws of nature (scientists); hence, it is mainly science that can help make the "good decisions": those that mimic the functioning of ecosystems or in softer versions are inspired by them or propose "close to nature management." For example, the ecosystem management of forests is based on this philosophy: to know scientifically, in order to do like nature, and avoid mistakes. Forest ecosystems and the self-regulated functioning thereof, without human intervention, are thus central. Threatened species must be protected at all costs because if they disappear, an entire ecosystem collapses. Ethics is consequentialist (avoiding negative consequences on ecosystems), and it leads to naturalizing humanity. Good practices are based on the laws of nature and the precautionary principle allows determining "what not to do" when scientific information is insufficient. Many representatives of environmental movements avail themselves of such an ethics to defend primary forests, for example, by invoking their exceptional and fragile biodiversity and that it is therefore necessary to protect them from any human intervention.

It is clear that there is no social unanimity on either of these ethical reasoning. Globally and with all the exceptions that such a generality could imply, the first reasoning is that of the economic world, the second is that of those identified as actors of deep ecology, and the last is that of ecologists or scientists who amalgamate science and ethics. The notion of sustainable development is anthropocentric: it is for future generations that we should think differently about our activities related to nature today. Even if sustainable development has been a unifying force, it cannot deliver on its promises for the proponents of a biocentric ethics or even to a lesser extent ecocentric.

These ethical reasoning fail to understand the well-founded decisions in cultures that do not share the idea that man and nature are separate entities. In the mythical stories of North American Native people, human and nature have always coexisted; there is no creative God; and it is the interaction between the human, the beaver, the tree, and the summer that creates the world (Savard 2004). They are (were) animist in the sense that Descola gives to this term. Their ethics in relation to nature is therefore neither anthropocentric, nor biocentric, nor ecocentric. Moral rules obviously exist as in any culture, but they may seem very strange to those for whom nature and human are separate entities. For example, hunter–gatherers Ilnu honored the spirits of the animals they hunted. These rituals in particular allowed knowing that things were done in a "good" way. The compensation of what is taken is necessary, so there is a duty of reciprocity, and it is important to maintain the balance between the visible and invisible worlds, between nature, spirits, and humans.

4.4 ANTHROPOCENE

4.4.1 NEW ERA

Environmental ethics in naturalist ontology is based on the idea that nature exists independently of human will: according to some, it is a creation of the divine, and others view it as a spontaneous development, but either way, humans have nothing to do with it. Not only the wind, the ocean, the stars, but also mountains and forests are not a human creation. This certainty is related with and specific to naturalist ontology.

However, current environmental issues, such as climate change, nuclear waste, and the heaps of other waste produced by humans, challenge the certainty that there is a stark distinction between human and nature. Human presence permeates nature in every corner of the planet. Should we consider that existing European forests are not "natural" because they bear the mark of secular interactions between humans and nature? To use the term by Latour (2015), those forests are the very image of a cocreation. The human is everywhere in today's nature, and nature is everywhere in the human's life. Today's climate is a cocreation and so are the deserts or the biodiversity of Alpine pastures and calcareous grasslands.

The concept of Anthropocene (Bonneuil and Fressoz 2013) provides new philosophical and ethical points of reference for understanding the place that humans occupy today in nature. This term has been readily accepted in the social sciences. Geologists have acknowledged that there seems to be enough evidence to back the concept of Anthropocene, and they are raising the question of when it began. At the 35th International Geological Congress held in 2016, a group of international experts established in 2009 tabled its conclusions regarding the possibility of designating a new period of the Earth's history. They consider that the permanent evidence of human activity is observable in geological strata throughout the world. History is therefore becoming geohistory, and we would have gone from the Holocene to the Anthropocene era. Both humans and nonhumans not only adapt to their environment but also create it. Stromatolites changed the composition of the atmosphere several billion years ago, and today, humans influence climate. By virtue of its choices and actions, humanity has become a de facto cocreator of the planet's fate. Unlike stromatolites, humanity acts today consciously or freely, which implies unprecedented responsibility. This begs the question of which ethical reasoning can justify human decisions so as to enable "good" action relation to nature in such a context or "good" participation in the geohistory of the planet (Latour 2015).

4.4.2 Partnership with Nature

The Anthropocene idea calls for the development of a renewed environmental ethics that is already germinated in social practices but remains hidden in a jumble of amalgamated scientific, economic, ethical, aesthetic, technical, and symbolic discourses. This is why, drawing on Serres (1990), Morin (2007), and Latour (2015), Huybens (2010) has put forward the proposal of a multicentric ethical reasoning. It is a metamorphosis of environmental ethics that articulates what is best in the different existing reasoning to invent a form of partnership between today's nature and today's humanity so that the potentialities of the one and the other are actualized in responsible projects for today and for tomorrow. Human responsibility in multicentric ethics therefore calls for addressing the world that will be left to the humans and nonhumans of tomorrow, with the previous generations.

Multicentric ethics is practiced (reasoning "in the circumstances"), humanistic (promotes human consciousness and kindness for humans and for nature), and based on values (freedom and responsibility for humans) and enjoins consider the most positive consequences for humans and nonhumans at the same time. This last point is important as other environmental ethical reasoning tend to privilege positive consequences for humans while minimizing negative consequences for nature (anthropocentric) and, conversely, privilege positive consequences for nature while minimizing negative consequences for humans (ecocentric). Multicentric ethics also contains the idea of ethical dialogue (developed below) as a moral prescription for allowing humans with different conceptions and assumptions to make the "good" decisions in given circumstances. Multicentric ethics also takes up the idea of restitutionary justice.

The ethics of partnership with nature allows choosing actions that have positive consequences for both humans and nature. The ethical injunction could be stated as follows: "we have to maximize our positive impact on nature for all species of the planet, including our own". For example, in terms of actions to be taken to protect the woodland caribou in Quebec, one might ask how forest management can help the species survive and thrive, and not just how to prevent the negative impacts of logging on this species, by creating protected areas of any human intervention where the naturalness of nature could express itself for the better" (Huybens and Lord 2016). The desired future should guide decision-making: "It is more important that a correct understanding of the future guides our polices, practices, customs and habits than that we have the calculus to add up the real consequences of innumerable future generations on many different value scales" (Macqueen 2005). In the field of forestry, the question "what desired forest are humans able to co-construct with nature today and for what desired world?" (Huybens and Henry 2013) is not the same as "how to leave nature as much as possible in the state in which it puts itself without us?" (ecocentrism) or "how to exploit natural resources to the best of our interests by minimizing the negative consequences for nature?" (anthropocentrism).

Nature as partner is obviously wordless in Chinese, English, or any other language. Scientific knowledge allows decoding its "language." Moreover, Ki-Zerbo and Beaud-Gambier (1992) warn against the pitfall of turning science and myths into antagonistic realities. Throughout time and space, human myths and beliefs have also revealed partnerships with nature, some of which could be of symbolic order. For example, the omnipresent ties between humans, the visible world, and the spiritual world in animistic cultures can be a source of inspiration for developing ethics of partnership: "(. . .) we ought to extend moral consideration to trees. Indeed, we ought to extend it to all the species within forests. In part, we should extend this moral consideration because forest ecosystems contribute to satisfying human aspirations. In part, it is because trees are intrinsically valuable—however slight that value might be in comparison with humans and their aspiration. In part, it is because reverential stewardship is a necessary attitude to the survival of our living—and many argue, spiritual—ecosystem" (Macqueen 2005).

For example, this is the case in development cooperation projects, where decisions cannot only address the increase of welfare and the eradication of poverty (a desirable world based on a particular idea of development associated with material well-being only). It is necessary to introduce not only the nature and its intrinsic value but also a real multicultural dialogue that allows honoring the assumptions of local populations in relation to the forest or the trees even if they are symbolic, imaginary, or spiritual. And it should be also the case when addressing the needs of the citizens of today's urban societies. "There are currently inquantifiable values placed on forests linked to broader human aspiration beyond material wealth. We need a forest ethics that reflects this reality rather than pretending it does not exist" (Macqueen 2004).

Multicentric ethics leads to humanizing humanity in its relationship with nature and as a precondition, in promoting dialogue between humans. Coconstructed decisions are complex, and no single culture or group has such in-depth knowledge required to offer the only good answer. The dialogue allows including the wisdoms of diverse cultures, scientific, traditional, and technical knowledge as well as the views of representatives of groups of interest.

4.5   ETHICAL DIALOGUE: DEVELOPING COMMUNICATION SKILLS

To better grasp the meaning of ethical dialogue, we will distinguish on the one hand the "dialogical dialogue," term borrowed from Raine (2005) in order to specifically explain the meaning of "coming to an agreement" and the concept of "ethics of dialogue" (Segers 2014) in order to frame the implementation of a process of coconstruction on the other hand.

4.5.1   DIALOGICAL DIALOGUE: "COMING" TO AN AGREEMENT

Coming to an agreement is different from agreeing. Dialogical dialogue presupposes disagreements, which are not due to partners lacking information. Incomprehension and misunderstanding in the dialogical paradigm are not exceptions but rather the starting point of any exchange of views (Patenaude 1997). Agreements and trust between interlocutors are seldom prerequisites: they will be the result of the dialogue.

The links of humans to the forest are not only economic, and denying the other components of these links does not make them disappear: they express themselves in particularly amalgamated discourses that make dialogue difficult. Dialogical dialogue allows honoring these divergent views (Stengers 2009) and certainly not to discriminate between who is right and who is wrong or between the winner and the loser.

Dialogic dialogue is a shared thought, not an increasingly sophisticated argumentation in the course of discussions between actors with diverging interests. Competition is a brake. Coconstruction between actors considered equal and legitimate allows the development of solutions that are not necessarily those that the actors carry at the beginning of the process. Dialogic dialogue does not consist, or at least not only, of "improving" a project carried by an actor who seeks to rally others to his/her point of view: "Many forest development professionals have co-opted the terms bottom-up development, participation and decentralization. But while these may appear to be endogenous approaches, in reality there is only recognition that local values are a desirable ingredient in the development process" (Studley 2007).

Entering into a dialogue does not entail reducing all values to the economic component: "Trying to measure each of these value categories (security, social affiliation, creativity, aesthetics and identity) on a financial scale would be like trying to measure income on a scale of aesthetic beauty or profit on a scale of friendship. We can do it, but the results are without meaning" (Macqueen 2005). Collective aspirations are not always and not only related to economic growth as a means or an end to development: "The priority is not therefore to engage in dialogue, but rather to engage in ethical dialogue where the establishment of a collective or global forest ethic is the end in view rather than the immediate resolution of conflict between individual (or sovereign state) aspirations" (Macqueen 2004).[3] Thus, nature and the nonhuman's intrinsic value and humans' spiritual and emotional values are as important as the commercial value of fiber. Beliefs, assumptions, myths, and emotions are therefore not belittled; on the contrary, they become explicit and integrated into the issue as elements and, hence, as sources of solutions.

> Forestry (. . .) appears now to offer an approach that both enhances the forests and the well-being of the people who depend upon them. This approach provides a platform to address multiple-aim forest management, sustainability, stakeholder needs, plural behaviour patterns, local and indigenous forest values, knowledge equity, and synergy between formal and customary forestry knowledge systems. (Studley 2007)

Engaging in dialogue promotes the expansion of knowledge of all parties with a view of coconstructing an issue that has become common[4] (Callon 2001) and solutions to bring together (Kaner 2007). Dialogical dialogue allows acknowledging diversity with curiosity as a way of developing a common creativity and a "shared framework of understanding" (Kaner 2007), all the more realistic as it is well rooted in the deep certainties of each. Obviousy, seemingly insurmountable differences of opinion due to the "Groan zone"[5] (Kaner 2007) are uncomfortable, frustrating, and even exasperating, but they are not a good reason to give up. "Since the greatest complexity contains within itself the greatest diversity, independence, freedom, and the greatest danger of fragmentation, solidarity, friendship, and love are vital bonds to human complexity" (Morin 2004). The link to find and maintain in the zone of turbulence of dialogical dialogue processes is made by love in the broadest sense, manifesting itself in the form of empathy, patience, kindness, and willingness to listen, and to articulate together the information that is presented as conflicting. Divergences become a force for the creativity of complex and coconstructed solutions.

4.5.2   ETHICS OF DIALOGUE

While dialogical dialogue includes free debate and some improvisation (Saint Arnaud 2009), it is desirable to frame it within the ethics of dialogue. Thus, some operating rules of the groups are to be explained and recalled. Procedures and argumentation are important, but not exclusively and not all the time. Rational conduct is expected, but not only it. Respect for others is essential. It is also important to guarantee the equality and freedom of members and to promote the expression of the assumptions held by other members of the group as much as one's own (Raine 2005).

The accompanist of dialogical dialogue processes has the responsibility of promoting and even instructing conducts that will enable listening, openness, and appreciation for otherness (Segers 2014). According to Raine (2005), this is a mediator and sometimes a member of the group whose legitimacy is accepted by others on the basis of acquired credibility in the course of the dialogue process. Studley (2007) lists the terms *go-between, intercessor, intermediary, intermediate, intermediator, mediator, middleman,* and above all *broker* to designate this person. "A broker is someone who acts as an intermediate agent in a transaction or helps to resolve differences" (Studley 2007). To become a mediator in a dialogical dialogue, the actor has acquired intercultural communication skills, he/she is thus more aware of his/her own paradigms and is able to explain without judging the different foundations of the various ethical, symbolic, cultural, or scientific assumptions held by actors. "In order to become knowledge-brokers, forest development professions need to learn about the cultures and paradigms of the knowledge holders" (Studley 2007).

Dialogue cannot be established on the basis of adding more rules or of the hope of accrued financial profit, but around an ideal that becomes common and shared values stemming from the dialogue. Mediator's skills in the context of a dialogical dialogue include strong awareness of the importance of all representations of the world in order to make forest-related decisions today: "For the forest development professional, it is therefore essential to know how to bridge between these different types of knowledge, ways of knowing and perceptions" (Studley 2007). Often, Western decision makers in particular are not used to making decisions on the basis of explicit symbolic grounds. Explaining the assumptions of all involved parties is therefore essential to dialogical dialogue, and this requires skills that must be learnt in theory and in practice concurrently.

4.6 CONCLUSION

There are a multitude of wisdom traditions that would contribute to enrich the ethics of the environment for today's human in today's nature. Let Descola (2010) conclude this chapter. ". . . Even if the solution we would like for the future, a different solution to live together both between humans but also between humans and non-humans, does not exist yet, we have at least the hope, since others have done before us in other civilizations and other societies, to be able to invent original ways of inhabiting the earth. Since all these solutions have been imagined by humans, it is not forbidden to think that we could imagine new ways of living together, and perhaps even better."

REFERENCES

Ballet, J. and F. De Bry. 2001. *L'entreprise et L'éthique*. Paris: Seuil and Economie.
Beauchamp, A. 1993. *Introduction à L'éthique de L'environnement*. Montréal: Éditions Paulines, Médiaspaul.
Blondel, E. 1999. *La Morale*. Corpus. Paris: Flammarion.
Bonneuil, C. and J.-B. Fressoz. 2013. *L'Événement Anthropocène*. Paris: Seuil.
Bonvoisin, D. 2017. Les jeux vidéo sont-ils plus verts que nature? Seminar presented at Point Culture, Louvain La Neuve, November 22, 2017.
Callon, M., Lascoumes, P., and Y. Barthe. 2001. *Agir Dans un Monde Incertain: Essai sur la Démocratie Technique*. Paris: Seuil.
Camerini, C. 2003. *Les Fondements Epistémologiques du Développement Durable: Entre Physique, Philosophie et Ethique*. Paris: L'Harmattan.
Descola, P. 2005. *Par-delà Nature et Culture*. Paris: Gallimard.
Descola, P. 2010. *Diversité des Natures, Diversité des Cultures*. Paris: Bayard.
Habermas, J. 1992. *De L'éthique de la Discussion*. Paris: Editions du Cerf.
Huybens, N. 2010. *La Forêt Boréale, L'éco-conseil et la Pensée Complexe: Comprendre les Humains et Leurs Natures Pour Agir Dans la Complexité*. Sarrebruck: Éditions Universitaires Européennes.

Huybens, N. and P. Henry. 2013. La Forêt Souhaitée: Une Réponse à la Non-acceptabilité Sociale des Activités de Coupe en Forêt Boréale? Réflexions Préliminaires. Saguenay: Université du Québec à Chicoutimi http://constellation.uqac.ca/2624/ (Accessed September 10, 2017).

Huybens, N. and D. Lord. 2016. La forêt souhaitée pour penser dans la complexité la protection du caribou forestier au Québec. In *Forêts, Savoirs et Motivations*, eds. Farcy, C. and N. Huybens, 329–346. Paris: L'Harmattan.

Kaner, S. 2007. *Facilitator's Guide to Participatory Decision-Making*. New York: Jossey-Bass.

Ki-Zerbo, J. and M. J. Beaud-Gambier. 1992. *Compagnons du Soleil: Anthologie de grands Textes de L'humanité sur les Rapports Entre L'homme et la Nature*. Paris: La Découverte, UNESCO.

Larrère, C. 2006. Ethiques de l'environnement. *Multitudes* 24:75–84.

Larrère, C. 2009. La justice environnementale. *Multitudes* 36:156–162.

Latour, B. 2015. *Face à Gaïa: Huit Conférences sur le Nouveau Régime Climatique*. Paris: La Découverte.

Legault, G. A. 2006. P*rofessionnalisme et Délibération Ethique: Manuel D'aide à la Décision Responsable*. Québec: Presses de l'Université du Québec.

Macqueen, D. 2004. *Forest Ethics: The Role of Ethical Dialogue in the Fate of the Forests*. London: International Institute for Environment and Development (IIED).

Macqueen, D. 2005. *Time and Temperance: How Perceptions about Time Shape Forest Ethics and Practice*. London: IIED.

Morin, E. 2004. *La méthode VI. Éthique*. Paris: Seuil.

Morin, E. 2007. *Vers L'abîme?* Paris: Edition de L'Herne.

Patenaude, J. 1997. Le dialogue comme compétence éthique. PhD Dissertation, Quebec City: Université de Laval.

Raine, P. 2005. *Le Chaman et L'écologiste*. Paris: L'Harmattan.

Reeves, H. 1986. *L'heure de S'enivrer: L'univers a-t-il un Sens?* Paris: Seuil.

Russ, J. 1995. *La Pensée Ethique Contemporaine*. Paris: Presses universitaires de France.

Saint Arnaud, Y. 2009. *L'autorégulation Pour un Dialogue Efficace*. Montréal: Presses universitaires de Montréal.

Savard, R. 2004. La Forêt Vive: Récits Fondateurs du Peuple Innu. Montréal: Boréal.

Sandström, C., Carlsson-Kanyama, A., Beland Lindahl, K. et al. 2016. Understanding consistencies and gaps between desired forest futures: An analysis of visions from stakeholder groups in Sweden. *Ambio* 45:100–108.

Segers, I. 2014. Éthique, dialogue et développement durable pour la pratique de l'éco-conseil. Master Thesis, Saguenay: Université du Québec à Chicoutimi.

Serres, M. 1990. *Le Contrat Naturel*. Paris: François Bourin.

Stengers, I. 2009. *Au Temps des Catastrophes: Résister à la Barbarie Qui Vient*. Paris: La Découverte.

Studley, J. 2007. *Hearing a Different Drummer: A New Paradigm for the "Keepers of the Forest"*. London: IIED.

ENDNOTES

1. Of course, "there is no uncompromising good (the angelical) and bad (the diabolical)" (Blondel 1999). The true and the false are not either absolute.
2. Science tells a possible and human "true," and it is not universal but child of time. Each science reveals the "true" of its discipline but the reality is not disciplinary. "The true significance of the epistemological revolution stemming from relativity and quantum mechanics is to no longer consider scientific rationality as synonymous with certainty

and therefore to deny the possibility of comprehensive, objective, and definitive knowledge of the universe" (Camerini 2003). Equally, "it would be better to have imperfect predictions than no predictions at all provided that the epistemic status of scientific prediction is clear" (Camerini 2003).

3. In this sense, ethical dialogue is equivalent to dialogical dialogue.

4. A problem becomes common when its different facets are integrated in a definition coconstructed by disagreeing parties: the issue of woodland caribou protection in Quebec includes climatic changes, the fragmentation of the primary forests, the falling paper price, lack of scientific knowledge about its ecology, and decline in biodiversity worldwide. The issue is not limited to either of these elements and certainly not to economic or environmental aspects associated with the decline of caribou.

5. When the dialogue becomes mostly a debate/fight and is obscured by the arguments of each other.

Section II

Urbanization of the Society

5 Main Findings and Trends of Urbanization

*Tahia Devisscher, Lorien Nesbitt,
and Adrina C. Bardekjian*

CONTENTS

5.1 MAIN TRENDS

Urbanization is considered here as the increase in the proportion of population living in cities and settlements versus the population in rural areas (UN 2014). This involves a process in which rural, dispersed, and primarily agriculturally focused populations become concentrated in larger, more dense urban settlements mainly focused on industrial and service economies (National Research Council 2003).

Rapid global urbanization is a relatively new phenomenon. As recently as 1950, only 30% of the world's population was urbanized (Bettencourt and West 2010). In 2007, the global urban population exceeded the rural population for the first time (Figure 5.1; UN 2014). By 2030, 60% will live in urban areas and this proportion is forecasted to reach 70% by 2050, with the majority of growth expected in Asia and Africa (UN 2014, 2015). By 2050, 64% of Asia's population and 56% of Africa's are forecasted to be urban (Revi et al. 2014; UN 2014).

At present, cities are home to 55% of the global population. About half of the urban population is living in small settlements (less than 500,000 inhabitants), while one out of seven people is installed in a city with 1 million inhabitants or more (UN 2014).

The world's most urbanized regions include North America (82% urban), Latin America and the Caribbean (80%), and Europe (73%) (UN 2014). Asia (48%) and Africa (40%) are still mostly rural, but this is expected to rapidly change. In Africa, countries in the sub-Saharan region show the lowest proportion of urban population (32.8%) compared to North African countries (47.8%), which have more progressive urban development strategies (AfDB 2012).

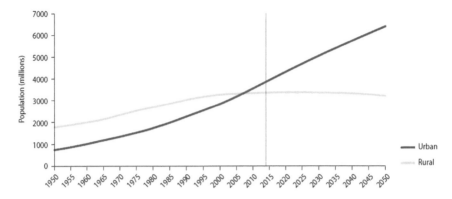

FIGURE 5.1 World population living in urban and rural areas, 1950–2050. (From UN, *World Urbanization Prospects: The 2014 Revision*, Department of Economic and Social, Population Division, UN, New York, 2014. Copyright © 2014 UN.)

In Asia, megacities of more than 10 million people continue to rise, such as China's Beijing (17 million) and Shanghai (24 million); Japan's Osaka–Kobe (12 million) and Tokyo (32 million); and the cities of Jakarta (23 million), Manila (16.5 million), and Seoul (16 million). However, most of Asia's urban areas are still composed of settlements with populations of 100,000–500,000 people (Fensom 2015).

In the second half of the twentieth century, upper-middle income countries[1] have undergone the most rapid urbanization process (UN 2014). This is particularly the case in Asia and Latin America. While the pace of urbanization in lower-middle income countries has been slower, these countries now host three-quarters of the world's urban population and many of the world's largest cities, such as Karachi, Dhaka, Lagos, Delhi, Cairo, and Kinshasa (Revi et al. 2014). In fact, as urbanization processes reach saturation in high- and upper-middle income countries, the fastest-growing urban areas are in medium-sized cities of Asia and Africa. As the world continues to urbanize, the greatest sustainable development challenges will thus likely be in cities located in lower-middle income countries (UN-Habitat 2003; Sheuya 2008; UN 2014).

Globally, 60% of the area expected to be urban by 2030 has yet to be built, creating new challenges and opportunities for the future (Clos 2016).

5.2 URBANIZATION DRIVERS

The emergence of cities begins with the transition from "hunter–gatherer" societies to agricultural food production and the resulting establishment of villages (Mumford 1956), with the caveat that villages were also established by indigenous cultures in coastal Ecuador and the Pacific Northwest of North America in the absence of agricultural food production (Diamond 1999). The evolution of cities from villages was supported by improvements in plant cultivation and livestock rearing in the Neolithic age, which allowed for higher-volume production and food storage and a departure from subsistence agriculture (Mumford 1956). This fundamental change led to an increase in surplus human labor that would be employed in nonagricultural work in cities.

Since then, urbanization has been mainly driven by the requirements of economic development. For example, in Europe and North America, rapid urbanization in the late nineteenth and twentieth centuries was closely linked to the industrial revolution (UN 2014). Similar relationships with economic development have accompanied the urbanization processes that occurred in the late twentieth and early twenty-first centuries in some parts of Latin America, the Caribbean, and Eastern Asia (UN 2014; Figure 5.2).

Rural populations tend to be drawn to cities for education and employment opportunities in the industrial and service sectors (National Research Council 2003). In turn, this concentration of labor and education reinforces the trend by providing the necessary environment for businesses to grow. Population and business density also facilitate knowledge and information sharing, which supports innovation and thus further stimulates urban growth (UN 2014). In this century, an ever-growing service-oriented economy has been the main driver of urbanization in many regions. Urban growth may also be associated with improvements to public health in urban areas relative to rural areas, leading to declining urban death rates and thus urban population growth that is relatively faster than rural growth (de Vries 1990).

Urbanization is also often driven by crises and forced migration. This relates to situations of widespread threat to life, physical safety, health, or basic subsistence beyond the coping capacity of individuals and the communities in which they reside (Martin et al. 2013). Crises and associated migration to cities may result from political conflict and acute events such as natural and human-made hazards—earthquakes, severe droughts, wildfires, and nuclear accidents—or from gradual processes that can slowly cause environmental damage, political instability, malnutrition, and poverty (Dickson et al. 2012; Martin et al. 2013; ARUP 2014). These extreme events and more chronic processes frequently intersect, affecting people that are forced to migrate. Crisis migration is complex and diverse, and it can be permanent or temporary. Some more recent rural–urban migrations are found to

FIGURE 5.2   Aerial view of Seoul (South Korea). (Courtesy of Cecil Konijnendijk.)

be circulatory, with migrants fluctuating between the rural origin and the urban destination (WC 2011).

5.3 URBANIZATION IMPACTS

Rapid urbanization is exerting pressure on fresh water resources, the living environment, and public health. Accompanied by increased competition for land, unequal distribution of wealth, and often unplanned expansion of urban infrastructure and services, urbanization is also exposing more citizens and local economies to increasing hazards, both natural and human-made (Dickson et al. 2012; UN 2015).

Cities in low-income countries face increased risk because of rapid and unplanned growth and chronic problems with urban management. While all continents report more disasters, the rate of disaster increase from 1975 to 2006 has been the greatest in Africa and Asia, continents which are experiencing the highest rates of urban growth worldwide (UN-Habitat 2007). Examples of Asian and African countries where human settlements were recently affected by large natural and human-made disasters include Indonesia (2004 tsunami), Pakistan (2005 earthquake), Lagos (2006 oil pipeline explosion), and Japan (2011 triple disaster). According to the United Nations Office for Disaster Risk Reduction (UNISDR) (2015), annual losses due to disasters in low-income countries are five times higher than in high-income countries when expressed as a proportion of social expenditure.

In many low- and middle-income countries, a key driver of urban risk is highly unequal access to urban space, infrastructure, services, and security (UNISDR 2015). Inequality in many cities has continuously risen since the 1980s (UN-Habitat 2014). As a result, communities constrained by lower access to resources show higher levels of vulnerability within their constituencies. For example, poor communities living in periurban areas or informal settlements are more vulnerable because they reside in high-risk areas and unstable shelters, with limited access to basic safety nets and emergency services that can help them cope with chronic stresses or prepare for sudden shocks (Dickson et al. 2012).

In the future, urban risk is expected to increase given current urbanization trends and climate change interactions. Through changes in temperature, precipitation patterns, and sea level, climate change is expected to influence hazard levels and exacerbate disaster risks. For example, the UNISDR (2015) estimated that by mid-century, 40% of the global population would be living in river basins with severe water stress, particularly in Africa and Asia. Global climate projections also point toward an increase in the number, intensity, and frequency of heat waves affecting particularly European and North American cities in the second half of the twenty-first century (Meehl and Tebaldi 2004). In tandem with large-scale events, cities are likely going to face more frequent small-scale events, such as localized floods and severe droughts, although these are seldom systematically recorded and often ignored at the global level (UN-Habitat 2007).

As cities grow, ecosystem degradation also drives urban risk. Cities affect the ecology of the local landscapes through, for example, fragmentation of habitats, isolation of faunal and floral populations, or loss of natural areas for the benefit of infrastructure development (Nowak and Walton 2005; Pauchard et al. 2006;

Devictor et al. 2007; Shen et al. 2008). Moreover, cities can have disproportionate environmental impacts at regional and global scales well beyond their borders (Seto et al. 2012). For instance, there is mounting evidence that cities can influence regional weather patterns through complex interactions between land use, weather, and climate feedback (Grimmond 2011). Further, most of the ecosystem services consumed in cities are in fact generated by ecosystems outside of the cities themselves, including ecosystems half a world away (Rees and Wackernagel 1996). According to the United Nations (UN) (2015), cities occupy just 3% the Earth's land but consume 60–80% of the global energy and contribute to 75% of carbon emissions. This highlights the importance of considering cities in connection to at least their surrounding landscape to better understand and manage the urban–rural (tele)connections relevant to resilience building (Grimm et al. 2008).

With the current urbanization trend, cities are gaining political attention over rural affairs and institutions, particularly in highly urbanized countries. In postindustrial societies, urbanization has gradually induced the abandonment of rural areas, particularly among the younger generations, causing the erosion of productive lands. The aging populations remaining in rural areas have in some cases lost the capacity to generate value in extensive grazing and family subsistence farming, switching to cash crops when possible (Chapter 8). The rural exodus has led to the intensification of most productive lands and the abandonment of marginal lands (Barbero-Sierra et al. 2013) and loss of traditional cultural landscapes in many cases (Van Eetvelde and Antrop 2004).

While it is too early to state that ongoing urbanization is leading to a global social transformation from rural to urban culture (Scott et al. 2007), it is fair to say that contemporary rural and urban landscapes are rapidly changing, often resulting in citizens increasingly losing their direct connection with nature. This alienation from nature may influence citizens' perceptions of wildlife and nature in general (Chapter 7). Studies in Europe, for example, suggest that forest issues are not well understood outside the small forestry community and that there is a significant gap between the public understanding of forests and on-the-ground realities (European Commission 2009).

5.4 URBAN PLANNING

While cities may be drivers of potential disasters, they can also provide great solutions to humanity's problems (Bettencourt and West 2010). In general, cities can bring major advancements in terms of innovation, trade, culture, growth, or social development, but the challenge to achieve this potential is that they need to be built, developed, and maintained sustainably (Bettencourt and West 2010; UN 2015). This challenge has been recognized by the UN and included in the Sustainable Development Agenda under Goal 11 aimed to "make cities inclusive, safe, resilient and sustainable" (UN 2015).

Creating healthy, sustainable cities requires balancing urbanization pressures with institutional capacity in models that reflect and are relevant to local realities (UN-Habitat 2003; Sheuya 2008). Urbanization processes vary around the world, and thus, sustainable urban development outcomes take different forms in different places. However, regardless of location, sustainable urbanization demands investment in basic infrastructure and services, and inclusive planning models to ensure the equitable distribution of infrastructure and services as more people make cities their home (UN 2014).

When cities grow, their form and the well-being of their residents are shaped by the interactions among governments, businesses, the public, and other institutions. The process of city building calls for mechanisms that can balance the objectives and actions of both private interests and public responsibilities: private interests in the form of private citizens and businesses and public responsibilities in the form of various levels of government and associated agencies. With planned growth and high institutional capacity, governments have the opportunity to fulfill their public responsibilities while supporting private interests through the supply of municipal infrastructure, such as roads, clean water, and electricity, and services such as education and healthcare, thus guiding and shaping the development of cities (UN 2014).

While planned urban growth has many benefits, rapid and unplanned urban growth is still a common challenge around the world (UN 2014). Unplanned growth threatens sustainable urban development when necessary infrastructure is not developed or services are not equitably distributed among citizens (White 2013; UN 2014). Common effects of unplanned growth include congestion, lack of basic services, shortage of adequate housing, and declining infrastructure (UN 2015). In many instances, unplanned urban growth takes the form of informal land invasions and slum development, especially in periurban areas (White 2013). Land invasions are a conglomeration of informal settlements (i.e., outside of the formal housing market), which are created by illegally occupying land to live in as a primary residence or to subsequently sell on the informal land market. This process particularly occurs when local populations are financially unable to participate in the formal housing market (UN-Habitat 2003). Some negative aspects of informal or unplanned development include inadequate access to safe water, sanitation, electricity, and other basic infrastructure. This often occurs in the context of high income inequality and when urbanization outpaces the capacity of local institutions to guide growth and provide necessary infrastructure and services (Angel 2008; Sheuya 2008; Moser 2009).

An additional effect of unplanned growth is the increased vulnerability of cities to the effects of climate change and other risks. In recent years, cities are recognizing the urgency of changing approaches to deal with risk and have made considerable progress in strengthening early warning systems and mainstreaming disaster and climate change considerations into the planning and management of urban services (UNISDR 2015). Managing risk is slowly being integrated in the early stages of development planning, moving away from add-on disaster risk management. Most recently, the concept of resilience has been adopted in the urban planning sector as a way to bridge the gap between disaster risk reduction and climate change adaptation in and around cities (ARUP 2014).

As cities prepare to face the challenges of the twenty-first century, the role of urban forests is becoming increasingly recognized and urban forestry is gaining momentum, as denoted by the "Forest and Cities" theme of the International Day of Forests in 2018. Urban forests offer a range of environmental, physical, psychological, social, and economic benefits to urban residents (Ulrich et al. 1991; Kuchelmeister 2000; Thompson 2002; Konijnendijk et al. 2005, 2013; Heidt and Neef 2008; Poudyal et al. 2009; Kenney et al. 2011), which in turn help enhance cities' capacity to be resilient to multiple disaster risks (Nowak and Crane 2002; Konijnendijk et al. 2008; Paoletti 2009; Gómez-Baggethun and Barton 2013; Aronson et al. 2015; Avolio et al. 2016; Kardan et al. 2016; Ferrini et al. 2017). Ecosystem services, such as urban temperature regulation, water supply, runoff reduction, and food production, can be critical for the resilience of cities in the face of multiple disturbances, helping cities adapt and thrive in changing conditions (Gómez-Baggethun et al. 2013). Given the many benefits provided by urban forests, greening cities is gaining momentum among urban planners in both developed and developing countries, and studies on urban greenspace are on the rise (Gairola and Noresha 2010). However, urban forests are also undergoing increasing pressures due to new expectations, behaviors, and varied development demands from society (Chapter 6).

5.5 CONCLUSIONS

This chapter provided an overview of current urbanization trends and the main challenges and opportunities cities are facing in the twenty-first century. While urbanization trends are not the same in the different regions of the world, cities are generally facing common challenges linked to inefficient planning and management of resources, increasing diversity and competition of developmental interests and expectations, and potential climate-related risks. Undoubtedly, forests will have an important role to play in facing these global urban challenges. While forests in and around cities are undergoing increasing pressures due to varied developmental demands from society, they are also opening doors to new approaches encouraging individual and collective well-being, innovation, and new methods of urban design and planning with people–nature interactions in mind.

REFERENCES

AfDB. 2012. *Urbanization in Africa*. Abidjan: African Development Bank. https://www.afdb .org/en/blogs/afdb-championing-inclusive-growth-across-africa/post/urbanization-in -africa-10143/ (Accessed July 2017).

Angel, S. 2008. An arterial grid of dirt roads. *Cities* 25:146–162.

Aronson, M. F., Handel, S. N., La Puma, I. P., and S. E. Clemants. 2015. Urbanization promotes non-native woody species and diverse plant assemblages in the New York metropolitan region. *Urban Ecosystems* 18:31–45.

ARUP. 2014. *City Resilience Framework*. The Rockefeller Foundation and ARUP. https:// assets.rockefellerfoundation.org/app/uploads/20140410162455/City-Resilience -Framework-2015.pdf (Accessed March 2017).

Avolio, M. L., Pataki, D. E., Gillespie, T. W. et al. 2016. Tree diversity in Southern California's urban forest: The interacting roles of social and environmental variables. In *Urban Forests: Ecosystem Services and Management*, ed. Blum, J., 233–264. Oakville: Apple Academic Press Inc.

Barbero-Sierra, C., Marques, M. J. and M. Ruíz-Pérez. 2013. The case of urban sprawl in Spain as an active and irreversible driving force for desertification. *Journal of Arid Environments* 90:95–102.

Bettencourt, L. and G. West. 2010. A unified theory of urban living. *Nature* 467(7318):912–913.

Clos, J. 2016. The opportunity to build tomorrow's cities. https://www.weforum.org/agenda /2016/04/urbanization-dr-joan-clos/ (Accessed October 2017).

de Vries, J. 1990. Problems in the measurement, description, and analysis of historical urbanization. In *Urbanization in History*, eds. van der Woude, A., Hayami, A., and J. de Vries, 43–60. Oxford, UK: Clarendon Press.

Devictor, V., Julliard, R., Couvet, D., Lee, A., and F. Jiguet. 2007. Functional homogenization effect of urbanization on bird communities. *Conservation Biology* 21:741–751.

Diamond, J. 1999. *Guns, Germs, and Steel*. New York: W. W. Norton.

Dickson, E., Baker, J. L., Hoornweg, D., and A. Tiwari. 2012. *Urban Risk Assessments: Understanding Disaster and Climate Risk in Cities*. Urban Development Series. Washington, DC: World Bank.

European Commission. 2009. *Shaping Forest Communication in the European Union: Public Perceptions of Forests and Forestry*. Brussels: DG Agriculture and Rural Development.

Fensom, A. 2015. *Asia's Urbanization Just Beginning*. The Diplomat. http://thediplomat.com /2015/01/asias-urbanization-just-beginning/ (Accessed July 2017).

Ferrini, F., Konijnendijk van den Bosch, C., and A. Fini. 2017. Routledge Handbook of Urban Forestry. Abingdon: Taylor & Francis.

Gairola, S. and M. S. Noresah. 2010. Emerging trend of urban green space research and the implications for safeguarding biodiversity: A viewpoint. *Nature and Science* 8(7):43–49.

Grimm, N. B., Faeth, S. H., Golubiewski, N. E. et al. 2008. Global change and the ecology of cities. *Science* 319(5864):756–760.

Grimmond, C. S. B. 2011. Climate of cities. In *Routledge Handbook of Urban Ecology*, eds. Douglas, I., Goode, D., Houck, M., and R. Wang. Abingdon: Taylor & Francis.

Gómez-Baggethun, E., Gren, A., Barton, D. N. et al. 2013. Urban ecosystem services. In *Global Urbanization, Biodiversity and Ecosystem Services: Challenges and Opportunities*, eds. Elmqvist, Th., Fragkias, M., Goodness, J. et al., 175–251. Dordrecht: Springer.

Gómez-Baggethun, E. and D. N. Barton. 2013. Classifying and valuing ecosystem services for urban planning. *Ecological Economics* 86:235–245.

Heidt, V. and M. Neef. 2008. Benefits of urban green space for improving urban climate. In *Ecology, Planning, and Management of Urban Forests*, eds. Carreiro, M., Song, Y. and J. Wu, 84–96. New York: Springer.

Kardan, O., Gozdyra, P., Misic, B. et al. 2016. Neighborhood greenspace and health in a large urban center. In *Urban forests: Ecosystem services and Management*, ed. Blum, J., 59–90. Oakville: Apple Academic Press.

Kenney, W. A., Van Wassenaer, P. J. E., and A. L. Satel. 2011. Criteria and indicators for strategic urban forest planning and management. *Arboriculture and Urban Forestry* 37(3):108–117.

Konijnendijk, C. C., Nilsson, K., Randrup, T. B., and J. Schipperijn. 2005. *Urban Forests and Trees*. Berlin: Springer-Verlag.

Konijnendijk, C. C. 2008. *The Forest and the City: The Cultural Landscape of Urban Woodland*. Dordrecht: Springer.

Konijnendijk, C. C., Annerstedt, M., Nielsen, A. B., and S. Maruthaveeran. 2013. *Benefits of Urban Parks—A Systematic Review*. Copenhagen: International Federation of Parks and Recreation Administration.

Kuchelmeister, G. 2000. Trees for the urban millennium: Urban forestry update. *Unasylva* 51:49–55.

Martin, S., Weerasinghe, S., and A. Taylor. 2013. Crisis migration. *Brown Journal of World Affairs* 20:123.

Meehl, G. A. and C. Tebaldi. 2004. More intense, more frequent, and longer lasting heat waves in the 21st century. *Science* 305(5686):994–997.

Moser, C. 2009. *Ordinary Families, Extraordinary Lives: Assets and Poverty Reduction in Guayaquil, 1978-2004*. Washington, DC: Brookings Press.

Mumford, L. 1956. The natural history of urbanization. In *Man's Role in Changing the Face of the Earth*, ed. Thomas W. L., 382–398. Chicago, IL: University of Chicago Press.

National Research Council. 2003. Cities Transformed: Demographic Change and Its Implications in the Developing World. Washington, DC: National Academies Press.

Nowak, D. J. and D. E. Crane. 2002. Carbon storage and sequestration by urban trees in the USA. *Environmental Pollution* 116:381–389.

Nowak, D. J. and J. T. Walton. 2005. Projected urban growth (2000–2050) and its estimated impact on the US forest resource. *Journal of Forestry* 103(8):383–389.

Paoletti, E. 2009. Ozone and urban forests in Italy. *Environmental Pollution* 157:1506–1512.

Pauchard, A., Aguayo, M., Peña, E., and R. Urrutia. 2006. Multiple effects of urbanization on the biodiversity of developing countries: The case of a fast-growing metropolitan area (Concepción, Chile). *Biological Conservation* 127:272–281.

Poudyal, N. C., Hodges, D. G., and C. D. Merrett. 2009. A hedonic analysis of the demand for and benefits of urban recreation parks. *Land Use Policy* 26(4):975–983.

Rees, W. and M. Wackernagel. 1996. Urban ecological footprints: Why cities cannot be sustainable—And why they are a key to sustainability. *Environmental Impact Assessment Review* 16(4–6):223–248.

Revi, A., Satterthwaite, D. E., Aragón-Durand, F. et al. 2014. Urban areas. In *Climate Change 2014: Impacts, Adaptation, and Vulnerability—Part A: Global and Sectoral Aspects*. Contribution of Working Group II to the Fifth Assessment Report of the Intergovernmental Panel on Climate Change, eds. Field, C. B., Barros, V. R., Dokken, D. J. et al., 535–612. New York: Cambridge University Press.

Scott, A., Gilbert, A., and A. Gelan. 2007. *The Urban-Rural Divide: Myth or Reality?* Policy Brief Series 2. Aberdeen: Macaulay Institute.

Seto, K. C., Güneralp, B., and L. R. Hutyra. 2012. Global forecasts of urban expansion to 2030 and direct impacts on biodiversity and carbon pools. *Proceedings of the National Academy of Sciences* 109(40):16083–16088.

Shen, W., Wu, J., Grimm, N. B., and D. Hope Shen. 2008. Effects of urbanization-induced environmental changes on ecosystem functioning in the Phoenix metropolitan region, USA. *Ecosystems* 11:138–155.

Sheuya, S. A. 2008. Improving the health and lives of people living in slums. *Annals of the New York Academy of Sciences* 1136:298–306.

Thompson, C. W. 2002. Urban open space in the 21st century. *Landscape and Urban Planning* 60(2):59–72.

Ulrich, R. S., Simons, R. F., Losito, B. D., Fiorito, E., Miles, M. A., and M. Zelson. 1991. Stress recovery during exposure to natural and urban environments. *Journal of Environmental Psychology* 11(3):201–230.

UN (United Nations). 2014. *World Urbanization Prospects: The 2014 Revision*. New York: Department of Economic and Social Affairs, Population Division.

UN. 2015. *Sustainable Development Goals. Goal 11: Make Cities Inclusive, Safe, Resilient and Sustainable*. New York: UN. http://www.un.org/sustainabledevelopment/cities/ (Accessed 10 October, 2017).

UN-Habitat (United Nations Human Settlements Programme). 2003. *The Challenge of Slums—Global Report on Human Settlements 2003*. London: Earthscan Publications on Behalf of UN-Habitat.

UN-Habitat. 2007. *Enhancing Urban Safety and Security: Global Report on Human Settlements 2007*. London: Earthscan.

UN-Habitat. 2014. *Urban Equity in Development—Cities for Life*. Concept Paper. Nairobi: UN-Habitat.

UNISDR (United Nations Office for Disaster Risk Reduction). 2015. *Making Development Sustainable: The Future of Disaster Risk Management*. Global Assessment Report on Disaster Risk Reduction. Geneva: United Nations Office for Disaster Risk Reduction.

Van Eetvelde, V. and M. Antrop. 2004. Analyzing structural and functional changes of traditional landscapes—Two examples from Southern France. *Landscape and Urban Planning* 67(1):79–95.

WC (Wilson Center). 2011. *Migration, Urbanization, and Social Adjustment*. Washington, DC: WC. https://www.wilsoncenter.org/publication/migration-urbanization-and-social-adjustment (Accessed August 2017).

White, T. 2013. Advances in the evaluation of informal settlement upgrading in Brazil. PhD diss., University of British Columbia.

World Bank. 2015. *Country Classification by 2015 Income*. Washington, DC: World Bank https://blogs.worldbank.org/opendata/new-country-classifications-2016 (Accessed 9 October 2017).

ENDNOTE

1. Following the World Bank's classification (2015): Low-income countries are where the gross national income (GNI) per capita is of US$1,000 or less. Lower-middle income countries are where GNI per capita is between US$1,000 and US$4,000. Upper-middle income countries are where GNI per capita is between US$4,000 and US$12,500. High-income countries are where GNI per capita is of US$12,500 or more.

6 Urban Lifestyles
Forest Needs and Fears

Cecil Konijnendijk van den Bosch

CONTENTS

6.1 INTRODUCTION

With most humans now living in urban areas and further urbanization being eminent, urban lifestyles have become dominant. Lifestyles are defined as the typical way of life of an individual, group, or culture (Merriam-Webster 2017). They comprise specific habits, attitudes, tastes, economic levels, and so forth. Habits, attitudes, and the like all impact people's preferences, perceptions, and behaviors, for example, toward nature and forests.

This chapter discusses how urban lifestyles have resulted in changing relations between people and forests and with emphasis on woodland and other nature situated in or near cities and towns. Cities and forests have had a long and complex history, covering the spectrum from cities growing at the expense of forests, to cities conserving forests for the many benefits they provide, from fuelwood and timber to providing recreational areas (Konijnendijk 2018).

As we will see, urban lifestyles have, over time, helped foster the emergence of the conservation movement, with concerned urbanites often taking the lead in protests over the loss of nature, both nearby and more remote. These lifestyles have on the other hand frequently resulted in a distancing from nature or at least from more in-depth understanding of natural processes as well as the management of forests and other resources. The latter is problematic, as forests, trees, and other natural components are much needed also in urban areas. For example, noncommunicable diseases such as stress and cardiovascular diseases are on the rise and access to nature can have a positive impact of public health and well-being (World Health Organization Regional Office for Europe 2016). But we also need forests and green spaces for the many ecosystem services they provide, such as mitigating the impacts of climate change on cities, reducing air pollution, and regulating water.

The chapter starts with placing the relationships between cities and their residents on the one hand and forests on the other in a historical perspective. It discusses the love–hate relationship between cities and forests and describes how urban residents have played a crucial role in the rise of the conservation movement in Europe and elsewhere. Next, the proliferation of (changing) urban lifestyles is discussed, especially in relation to changing uses and images of nature and forests. The current importance and multiple roles of urban forests are then introduced, for example, in terms of provision of wide range of ecosystem services that contribute not only to people's health and well-being but also to, for example, social cohesion. This is followed by an analysis of barriers that restrict urbanites from using their local urban forests. In many cases, the full benefits of urban forests are not always realized, for example, due to ambivalent feelings urban dwellers have toward nature and through what could be coined "ecosystem disservices." Fears not only of wild animals, diseases, confrontations with the forces of nature but also of other people can be a real obstruction for visiting natural areas. With cities becoming more culturally diverse, an increasing range of both positive and negative relations with nature has emerged. Approaches to removing these barriers and dealing with increasingly diverse urban populations are the topic of the subsequent section in which urban forestry is introduced as an interdisciplinary field that can provide a socially inclusive framework for these approaches. The chapter ends with conclusions and some perspectives for the future.

6.2   URBANIZATION AND FORESTS FROM A HISTORICAL PERSPECTIVE

In recent times, humankind has moved from being primarily rural to being predominantly urban. By the year 2030, 60% of all humans will live in cities (United Nations 2015). The growing importance of urban areas as human living environments is reflected in the 2015 United Nations Sustainable Development Goals (United Nations 2015). One specific goal, goal number 11, highlights urban areas, under the heading "Making Cities Inclusive, Safe, Resilient and Sustainable." It stresses the importance of access to green spaces for all segments of the population, including often marginalized groups such as children and the elderly.

The urbanization of society and human habitats has a long history. Moreover, modern relationships between nature, forest, people, and culture have largely been shaped in cities which testify "to our ability to reshape the natural environment in the most profound and lasting ways" (Kotkin 2005). Experts have identified urbanization as the most important environmental influence on behavior, as well as the most important behavioral influence on the environment (Gallagher 1993).

Trees and forests have always been a part of cities, initially as sources of food, fodder and wood, and sometimes shade. However, Lawrence (2006) describes how more organized planting and management of trees emerged in the sixteenth century in Europe. Trees started to be used in two main public settings of cities. First of all, these included spaces used for public activities enhanced by the presence of trees, such as parade routes, areas for recreation, and public spaces that functioned as squares of parks. Secondly, trees have been used as extensions of the private garden and mostly often as street trees in front of houses. Prior to the sixteenth century, trees primarily belonged to monastery gardens and some private gardens.

Lawrence (2006) distinguishes three key dimensions of the presence of city trees during history: aesthetics, power and control, and national identity. Over time, changes in aesthetics of architecture, garden design, and urbanism influenced the ways trees were planted in cities. Changes in recreational activities also had an impact. Gradually, city trees became symbols of pleasure. The dimension of power and control was reflected in, for example, the design of green spaces, species choice, and (limited) access to parks and gardens. Finally, from the perspective of national identity and fashions, the establishment and use of trees can be characterized as the gradual replacement of national styles of urban landscapes with a cosmopolitan set of urban landscape forms.

However, forests preceded urban trees. Other work has described not only the conflicts between cities and forests over time but especially also their joint development [e.g., the book by Konijnendijk (2018); see also Figure 6.1]. As cities developed, many were heavily dependent on surrounding woodland for the provision of construction wood, fuelwood, as well as fuel and raw materials for various industries (Perlin 1989). This often led to overexploitation and clearing of forests, although Bell et al. (2005) mention how differences in the level and speed of urbanization contributed to the emergence of regional "forest cultures" in Europe. While cities under the Nordic and central forest cultures were more successful in conserving at least some of the surrounding forests, in some cases not only due to later industrialization and urbanization but also because of a culture in which forests have played a more central role, urbanization has had dramatic impacts on the surroundings of cities in Europe's northwestern and southern countries. In Nordic and other countries,

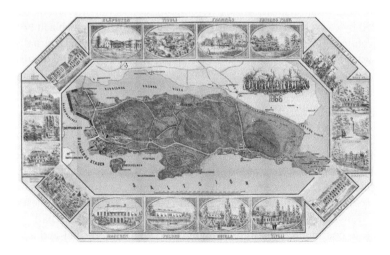

FIGURE 6.1 Djurgården woodland park and Stockholm (map of Södra Djurgården in 1866). (Courtesy of S. Leczinsky and Tr. with Ph. H. Mandel—Wikimedia Commons.)

cities were gradually "built into" the forest. In these countries, as well as under the central European forest culture, woodlands became appreciated for their many benefits and considered an important part of national and local cultures and identities.

The relationship between cities and forests has been complex. On the one hand, both have been at odds, as urbanization and the need for agricultural land led to clearing of forestlands. On the other hand, however, forests provided cities with a range of essential products and services, although this often led to overexploitation. During medieval times, the large class of peasants living near European cities as well as the urban poor was dependent on nearby forests for fuelwood, fodder, and grazing cattle (Konijnendijk 2018). From "wildwood" which was not named, owned, or managed, various forms of cultural and managed woodlands emerged (Rackham 2004). In many cases, forests near cities were owned by the nobility and sometimes the clergy, but forests could also be part of common lands, used and managed by a wide-ranging mix of commoners. These wooded commons were gradually appropriated by rulers and the upper classes, which resulted in social conflict. Wooded areas near cities and towns were often popular hunting domains for kings and nobility. A positive aspect of this was that these wooded areas were strictly protected by their owners and survived intensifying urbanization.

Gradually, the bourgeoisie and private entrepreneurs started to purchases city forests, not only for recreation and prestige reasons, but also for wood production purposes (Konijnendijk 2018). The city governments also became involved, not in the least because of concerns of securing a steady supply of wood. Municipal forest ownership and municipal forestry became widely established, and many cities started a centuries' long relationship with "their" local forests. Over time, of course, the priorities for municipal forestry changed. Urbanization led to new demands in terms of leisure activities, while the industrialization meant that fuels other than wood came into use. Moreover, other forest uses gradually came into focus, such as the provision of drinking water and the protection of nature.

Cities were often also the places were nature conservation efforts emerged, as illustrated by cases such as the Vienna Greenbelt and the Fontainebleau Forest near Paris (Konijnendijk 2018). Influential citizens, artists, writers, and journalists often led the way in raising awareness about the importance of conserving natural areas, also in and near cities.

6.3   URBAN LIFESTYLES AND CHANGING PEOPLE–FOREST RELATIONSHIPS

Most of us today have urban lifestyles, even when we do not live in cities. Our habits, attitudes, tastes, etc., are all strongly colored by our current urban society. Urbanization has had dramatic impacts on, for example, the way society is organized and on social bonds. In the late nineteenth century, Tönnies (1887) described an ongoing transition from "Gemeinschaft" (community) to "Gesellschaft" (society), where close community ties were being replaced by more formal and anonymous societal arrangements. Zukin (1998) mentions how during the past decades, the meaning of "urban lifestyles" has changed from a "fairly table prerogative of social status" to an aggressive pursuit of cultural capital. As immigrants, racial, ethnic, and sexual minorities have become more visible actors in both public spaces and cultural fields, they have made a variety of "alternative" lifestyles more visible, especially in the big cities where they are concentrated. In Canada, for example, immigration to larger metropolitan areas is almost the sole reason for continuing population increase. Statistics Canada predicts that foreign-born populations in both Toronto and Vancouver will be over or close to 50% by 2031 (Simmons and Bourne 2013).

"Urban" cultures have been defined and redefined and are often related to consumption. As a result, cities in the postmodern era are no longer landscapes of production (as they were during starting from the Industrial era), but rather landscapes of consumption.

One common denominator for different urban lifestyles is the potential to lose daily contact with nature. Cox et al. (2017) stress that the variation in experience of nature across populations is poorly understood. They surveyed over 1000 urban residents in the United Kingdom to measure four distinctly different nature interactions: indirect (viewing nature through a window at work and at home), incidental (spending time outside at work), intentional (as related to time spent in private gardens), and a second type of intentional (in this second case as time spent in public parks). These interactions have all been found to be beneficial, ranging from increased psychological well-being and reduced stress at work, to reduced mortality from cardiovascular disease resulting from intentional interactions through, e.g., recreational use of parks. Study findings showed that 75% of nature interactions were experienced by half the population. Moreover, only about one-third of the population experienced the large majority of all interactions that took place in nature itself, rather than just viewing nature. Experiencing nature regularly appears to be the exception rather than the norm, with a person's connection to nature being positively associated with incidental and intentional experiences.

There is not a single urban lifestyle, and it is important to recognize the cultural and sociodemographic diversity in urban populations. Buijs et al. (2009) discuss cultural differences in the meanings attached to nature, something which is important in a time of increasing migration and cultural diversification. People of different cultural backgrounds have different beliefs and preferences regarding nature and different normative views about the relationship between people and nature. The authors distinguish different images of nature as cognitive reflections of prior experiences with and discourses about nature. Based on other literature, they present three distinct images of nature, namely, the wilderness image (which is ecocentric), the functional image (which is anthropocentric), and the inclusive image (which is both ecocentric and representing an intimate relationship between humans and nature). In a study among immigrants from Islamic countries and native citizens of the Netherlands, the authors found confirmation in the thesis that immigrants in Western countries often have a lower preference for nonurban landscapes, and especially for wild and unmanaged landscapes. Native Dutch, on the other hand, expressed a strong preference for what they perceived to be "wilder nature," with over half of the respondents endorsing the wilderness image.

Also in the Netherlands, Kloek (2015) stresses that one should be careful to generalize, as, e.g., ethnic groups are not homogenous entities and harbor cultural diversity also within them. The authors interview young Dutch adults with Chinese, Turkish, and nonimmigrant backgrounds, finding that personal identities and age also inform recreational behavior and nature views, just as ethnicity. They highlight the importance of a multiplicity of identities, both collective and personal, with the latter relating to an individual's unique biography, experiences, and personal characteristics.

6.4   IMPORTANCE OF URBAN FORESTS TO URBAN DWELLERS

Urban forests, ranging from single trees in private gardens to periurban woodland, are a key component of green infrastructure and urban nature. In the previous section, several of the benefits of both direct and indirect interactions with nature (such as trees and forests) were highlighted, for example, in terms of promoting public health and human well-being. But urban forests and other urban nature provide a wide spectrum of benefits, resulting from different ecosystem services. Urban trees and urban green can help cool cities, reduce the impacts of air and water pollution, and assist cities in dealing with floods and other extreme weather events. They can provide food and fuelwood. Moreover, urban nature provides meeting places, inspiration, and opportunities for learning, while also stimulating creativity [e.g., the book by Miller et al. (2015) and the studies Plambech and Konijnendijk (2015) and Roy and Byrne (2014)].

The book *The Forest and the City* (Konijnendijk 2018) explores the many important roles of urban woodland. These includes, for example, woodland as areas for recreation, more adventurous play, settings for nature education, and places of spiritual and creative stimulation. Forests, parks, and other green spaces can also contribute to social cohesion, providing democratic and nondemanding meeting places and building place identity among communities. The contributions of forests and other natural environments have been highlighted in a study of Danish artists and creative professionals (Plambech and Konijnendijk 2015). Exposure to nature was especially important during the first parts of the creative process, for example, when ideas for new projects were incubated.

Children are an important group where contacts with forests in urban settings are concerned (Figure 6.2). Lerstrup (2016) studied the role of woodlands as settings for Danish forest kindergartens, i.e., all-day schools for preschool children that spend substantial parts of their time in forest environments. Lerstrup used Heft's affordance framework to analyze how specific elements and aspects of woodlands stimulated children play and activities. She highlighted the importance of, for example, ditches, loose branches, shrubs, and animals. Although her study did not provide conclusive evidence on this matter, comparison between the forest environment and a more conventional school playground did provide insights into differences in children's activities, with the forest environment stimulating more collaborative play and a wider range of activities.

Learning from nature is not limited to children, as highlighted in *The Forest and the City* (Konijnendijk 2018). Many universities and colleges have had demonstration areas within local urban forests, while many cities (and especially those with universities) also host arboretums. Public forest managers as well as not-for-profits use local forests to educate residents about nature, forest conservation, and forest management.

FIGURE 6.2 Urban woodland is essential for keeping children in contact with nature and for enhancing their play, creativity, and learning. (Photo by the author.)

While urban forests are primarily recognized for their benefits, we should also be aware that they can also be a source of various types of harm and nuisance. Lyytimäki (2017) speaks of ecosystem disservices, which can be defined as the functions, processes, and attributes generated by the ecosystem that result in perceived or actual negative impacts on human well-being (Shackleton et al. 2016). Occasionally, they have also been referred to as missed opportunities to enjoy ecosystem services. Disservices can relate to not only aesthetic issues (e.g., "ugly" trees) and health issues such as vector-borne diseases and allergies, but also safety and security issues, such as physical risks, and fear for wildlife or because of (perceived) human misconduct.

These are not only a nuisance, but can also result in social controversies and conflicts related to urban green management. In the next section, we will look into urban forests as ambivalent landscapes.

6.5   URBAN FORESTS AS AMBIVALENT LANDSCAPES

As we have seen, nature and landscapes are generally considered beautiful and beneficial (Van den Berg and Konijnendijk 2017). However, as also described through the introduction of the concept of ecosystem disservices, people's reactions to forests and nature are not always positive. At least some natural landscapes evoke a mixture of positive and negative feelings and thoughts (Bixler and Floyd 1997), with this ambivalence seemingly most pronounced for "wilder" landscapes with a low degree of human influence or presence.

William Cronon has written extensively about our ambivalent relationships with wilder nature, as expressed in the concept of wilderness [e.g., Cronon (1996)]. Although "wilderness" has been defined in many ways, it has generally been used as referring to natural areas untouched (or unmanaged) by humans. For most of Western history, wilderness has been considered a place to fear and avoid, being associated with the deserted, savage, desolate, and the barren, and places where one can lose oneself in "moral confusion and despair" (Cronon 1996).

Over time, a more positive approach to wilderness emerged, especially after the Enlightenment in Europe. During the era of Romanticism, wilderness even became sacred and associated with the deepest core values of the culture that created and idealized it (Cronon 1996). Closely associated to wilderness was the evolving concept of the "sublime," i.e., a sense of awe and reverence, mixed with elements of fear (Burke 1958). Countries such as the United States built national identity and pride around wilderness and national parks. Later, theories such as Biophilia (Wilson 1984) further strengthened the love of nature, also in its "wilder" forms. Despite this, negative perceptions of wilderness also prevail.

As discussed by Van den Berg and Konijnendijk (2017), the literature shows that "wilder" areas such as unmanaged forests can evoke positive and negative feelings but that highly managed natural settings that are strongly controlled by humans may also evoke negative thoughts and feelings. Such settings are often perceived as overly formal and excessively tidy, and, thereby, unnatural (Özgüner and Kendle 2006).

Forests have often been associated with both positive and negative feelings and perceptions. Porteous (2002) introduces the rich heritage of folktales and myths related to forests. Many of the ancient myths and tales held a note of warning and were associated with people's fears. In his German perspective of forest–people relationships, Lehmann (1999) explains the association of forests and fear from the perspective of the forest being a dark border area, lurking at the fringe of civilized society. Interviews with German forest users—many of whom urban—indicated that people often have fears of being left behind in the forest, for example, based on childhood memories. This fear, like the fear of getting lost, is a recurring theme.

Van den Berg and Ter Heijne (2004) have studied impressive nature experiences in relation to gender and sensation seeking. Based on quantitative analyses of people's personal encounters with natural threats, they identified four clusters of situations that tend to evoke both fear and fascination in people: (a) close encounters with wild animals, (b) confrontations with the forces of nature (e.g., a storm or an earthquake), (c) overwhelming situations (e.g., being intimidated by the greatness of a forest), and (d) disorienting situations (e.g., getting lost in the woods).

Jones and Cloke (2002) mention that "places of trees" can be places of fear as well as of exclusion. They provide examples of groups of people "claiming" green spaces, even in the best intention, but then also excluding other groups who do not adhere to the group's norms. Byrne (2012) shows how Latino inhabitants of Los Angeles often did not use nearby urban national park as they thought that other (White) users did not want them there and would even throw them out if they would visit.

The role of forests as "places of fear" has its roots in some of the primeval fears for forests, related to, e.g., getting lost or the dangers associated with wild animals. Schama (1995) writes about the ambiguity in our relations with forests over time, stating that forests represent two kinds of arcadia, namely, "shaggy and smooth, dark and light, a place of bucolic leisure and a place of primitive panic." Some of our other fears of forests, especially in urban settings, are related to fear of crime. Sreetheran and Konijnendijk van den Bosch (2014) review the literature on fear of crime in green spaces, identifying a wide range of factors that impact fears, ranging from individual factors such as gender and ethnicity to social factors related to other users and "stranger danger." Fear can also be associated with characteristics of the forest or green space itself, such as an unkempt or unmanaged appearance, dense vegetation that limits social control, and wildlife. Lehman (1999) describes how woodlands in and near cities have been regarded as less-controlled "fringe" environments for those who want to escape from the rule of law of from societal norms. Although statistics show that woodlands generally do not have high rates of violent or other crimes, there is a long and sometimes colorful history of criminals, robbers, outlaws, political refugees, and even terrorists turning to urban forests as hiding places. Stories range from those involving fairy tale figures such as Robin Hood, to outlaws, such as Dick Turpin, and terrorists such as the members of the Rote Armee Fraktion in Germany (Konijnendijk 2018). Fear of crime associated with woodland can hamper their recreational use, as demonstrated by Jorgensen and Anthopoulou (2007). Urban woodlands are not seldom seen as fraught with dangers including intimidation from groups of young people and physical or sexual assault.

As mentioned, levels of fear differ between individuals of different cultural and sociodemographic backgrounds. Women, for example, are more often afraid of going into woodlands by themselves than men [e.g., the studies by Jorgensen et al. (2006) and Sreetheran and Konijnendijk van den Bosch (2014)]. Skår (2010) states that a fear of the outdoors affects women's quality of life, but not only because so many people are denied the healing benefits of nature. But it is also problematic because they are also denied of the empowering opportunities that participation in outdoor nature recreation provides.

Although very few urban forests provide threats from wild animals such as predators, Konijnendijk (2018) does present examples of, e.g., mountain lions attacking joggers in the periurban forests of California. Lehmann (1999) also mentions people's fear of certain animals and of the increasing risks associated with vector-borne

FIGURE 6.3 "Wilder" urban nature can evoke ambivalent feelings, although people with a more ecocentric worldview are often more appreciative. Lighthouse Point Park in West Vancouver. (Photo by the author.)

diseases such as Lyme disease (transmitted by ticks). As early as in the early 1990s, between 40,000 and 80,000 Germans were infected each year with Lyme disease (Lehmann 1999).

Even trees can evoke ambivalent meanings and emotions, as illustrated by Lehmann (1999) and Bonnes et al. (2011) who refer to accidents caused by, e.g., falling limbs and trees and larger trees as hiding places for attackers.

Van den Berg and Konijnendijk (2017) place our ambivalence toward nature in a wider perspective, relating it to different nature views and nature images. Earlier, I referred to the work of Buijs et al. (2009). The authors state that an ecocentric view is associated with a wilderness image in which natural settings that are untouched by humans are highly preferred and considered beautiful and "good" examples of nature (Figure 6.3). An anthropocentric view of nature, on the other hand, is associated with a more functional image with a strong preference for intensively managed settings that are useful for humans. The different images of nature can then also be related to fears, as, for example, "wilder" nature will be much less appreciated by people with an anthropocentric view (who might be looking for cues of management and get disturbed when these are absent).

6.6 OVERCOMING BARRIERS TO URBAN NATURE USE: THE ROLE OF URBAN FORESTRY

Not only are we faced with the challenge of overcoming the barriers that restrict people from using their looking forest or nature area. We also have to take into account that with increasing urbanization, people's images or views of how forests are or should be managed have changed. Rametsteiner and Kraxner (2003) discuss some of the challenges forestry is facing due to changing public perceptions. In general, forests are associated with positive roles such as being "green," providing fresh air, and acting as symbols of nature, but active forest management practices are not always understood or supported (Konijnendijk 2018). Forestry has had difficulties in fighting off negative public perceptions of clear-cutting and overexploitation of forests, especially as more people have "moved away from the land."

During the past 50 years or so, the interdisciplinary field of urban forestry emerged from a need to develop an urban approach to managing forests and trees. Traditionally, most forestry educations have had very little focus on urban settings, and little attention was given to aspects related to urban governance, intensive communication with local residents, recreational use, the integration of forests into wider green infrastructure, and how best to grow trees in harsh urban environments. Initially coined in North America (during the 1960s), urban forestry involves the planning and management of woodlands, trees, and associated green spaces in urban settings [e.g., the study by Konijnendijk et al. (2017)]. Urban forestry combines theories, methods, and practices from a range of disciplines, including not only arboriculture, horticulture, and landscape architecture, ecology, but also planning and social science. Obviously, forestry as the management of forest ecosystems for the multiple benefits these provide is also an important contributor. It brings in its long-term perspective on natural resource management, the need for regular resource assessments, and a sustainability perspective. Forestry's work with providing multiple benefits from the same resource has also been influential in urban forestry. As mentioned, urban forestry is one of the fields which links the natural sciences with the social sciences and humanities. Moreover, it is just as much about people as about trees. Many definitions of, and scholarly works on urban forestry, have stressed its socially inclusive nature. This normative perspective recognizes that urban forestry can only be truly successful when it is done in close collaboration with local (urban) communities. This explains why governance and public engagement are often given considerable attention.

The human focus of urban forestry also includes consideration of the discussed diverse and changing perceptions of forests and nature in urban settings. As urban forestry aims to be socially inclusive and wants to provide a range of benefits to all members of local communities, it is challenged by people's ambivalent perspectives of forests, fear or crime and the like. Moreover, recreational and other uses of urban forests can be restricted due to lack of information, education, and awareness. Enhancing the latter is a task for professional urban foresters and their partners such as not-for-profits.

Cox et al. (2017) highlight the trend of declining nature experiences, at least for some sociocultural groups. Given the many benefits associated with nature experiences, for example, in terms of enhanced human health and well-being, it is important to reverse this trend. Several approaches can be applied, including expanding green infrastructure and urban forests, and bringing more green space and trees into people's living and working environments. However, enhancing supply of green space and offering opportunity are not enough. The design and management of urban forests and other green space needs to focus on increased orientation toward nature for a larger group of people. Cox et al. (2017) mention the importance of taking into account a wide range of backgrounds, including people's past nature experiences and their history of nature play during their childhood. Nature-based activities can also help enhance people's interactions with nature.

In its endeavors to connect more people with urban nature, urban forestry needs to recognize the wide range of individual, social, and cultural characteristics. Skår (2010) stresses that forest experiences are not uniform nor stable, but rather dynamic, socially constructed and changeable, and shaped by people's lived experiences. Thus, we need to understand the emotional and other relationships between different people and different places.

Earlier, we mentioned women as a group who is often hindered in obtaining nature experiences by fear of crime. The barriers are even more several for women of different ethnic backgrounds and/or recent immigrants. In a project in the United Kingdom, local foresters tried to break this barrier by bringing 13 women's groups from different cultures on guided forest walks in woodlands near London and Nottingham (Burgess 1995). The groups included Asian and Afro-Caribbean women. The women felt safer in a group and being accompanied by a local forester. They also said that they noticed that the forest does not have to be a dangerous place. Suggestions for improving forest use by all groups include improving signage and information, maps, community liaisons, and policing of forests.

The framework presented by Sreetheran and Konijnendijk van den Bosch (2014) can assist with understanding fear of crime and finding ways of overcoming it. Different contributing factors can be addressed by urban forests, from engaging with people from different background to enhancing mutual understanding between different groups. Much can also be done from a woodland design and management perspective, although there are trade-offs, for exampl,e when removing vegetation to enhance social control, which can lead to somewhat lower biodiversity values and less of a "wilderness" experience. It is important that urban foresters work closely together with local communities in their efforts to enhance the use and appreciation of local forests and green space.

Skår (2010) mentions two main ways in which fear can be mitigated. First of all, landscapes can be designed in ways that do not trigger fear responses in people, e.g., by designing and placing natural elements or lighting in accordance with specific preferences that enable both clear visibility and refuge. Nasar and Jones (1997) highlight the importance of "cues of care" that show that an area is appropriately managed. Second, fear can be reduced through the minimization of intrapsychological constraints such as stress, anxiety, and perceived self-skill. Related to the former way of mitigating fear, C. R. Jeffery coined the term *crime prevention through environmental design* in the early 1970s [cited in the study by Khurana (2006)]. It is based on the idea that proper construction and layout of the environment can help inhibit criminal activity and improve community living.

Children are another important demographic. The author Richard Louv (2005) had a major impact with his book *Last Child in the Wood* in which he presents the decreasing exposure of children to nature and the negative consequences of this. He even coined a new "disease," Nature Deficit Disorder, to highlight the impacts of less nature contact on children's development, play, learning, and so forth. In the Nordic countries and increasingly elsewhere, forest schools are a common phenomenon to make children at ease in the forest. In these schools, children spent part of their biology and sometimes also other lessons in the forest (O'Brien and Murray 2007).

Dealing with fears primarily caused by animals is another area to look at. Ongoing efforts to reduce diseases transferred by animals, for example, by controlling animal populations, is one area of action. Proper wildlife management is also needed. A balance will have to be found between bringing the forest and "wild" nature all the way to people's doorsteps and dealing with the associated risks of this, such as increases in tick and mosquito-borne diseases. Urban nature, by its very characteristics, should be allowed to keep its "wild edge" and its role as adventurous space, offering exciting and less controlled opportunities for play and recreation.

When parks of forests have become "no-go zones," for example, because of a rise in criminal activities, efforts are needed to "reclaim" the green space and enhance social and management control. The local community will have to be mobilized, while the visible presence of managers can deter criminal behavior.

Activities to prevent and reduce crime can also have a downside, as discussed in a British report on the social values of public spaces (Worpole and Knox 2007). People run the risk of being "designed out" of the public realm. The authors state that the British government's emphasis on crime and safety in public spaces is depriving these spaces of their historic role as a place where differences of lifestyles and behavior is tolerated. Moreover, people considered "abusers" or nuisances by others are also people with a right to use green spaces.

Earlier, we mentioned the concept of ecosystem "disservices." Lyytimäki (2017) provides suggestions for how to deal with these and minimize the negative impacts of urban forests and other ecosystems. He states that early-phase participation by stakeholders can decrease the risk of misunderstandings, increase the possibilities for efficient uptake of research results, and give important possibilities for incorporating local tacit knowledge into the assessment. In general, the author calls for the inclusion of lay knowledge as a way of enhancing trust and social cohesion. Lyytimäki (2017) also stresses that all understandings of ecosystem services are inherently value-based and often individual, meaning that what is harmful may widely differ depending on the context and the persons involved. As with fear of crime, age, gender, personal experiences, knowledge level, social settings, and cultural background all influence people's preferences. Successful urban forestry must be sensitive to this. Ecosystem services and disservices can also be linked, for example, when less intensive urban forest management results in higher biodiversity and the appreciation of some people, while other residents will see the uncontrolled growth of natural vegetation as undesirable.

6.7   CONCLUSION AND PERSPECTIVE

This chapter has presented the long-standing relationship between cities and their residents on the one hand and forests and trees on the other. As seen, forests have been a crucial component of cities and towns, as they have provided urban communities with a wide range of essential goods and services. Where forest products such as timber, fodder, fuelwood, and game where the focus of earlier forest–city relationships, attention gradually shifted toward cultural ecosystem services such as providing settings for recreation. The role of forests in, e.g., adapting cities to ongoing climate change has also come to the fore. But perhaps the most crucial contributions of forests in our current urban era relate to their positive contributions to public health and to keeping urban residents, from children to ethnic minorities, in contact with nature.

We not only need forests and trees, but we also have to recognize that our feelings toward forests and trees are ambivalent. Ongoing urbanization and our gradual distancing from the land have enhanced the focus on some of the negative images and perceptions and to the emergence of the concept of ecosystem disservices. Much of the negative attitudes toward forests among at least some segments of urban populations can be addressed through, e.g., raising awareness, enhancing people's engagement with forests and forestry and more sensitive design, planning and management of our urban forests. Here the profession of urban forestry has an important role to play, as it has social inclusiveness in its very core and can draw upon the perspectives and expertise from a wide range of disciplines. Yet even urban foresters are not always appropriately trained to take into account the diverse and ambivalent relations between urban people and urban forests.

In other work [e.g., the book by Konijnendijk (2018)], I have argued applying a place-based perspective to urban forestry, recognizing that our relationships with forests, trees, and other nature are shaped based on our individual and social engagement and relation-building with the local environment. Other authors have argued for both "place making" and "place keeping" as key approaches to managing well-functioning urban green space. This will provide us with a framework for addressing the forest needs and fear of urban populations in the future.

REFERENCES

Bell, S., Blom, D., Rautamäki, M., Castel-Branco, C., Simson, A., and I. A. Olsen. 2005. Design of urban forests. In *Urban Forests and Trees—A Reference Book*, eds. Konijnendijk, C. C., Nilsson, K, Randrup T. B., and J. Schipperijn, 149–186. Berlin: Springer.

Bixler, R. D. and M. F. Floyd. 1997. Nature is scary, disgusting, and uncomfortable. *Environment and Behavior* 29:443–467.

Bonnes, M., Carrus, G., and P. Passafaro. 2011. The ambivalence of attitudes toward urban green areas: Between proenvironmental worldviews and daily residential experience. *Environment and Behavior* 43(2):207–232.

Buijs, A. E., Elands, B. H. M., and F. Langers. 2009. No wilderness for immigrants: Cultural differences in images of nature and landscape preferences. *Landscape and Urban Planning* 91:113–123.

Burgess, J. 1995. *"Growing in Confidence"—Understanding People's Perceptions of Urban Fringe Woodlands*. Technical Report. Cheltenham: Countryside Commission.

Burke, E. 1958. *A Philosophical Enquiry in the Origin of Our Ideas of the Sublime and the Beautiful*. First published in 1756. London: Oxford University Press.

Byrne, J. 2012. When green is white: The cultural politics of race, nature and social exclusion in a Los Angeles urban national park. *Geoforum* 43:595–611.

Cox, T. C., Hudson, H. L., Shanahan, D. F., Fuller, R. A., and K. J. Gaston. 2017. The rarity of direct experiences of nature in an urban population. *Landscape and Urban Planning* 160:79–84.

Cronon, W. 1996. The trouble with wilderness, or, getting back to the wrong nature. In *Uncommon Ground: Rethinking the Human Place in Nature*, ed. Cronon, W., 69–90. New York: Norton.

Gallagher, W. 1993. *The Power of Place: How Our Surroundings Shape Our Thoughts, Emotions and Actions.* New York: Harper Collins.

Jones, O. and P. Cloke. 2002. *Tree Cultures—The Place of Trees and Trees in Their Place.* Oxford: Berg.

Jorgensen, A. and A. Anthopoulou. 2007. Enjoyment and fear in urban woodlands—Does age make a difference? *Urban Forestry and Urban Greening* 6(4):267–278.

Jorgensen, A., Hitchmough, J., and N. Dunnet. 2006. Woodland as a setting for housing-appreciation and fear and the contribution of residential satisfaction and place identity in Warrington New Town, UK. *Landscape and Urban Planning* 79(3–4):273–287.

Khurana, N. 2006. Is there a role for trees in crime prevention? *Arborist News*, August 2006: 26–28.

Kloek, M. E. 2015. Colourful green: Immigrants' and non-immigrants' recreational use of greenspace and their perceptions of nature. PhD Dissertation, Wageningen: Wageningen University.

Konijnendijk, C. C. 2018. *The Forest and the City: The Cultural Landscape of Urban Woodland.* 2nd, revised edition. Berlin: Springer.

Konijnendijk van den Bosch, C., Ferrini, F., and A. Fini. 2017. Introduction. In *Routledge Handbook of Urban Forestry*, eds. Ferrini, F., Konijnendijk van den Bosch, C., and A. Fini, 1–13. London: Routledge.

Kotkin, J. 2005. *The City: A Global History.* London: Weidenfeld & Nicolson.

Lawrence, H. W. 2006. City Trees: A Historical Geography from the Renaissance through the Nineteenth Century. Charlottesville, VA: University of Virginia Press.

Lehmann, A. 1999. *Von Menschen und Bäumen. Die Deutschen und ihr Wald.* Reinbek: Rowohtl Verlag.

Lerstrup, I. 2016. Green settings for children in preschools—Affordance-based considerations for design and management. PhD Dissertation, Copenhagen: University of Copenhagen.

Louv, R. 2005. *Last Child in the Woods: Saving Our Children from Nature-Deficit Disorder.* New York: Algonquin Books.

Lyytimäki, J. 2017. Disservices of urban trees. In *Routledge Handbook of Urban Forestry*, eds. Ferrini, F., Konijnendijk van den Bosch, C., and A. Fini, 164–175. London: Routledge.

Merriam-Webster. 2017. Online dictionary, entry for "lifestyle." https://www.merriam-webster.com/dictionary/lifestyle (Accessed March 24, 2017).

Miller, R. W., Hauer, R. J., and L. P. Werner. 2015. *Urban Forestry: Planning and Managing Urban Greenspaces.* Third edition. Long Grove: Waveland Press.

Nasar, J. and K. M. Jones. 1997. Landscapes of fear and stress. *Environment and Behavior* 29(3):291–323.

O'Brien, L. and R. Murray, R. 2007. Forest school and its impacts on young children: Case studies in Britain. *Urban Forestry and Urban Greening* 6(4):249–265.

Özgüner, H. and A. D. Kendle. 2006. Public attitudes towards naturalistic versus designed landscapes in the city of Sheffield (UK). *Landscape and Urban Planning* 74:139–157.

Perlin, J. 1989. *A Forest Journey: The Role of Wood in the Development of Civilization.* Cambridge, MA: Harvard University Press.

Plambech, T. and C. C. Konijnendijk van den Bosch. 2015. The impact of nature on creativity—A study among Danish creative professionals. *Urban Forestry and Urban Greening* 14(2):255–263.

Porteous, A. 2002. *The Forest in Folklore and Mythology.* Reprint of 1928 original. Mineola, NY: Dover Publications.

Rackham, O. 2004. Trees and Woodland in the British Landscape: The Complete History of Britain's Trees, Woods and Hedgerows. Revised edition. New York: Phoenix Press.

Rametsteiner, E. and F. Kraxner. 2003. *Europeans and Their Forests. What Do Europeans Think about Forests and Sustainable Forest Management?* Vienna: FAO and UN-ECE Forest Communicators Network.

Roy, S. J. and A. Byrne. 2014. A systematic quantitative review of urban tree benefits, costs, and assessment methods across cities in different climatic zones. *Urban Forestry and Urban Greening* 11:351–363.

Schama, S. 1995. *Landscape and Memory.* London: HarperCollins Publishers.

Shackleton, C. M., Ruwanza, S., Sinasson Sanni, G. K. et al. 2016. Unpacking Pandora's Box: Understanding and categorising ecosystem disservices for environmental management and human well-being. *Ecosystems* 19(4):587–600.

Simmons, J. and L. S. Bourne 2013. *The Canadian Urban System in 2011: Looking Back and Projecting Forward.* Research paper 228. Toronto: Cities Centre, University of Toronto.

Skår, M. 2010. Forest dear and forest fear: Dwellers' relationships to their neighbourhood forest. *Landscape and Urban Planning* 98:110–116.

Sreetheran, M. and C. C. Konijnendijk van den Bosch. 2014. A socio-ecological exploration of fear of crime in urban green spaces—A systematic review. *Urban Forestry and Urban Greening* 13(1):1–18.

Tönnies, F. 1887. *Gemeinschadt und Gesellschaft.* Leipzig: Fue's Verlag.

United Nations. 2015. *Sustainable Development Goals.* New York: United Nations http://www.un.org/sustainabledevelopment/sustainable-development-goals/ (Accessed April 28, 2016).

Van den Berg, A. E. and M. Ter Heijne. 2004. Angst voor de natuur: een theoretische en empirische verkenning. *Landschap* 3:137–145.

Van den Berg, A. and C. C. Konijnendijk. 2017. Ambivalence towards nature and natural landscapes. In *Environmental Psychology: An Introduction*, eds. Steg, L., Van den Berg, A. E., and J. I. M. de Groot, 67–76. Chichester: BPS Blackwell.

Wilson, E. O. 1984. *Biophilia.* Cambridge, MA: Harvard University Press.

World Health Organization Regional Office for Europe. 2016. *Urban Green Spaces and Health: A Review of the Evidence.* Bonn: European Office of the World Health Organization.

Worpole, K. and K. Knox. 2007. The social values of public spaces. York: Joseph Rowntree Foundation.

Zukin, S. 1998. Urban Lifestyles: Diversity and standardisation in spaces of consumptions. *Urban Studies* 35(5–6):825–839.

7 Toward a Social Representation of Forests by Western Urbanized Societies

Christine Farcy, Sylvie Nail, Julie Matagne,
Anne-Marie Granet, Olivier Baudry,
and Ewald Rametsteiner

CONTENTS

7.1 INTRODUCTION

For city dwellers, urbanization induces an increasing separation, both physical and psychological, from the land and the processes and resources that sustain it (MacCleery 2011). Although the linkage between this separation and a global social transformation from a rural to an urban culture is controversial (Scott et al. 2007), the evolution in the perception that urban populations can have on forests is demonstrated in various studies undertaken in the Western context.

For Eizner (1995), forests constitute both a prototype and a paradigm of nature for urban dwellers. While emotional, spiritual, or even mystical values seem to be on the rise, few references are made to wood production, and little awareness exists on the economic use of forests (Schmithüsen 1999). Feelings and virtual knowledge increasingly replace background and daily concrete experiences of natural processes. When reviewing various studies on perceptions and attitudes toward forests in Europe, Schmithüsen (1999) observes that "there are indications that the population shares to some extent the perceptions evoked by city dwellers": a symbol of nature and cost-free space for recreation, forests symbolize the free play of natural forces with little or even no human influences, contrasting with inhabited and intensively exploited areas such as agricultural land.

Thus, at the dawn of the millennium, a rather homogeneous and urban-driven vision of forests based on distance, virtuality, and immateriality seems to emerge in Western countries and, with it, a growing difficulty for society to understand what the essence of professional forest ethics has been for decades (see Chapters 2 and 3).

This chapter aims to follow up on the issue and even to consolidate intuitions mentioned above. To overcome the individual nature of the concept of perception and take it to society level, this chapter focuses on the concept of social representation introduced by Moscovici in the 1960s. Although little is used in forestry, its close links with communication processes deserve more interest than they have received so far, while allowing dynamic thinking and prospect. Buijs (2009) demonstrates the relevance of the concept when studying social representations of nature in the Netherlands, while Paré (2012) more precisely investigates the influence of militant media on the social representation of forests in Canada. Social representations will be addressed through triangulations, by cross-cutting results of some of the more recent studies made in the Western context.

The chapter will focus on the social representations of the object "forest." How can social representations inform us about the new connection between people and forests in today's urbanized societies? Recent studies have been undertaken by scientists and experts of the forest community, on perceptions of forest owners or forest managers or on the acceptability of forest management by society. Implicitly, their main goal is following a wish already expressed by Westoby in 1987, to find out how society understands forests and forestry, what actions are acceptable, and how to influence such perceptions (Rametsteiner and Kraxner 2003). This chapter intentionally takes the opposite stance and changes the centre of gravity of the analysis, i.e., it aims at offering an external viewpoint on forests and at hearing views expressed outside the forestry sector. Hence, the concept of social representation proposed in this chapter does not deal with forestry actors and remains upstream, considering people and society.

7.2   SOCIAL REPRESENTATION

In social sciences, social representation is commonly seen as common-sense knowledge about an object which is shared by a community or a social group and whose emergence and transmission rely on communication processes.

Forged in the 1960s within the academic community in the realm of social psychology, social representation is defined as "the elaboration of a social object by the community for the purpose of behaving and communicating" (Moscovici 1963). It is considered as a bridge between the individual and the social world. It differs from individual perceptual processes related to human physiology, on the one hand, and from the concept of collective representations developed by Durkheim (1898) and considered as shared by the whole society, on the other hand.

In the historical overview of their handbook on social representations, Moliner and Guimelli (2015) stress that for Moscovici, social representations are the products of social groups that make up society. The coexistence of different social representations of the same object is thus common, and they consider this feature as crucial in changing societies. They note that his theory also highlights the role of communication processes in explaining the emergence and transmission of social representations. Through related processes such as influence, normalization, and compliance, individual beliefs can indeed become consensual, while collective beliefs can become the norm for others.

Moscovici (1969) expanded the concept, pointing out the twofold function of a social representation as a system of values, ideas, and practices: "first, to establish an order which will enable individuals to orient themselves in their material and social world and to master it and, secondly, to enable communication to take place among members of a community by providing them with a code for social exchange and a code for naming and classifying unambiguously the various aspects of their world and their individual group history."

The various functions of social representation are illustrated by Bauer and Gaskell (1999) with the romantic image of nature for environmental groups: "it may define group identity and membership, it may mythically legitimize activism on behalf of environmental preservation, or merely serve the attitudinal function of evaluating genetically modified crops."

Thus, representations develop thanks to experiences and practices, built into systems of cultural and social attitudes, as well as norms and are transmitted through tradition, education, and communication (Jodelet 1984). In a recent study from the unprecedented vantage point of advertising, a form of marketing communication that mainly uses dominant social representations and thus contributes to their reinforcement, Smajda et al. (2016) analyze how the image of trees and forests have been used in commercial advertising since the 1960s in a mainstream Belgian magazine. The authors observe a progressive, but clear shift toward a forest as a living reality in interaction with human behavior, rather than a passive object or a distant setting.

Representations are made up of cognitive elements such as theories, networks of ideas, images, or metaphors; they appear to those who share them as the objective reflection of an obvious and indisputable reality. They are called hegemonic when they dominate in a particular social group which is both structured and significant, and they prevail "in all symbolic or affective practices" [Moscovici (1988) cited by Höijer (2011)]. In his study of social representation and present communication, Höijer (2011) illustrates such a hegemony with the example of climate change conceived as a threat for human life and society.

An inspiring concept, social representation has been constantly enriched by theoretical contributions from various disciplines for over 50 years. A significant example lies in the so-called structural model developed by Abric (1976, 1993) and Flament (1989) in order to explain the different roles played by the various cognitive elements constituting social representations. Firstly, the model identifies a central core, a collective memory resistant to changes, the product of historical or symbolic determinisms, which constitutes the basis of the organization of all the elements and generates meaning and value to the members of the group. These central elements are characteristics that people consider as inseparable from the object. Secondly, the model encompasses a periphery, made up of elements related to experiences and individual history; those elements are less stable, negotiable, and even contradictory, and they act as bumpers and allow adaptation of the representation to take place, for example, in various social contexts (Moliner and Guimelli 1995). This dynamic conception explains how the development and adjustment of social representations are possible under new social circumstances or potentially threatening information about an object (Wagner 1998). Implementing Vergès's (1992) method to define the central core, Michel-Guillou (2006) studied the social representations of the environment among professionals in the agricultural sector in France.

Research on social representation has generated various methodological developments in social psychology, anthropology, sociology, ethnology, and communication sciences. Data come from observation, conversations, interviews, or selected texts and mass media excerpts. Analysis deals with content, narrative, discourse, or semiology. Methods are often based on the principles of grounded theory (Glaser and Strauss 1967) and have methodological triangulation and saturation as their main pillars. While considering the value of such a methodological pluralism, efforts are also made to better formalize approaches, for example, to design and conduct empirical research (Bauer and Gaskell 1999) or to identify the central core (Vergès 1992; Guimelli 1993).

In this chapter, following the triangulation approach, recent sources (mainly focusing on the Western context) are cross-cut with the intention of moving closer to a conceptual outline of the social representation of forests by urbanized societies. The high symbolic value of the forest and trees as noted, for example, by the Food and Agriculture Organization of the United Nations (2003) in different regions, cultures, and religions even allows to hope to identify key drivers or elements partaking of its central core.

7.3   ENVIRONMENTAL CONCERNS

Since the 1970s, the European Commission has monitored the evolution of European Union (EU) citizens' opinions on a regular basis, through a specific initiative called Eurobarometer. This allows it to prepare and assess its work on various issues and concerns. Since the 1990s, one of these has been agriculture and the Common Agricultural Policy (CAP). For the first time in 2015, a question related to forests was integrated in the CAP survey conducted in the 28 member states of the EU[1] (EU 2016): "What do you think are the most important benefits provided by forests?" Following Eurobarometer rules, the 27,822 respondents, coming from different social and demographic categories, were interviewed face to face at home, in their native language. They were given the opportunity to choose a maximum of three items out of seven predefined proposals[2] (Figure 7.1).

According to the trend illustrated in Figure 7.1, in 25 EU countries, including a poorly forested country such as the United Kingdom as well as a traditionally well-forested one like Austria, "absorbing carbon" and "habitat for animals" ranked

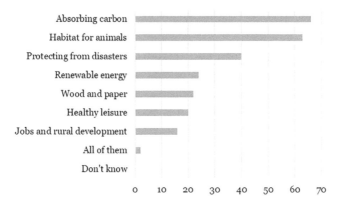

Absorbing carbon
Habitat for animals
Protecting from disasters
Renewable energy
Wood and paper
Healthy leisure
Jobs and rural development
All of them
Don't know

0   10   20   30   40   50   60   70

FIGURE 7.1 Eurobarometer: Europeans, Agriculture and the Common Agricultural Policy—Question 13 "What do you think are the most important benefits provided by forests?" Maximum of three answers, n = 27,822 from EU28, %. (Adapted from EU, *Special Eurobarometer 440 Europeans, Agriculture and the Common Agricultural Policy*, EU, Brussels, 2016. © EU 2016.)

number one or two by far. In 20 countries, "protecting from disasters" occupied the third position. These results reveal a vision mostly informed by environmental drivers.

In no EU country was the item "wood and paper" considered as the major benefit of forests. At best, it ranked third in the three Baltic countries and in Malta and fourth in Sweden and Finland, the most forested EU countries, with forest covers reaching more than 75% of the land cover and a contribution of the forestry sector to gross domestic product among the highest in Europe (between 3 and 5%). In Finland nevertheless, the item "renewable energy" occupies the first place, jointly with "absorbing carbon," while "habitat for animals" comes next. The item "healthy leisure" generally came fifth or sixth, except in Poland and Denmark where it came third, after "habitat for animals" and "absorbing carbon."

Despite the inevitable bias inherent to this kind of closed questionnaire, in particular, the possible implicit wish to conform to social norms, the trends expressed are significant, and the findings reinforce previous studies and surveys on the issue in Europe. Rametsteiner et al. (2009) in particular confirm the recent desire to move away from a focus on the use value of forests: "protection and the prevention of deforestation are key concerns regarding forests." They highlight the "increasing importance and greater public awareness on the issue of climate change and how it relates to forests" and bear out the higher importance granted to climate change by youth than by other age groups. Whereas environmental issues are closely associated with forests, the review shows that it is not the case either for ecosystem services or for recreational purposes. The same conclusions were recently reached in Slovakia (Dobsinska and Sarvasova 2016) and in France (Dobré et al. 2016) where despite the increasing use of forests for recreation, the environmental values of forest ranked first among respondents representing society at large. Those authors also confirm the distance between young people and the material and economic use of forests.

If the environmental drivers that inform the European vision are revealing of a positive role of forests, the main implicit feeling relates to concerns and anxiety, as observed from another large-scale survey, the Eurobarometer dedicated to "attitudes of Europeans toward biodiversity." For the 27,718 respondents, "the decline and possible extinction of animals and plant species, natural habitats and ecosystems" at global level is a serious concern, identified as very serious by 61% of them and as fairly serious by another 30% (EU 2015). Age is again an important factor, with young people (15–24 years old) apparently more worried about this decline and possible extinction (65%, instead of 58% for other age groups). It is noteworthy that respondents are less concerned about the biodiversity of their immediate surroundings, with 19% of them considering local decline as very serious, but 36% as fairly serious.

On the specific question addressing the rate of seriousness of five preidentified issues related to the loss of biodiversity, respondents considered "the degradation and loss of natural habitats like forests, meadows and wetlands" as the most pressing issue with 61% believing that it is a very serious issue, followed by "the loss of the benefits from nature such as crop fertilization, soil fertility, prevention of floods and droughts, climate regulation, clean air and water" (59%) and "the decline and disappearance of animal and plant species" (58%). "The disconnection from nature in urban areas and modern lifestyle" came next with 42%, while "the negative economic impacts of biodiversity degradation such as the loss of income from nature-oriented tourism or fisheries" came last with 36% (EU 2015).

Confirming the impact of media on the social representations of ecological issues (Gendron and Dumas 1991), a third specific Eurobarometer (EU 2014) allows to go more in depth by highlighting the dominant, albeit decreasing, role of television (65% among 27,998 respondents—72% in 2011) followed by the rising influence of the social media (41% against 30% in 2011) as a source of information on environmental issues; 62% of respondents felt well informed on environmental issues. On the question of the reliability of sources, the most trusted sources for environmental information are scientists, environmental organizations, and the television.

Thus, cross-cutting findings allows to confirm the dominant feelings of worry, threat, or sense of urgency shared by European citizens in relation to the environmental component of forests, in particular at global level. Forests, it seems, are seen as threatened and so become a cause to be defended on account of its existence and legacy values.

7.4 TREE CUTTING AND FOREST REMOVAL

While urbanization contributes to disconnecting everyday life from life cycles, the wish or request to explicitly prohibit the cutting of trees seems to represent an increasing trend in society, and activists are quite mobilized on the issue in all the regions (Demougin 1995; Eizner 1995; Perraud 1997; Dobré et al. 2006; Hunziker et al. 2013; McComb 2016). Neighborhood mobilization for preserving a specific wood or forest are also frequent. The media is playing a significant role in reinforcing those trends (Matagne 2017), as exemplified in an article in the British newspaper *The Guardian* published on April 7, 2017.[3] It refers to a recent mobilization by women in

Poland against so-called "massacre" of trees illustrated in a very symbolic way while far away from the issue at stake.

In the national survey periodically undertaken in France on the basis of a representative sample of 1,000 respondents, Granet and Amand (2016) observe a recent convergence of rural and urban views on the meaning of cutting trees; in 2004, 19% of rural people interviewed considered cutting trees as destroying forests, while 26% of the whole population shared the same view. In 2010, 30% of both categories shared this negative view, and the trend was confirmed in 2015.

Likewise, in 2005, Bodson (2005) asked a representative sample of 1,000 respondents of the Walloon and Brussels populations in Belgium, well-forested in its southern part with emblematic timber production forests such as the Ardennes, what activities should be prohibited in the forests. While there was a large consensus against the practice of using motorbike and all-terrain vehicles in the forests, 47% of the respondents were in favor of outlawing hunting and 40% considered that tree felling should be prohibited. This last option was approved by 68% of young people (15–24 years old). The same trend was recently also observed in France by Granet, Dobré, and Cordellier within the framework of the national survey 2015 (Survey 2015 Forest and Society, Office National des Forêts and University of Caen-Basse Normandie, to be published).

Calls to public officials were used, for example, by Hallé (2011) who requested in a pamphlet addressed to French local and national decision-makers for trees in the cities to be respected and for steps to be taken to insert specific paragraphs on trees in the Universal Declaration of Human Rights. Arguing for a link between trees and the full exercise of the human condition, he requested the integration of a principle according to which "destroying or illtreating trees" could not be decided unilaterally. Those wishes found an echo in the recurrent attempts by policy-makers in charge of environmental affairs to officially wrap an increasing proportion of the forest "in cotton wool."

Current social representations seem to reflect contradictions and paradoxes. Indeed, while cutting trees is an act that seems to activate intimate and deeply rooted drivers, the link with wood is often invisible or inexistent. Indeed, a large meta-study undertaken in 2007 by Rametsteiner et al. (2007) shows that Europeans generally have a clearly positive attitude toward wood: "It is a material considered as natural, warm, healthy, good-looking, easy to use and environmentally friendly by a majority of people across Europe. ... People associate forests as such and the wooden furniture at home very positively, but forget about or dismiss the link between both, namely harvesting and wood processing, which often have negative associations. Only a very small percentage seems to associate wood products with wood production in forests, harvesting or the loss of forest."

People like wood as a material, but they do not like trees to be felled. Likewise, many people may love a good steak, but do not make the connection with a slaughtered animal. Wooden material and steak have become objects in themselves in today's perception, disconnected or even completely cut off from their origins by the shaping they have undergone in people's perceptions. This raises the issue of consistency in perceptions or values in individual or social representation, leading to "cognitive dissonance" and individual/social strategies to deal with such dissonances.

7.5 WHAT ABOUT CHILDREN AND YOUTHS?

As shown in various studies mentioned before, young people seem to constitute a group expressing particular worry about the state and future of the forests.

In Finland, where forests are part of life, concerns were expressed on the erosion of the link between forest and youths due to urbanization (Heino and Karvonen 2003). When Bodson (2005) showed the impact of the age factor, with 68% of young Belgian people (15–24 years) being in favor of prohibiting the cutting of trees, he remained indecisive on interpreting the trend: was it an effect of the respondents' young age or were new trends being expressed by a new generation? Not enough evidence was available to decide, but recent studies on youths and children can contribute to fuel the debate.

7.5.1 YOUNG PEOPLE

By deepening the analysis of data collected every 5 years in France, Dobré and Granet (2007) identified a number of specificities in the vision of youths (15–24) on forests. They revealed in particular two different social representations based on specialized uses of forests: "(1) A forest "park" arranged for "Parisians" and families, possibly themselves when they go there to practice a sporting activity" and "(2) the forest to be preserved and to transmit, symbol of a nature which must not disappear, that those who desire it could frequent. It is not a forest sanctuary but a forest that does not really need man and where the few human interventions (e.g., security) must remain invisible."

In her PhD study in communication sciences, Matagne (2017) studies the cognitive autonomy and skills of audiences with respect to tree- and forest-related media messages, with a specific focus on youth. Among the various experiments undertaken, she confirms that young people feel concerned by forests and mainly see forests from an ecological perspective rather than leisure or an economic one. She also compares the level of agreement of 120 Belgian master's degree students on various statements related to forests. Without much surprise, she shows that students with a background in forestry are more inclined to understand the economic role of forests and have less doubt in accepting the usefulness of human action in this field than others. In response to the statement on the need to prohibit woodcutting to satisfy human needs, while only 3% of foresters agreed, 25% of nonforesters agreed and 24% had no opinion. Trends were less divergent on the social and ecological roles of forests, while nonforesters were more willing to consider human presence or action negatively and to refer to respect and sacred wordings. However, both groups agreed on the potential destructive role played by humans by mainly disagreeing on the statement that forests will survive whatever humans may do. The disagreement expressed by the foresters is nevertheless more marked (72% against 60%).

7.5.2 CHILDREN

Few studies have been undertaken on the perceptions or social representations of forests by children. Some preliminary and very exploratory findings were produced by Farcy et al. (2015), when more than 9,000 drawings were submitted by children (between 6 and 10 years old) from 23 countries of the EU as part of a "What Is the Forest for Me?" contest, organized in 2013 by Directorate-General for Agriculture and Rural Development of the European Commission. The study focused on a categorical content analysis (de Bonville 2006; Bardin 2007) of a sample of 100 drawings remaining after two selection filters applied to further identify the winners of the contest. Variables were thematically wide ranging: next to descriptive and graphical ones, some focused on land tenure (Le Roy et al. 1996) or on the relationship between people and forests, while others related to child psychology (Figure 7.2).

While being exploratory and subject to bias related to the constitution of the sample and the selection of the winners, some prevailing trends were identified, taking

(a)

(b)

FIGURE 7.2   Some drawings from children of the EU: (a) Agnès, 8 years old, Hungary; (b) Jan, 9 years old, Poland.   (*Continued*)

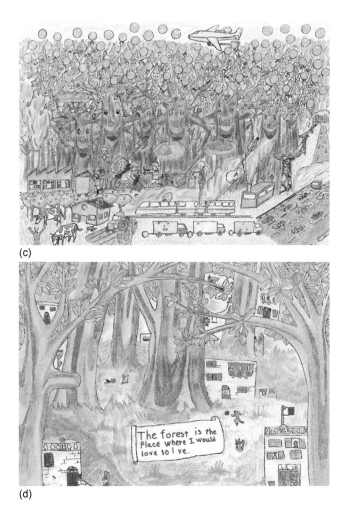

FIGURE 7.2 (CONTINUED)   (c) Daniel, 9 years old, Spain; (d) Kay, 8 years old, Malta. (Courtesy of EC–Directorate-General for Agriculture and Rural Development (DG AGRI), Forest Drawing Contest "What Is the Forest for Me?" 2013. © EC–DG AGRI.)

into account elements either recurrently present or missing in the drawings as representative of the content of social representations.

The conclusion is that the vision of children is positive, realistic, and systemic from an ecological point of view. Contrary to young people's feelings, only some drawings expressed a forest in danger, threatened with degradation or even annihilation by humans. We are also far from the traditional forests of children's literature, loaded with magic, mysteries, dangers, and anxieties. There are few traces of the fantastic and fictional audiovisual media, yet present on our screens. The functions of the forest, although only partially considered, are present, as are the contemporary stakes related to species, energy, and pollution.

What the children's representation lacks is essentially the fact that the present EU's forests are the result of human actions and that they require recurrent interventions of professionals and a permanent follow-up. However, these "forest agents" remain absent from the drawings and are even sometimes considered as harmful. Similarly, the function of wood production is only represented in a marginal way, whereas in reality, the European forest is predominantly productive.

A study of the role of forests and trees in the narrative underlying French-speaking board games for children in Belgium (Hubert et al. 2016) shows that players would be willingly involved in an opportunist relationship with forests by enjoying them without interfering in them rather than being involved in a hostile or aggressive relation with them.

Changes in the school systems and lifestyle could be seen as reinforcing this trend (Pergams and Zaradic 2008; Gubbels et al. 2011). Indeed, data from the United States show that the average outdoor time in 2002 was half that of that in 1982, while time at school was up by 20%, and passive leisure increased by 400% (Juster et al. 2004).

7.6   APPEAL OF NATURALNESS

Feeding on the idea of nature and the perceived naturalness of forest, antithesis of the urban, recreation in forests, including tourism, has been growing quite continuously since the 1970s (Kalaora 1997; Granet 2011; MacCleery 2011; Robert et al. 2012; Hunziker et al. 2013; Figure 7.3).

For instance, in the United States, while the recreational use on private and public forests has significantly increased, national forest recreational visits increased too, from 18 million in 1946 to 93 million in 1960 and to 350 million in 1996 (MacCleery 2011).

FIGURE 7.3   Nordmarka Forest in the surrounding of Oslo, peaceful place highly used by city dwellers. (Photo by Christine Farcy.)

If economic use is largely absent from public representations, forests are no longer perceived as sanctuaries or associated with wilderness without any kind of human intervention. In the United States, "since 1985, there has been a dramatic shift in the mission of federal and many other public lands, and the managing agencies are now largely focused on restoring and maintaining healthy ecological conditions and meeting recreational and amenity preferences of local and national stakeholders. National forest timber sales have declined by more than 80 percent since 1985 and now provide less than 2% of US timber consumption" (MacCleery 2011).

Why then does this trend toward a greater importance of the recreational function of forests not appear more clearly in the common agricultural policy (CAP) Eurobarometer survey (EU 2016)? With threats being heavily felt, is it difficult to dare express a personal pleasure in enjoying forests? Or is it that as observed in France in the 1980s to 1990s, environmental concerns are hiding or have replaced recreational activities, in terms of media communication and research at global level (Deuffic et al. 2004)? Another reason may be the global scope and supranational organization of environmental pressure groups, while the organizations focusing on the sociocultural dimensions and sports function mostly at the local level, with less visibility? Or still, is it that, in keeping with Moscovici's definition, social representations are not of the whole society but are products of social groups, some of which in this case are more attracted to forests than others?

In any case, even though they concern minorities, trends are significant. One of these has recently been, in the United Kingdom for instance, to purchase a woodland for the benefits of recreation, so as to reconnect oneself with its fauna and flora while explicitly "being away from it all," especially from electronic devices and social networks (Box 7.1).

BOX 7.1 "PEOPLE ARE BUYING WOODS (FOR LESS THAN YOU'D THINK) AS AN ESCAPE FROM HYPERCONNECTED SOCIETY"

Buying woodland has always been a refuge for the well-to-do classes: for land-owners in the 18th and 19th centuries, for investors wishing to benefit from government grant schemes for tree planting after the Second World War or for wealthy taxpayers willing to benefit from fiscal reductions through the owner-ship of woodlands in the late 20th century. So, this interest from the "happy few" in buying woodland in the early 21st century may be the mere continuation of a tradition and the title of the *Independent* article a rather misleading eyecatcher (Hooton 2016).

However, this trend, anecdotal in numerical terms, also points to a deeper-seated reaction: to cramped housing, crowded cities and a craving to reconnect with wildlife. This is at least the interpretation of the estate agency, which specialises in such transactions. A rapid analysis of the descriptions and photographs of the 40 properties on sale on their website (http://www .woods4sale.co.uk/) at the time of viewing enhances certain characteristics of the woodlands for sale, independently of the region where the woodland is located (England, Scotland, Wales), which are supposed to reflect the buyers' aspirations.

The photographs in each advert highlight the different seasonal delights (autumn colours, with mushrooms and colourful berries, carpets of blue-bells and unfurling ferns in the spring, long winter shadows and snow-laden branches, unkempt "romantic" aspect with mossy stones), as well as the mature trees and winding paths. The comments related to the woods for sale enhance both the accessibility of the woods from the hurly-burly of the nearby towns or cities and the peacefulness of the woods. The maturity of the trees is under-lined as well as, in many cases, the possibility of producing sustainable timber or at least firewood, although the only advert that places a strong emphasis on timber production is a conifer wood, traditionally not valued as much as deciduous woodland in the UK.

The texts also emphasise the great opportunity to make the most of the "great outdoors," through birdwatching, woodland craft, den-building or camping for instance. Interestingly, they also highlight whenever possible a connection with the past: the location of the woodland close to the site of a historic battle, the presence of a village nearby where the pub has served good, wholesome food for generations, the presence of a woodland in this very spot for centuries, with its native species. So, although not absent from the argu-ments, the economic interest of such an investment is quite limited, and what is for sale hinges on meanings and values linked to retreat from the bustle of urban life, on the one hand, and continuity, on the other hand. The continu-ity of the woodland itself through its long existence, but also of the practice of managing the land and, last but not least, of the human relationship with nature through outdoor pursuits in tree-covered environments: in a nutshell, the opposite of the characteristics of our urban societies described at the start with the terms distance, virtuality and immateriality.

Forests are places for sociability and emotions, as opposed to ordinary urban constraints, the precedence of the intellect and the "mental barriers of a tyrannical and hyperactive superego"; they can act as "safety valves … freeing one from social conventions and thus appeasing the mind" (Boutefeu 2008).

7.7 FORESTS, WELL-BEING, AND IDENTITY

Narrowing the scope on England, the first country in the world to become urban in its majority in the midnineteenth century and still one of the most urbanized ones today, but with low forest cover (less than 15%), allows us to better perceive the mainstream representations of forests. Copious academic literature on the benefits of forests for human health (physical and mental—see also Chapter 13), education, social integra-tion, and well-being more generally (O'Brien and Morris 2013) has progressively per-meated British society's representations of forests over the past 30 years.

The 12 Community Forests created around highly urbanized areas (three pilot areas in 1989 and nine community forests in 1991) have partly focused their agenda on these very issues, in conjunction with other institutions (prisons, hospitals, and schools, among others). One of the latest examples can be found in a series of videos made in late 2016 by the health sector to promote the development and adequate use of green and wooded spaces around hospitals.[4] In twenty-first century British society, wooded spaces are seen by most of the population as places associated with the benefits of nature at large.

In England, forests play an important part in the collective psyche, as remnants of a landscape heritage, reminiscent of the rural past, and therefore representative of shared values to be protected from change and the aggressions of urban life, the very values that Woods4Sale taps into (Box 7.1). They send back to a tradition in literature (folklore, fairy tales, and Shakespeare's plays, among others), in which the hero or heroine, willingly or by a whim of destiny, embarks on a journey leading to the discovery of self, through the trials of a long and solitary wandering through the forest. These journeys, among their characteristics, are composed of the isolation of the characters in the forest, the absence of social hierarchies, and an unusual solidarity between humans and nature, whether animals or trees (Nail 2008).

The protection of the so-called veteran trees, whether in Robin Hood's Sherwood Forest or isolated in the rural landscape, the term *unspoilt landscape*[5] related the New Forest National Park on its website or the very popular television series on "Remarkable Trees," are just three instances of these positive cultural representations of forests and trees (see also Chapter 9). The 2011 crisis also demonstrated it vividly to the incoming coalition government who wanted to dismantle the state forest sector: the overwhelming reaction of British people from all walks of life amply demonstrated that there was more to forests than economic issues. The report of the ensuing Independent Panel on Forestry permits to hold a mirror to the shared values associated with these landscapes, among them a close association between national identity and forests.

In a nutshell, the dominant representations of British forests today in the wide public, especially in England, hinge on two positive, although apparently contradictory, shared values: distant from where the majority of people live, they have acquired a quasi mythical power of conjuring up common national roots in a context of globalization; on many people's doorsteps as a result of active public policies over the last 30 years to develop urban forestry through the Community Forests Programme, they evoke everyday refuges to cultivate not only one's individual health, but also community cohesion and well-being through groups of protection of local woods (Forestry Commission 2017). But the paradox is only apparent, since the Community Forests themselves appeal to representations of the national past: the Community Forests' "extensive areas of woodland and open space will bring something of the magic of the first forests back to the edge of the city" (Johnston 2012).

Confirming this need, Dassié's (2016) analysis of the drivers of websites dedicated to virtual plantation or the adoption of trees shows that in practice, an invitation to plant trees which conjures up long life and eternity, often associated with the birth of a child, is less a way of defending trees than defending other causes. Indeed, "it is not so much about planting trees than about metaphorically rooting identities and an imaginary of crossborder social concord, through a project of cultural or identity hegemony."

7.8   SACREDNESS OF FORESTS

The aesthetic and emotional components of forests in literature and the cinema, as studied, for example by Schoentjes (2015) or Mottet (2017), confirm that emotions and sensations felt in the forests are visible manifestations of their symbolic function.

In that respect, the case of Germany provides interesting results. It is a widely held perception (particularly in some circles) that the German national identity and soul are strongly rooted in forests, and the romantic tradition according to which forest is a place of well-being, a way to flee the world, is still alive (Schama 1995). Germany is also the place where a renowned school of forestry developed. The PAC Eurobarometer (EU 2016) shows that this country follows the European trend described earlier ("habitat for animals": 70% and "absorbing carbon": 67%) while having "wood and paper" in the sixth position (19%) after "healthy leisure" (22%).

The outstanding success of the book *The Hidden Life of Trees: What They Feel, How They Communicate* by Wohlleben (2015) throws light on this. In his bestseller (sold at over 1 million copies in Germany in June 2017 and available in 35 countries), Wohlleben states that trees are not things but sentient and noble creatures able to communicate with each other and to establish relationships of friendship. He compares the fate of trees with that of the whales, both being noble creatures massacred by man. Although built on a scientifically controversial basis and an excess of anthropomorphism, the global success of such a discourse says a lot about the readers and without any doubt reveals major trends.

In an article of *Die Zeit* edited in *Courrier International*, Sussebach and Jänicke (2016) depict how Wohlleben has managed to crystallize the sacred nature of forests inherent to the social representation of city dwellers. The article shows his multiple facets: Wohlleben not only writes, but also has a sense of entrepreneurship in touch with the present mood (Box 7.2). Indeed, in 2003, he created a natural cemetery located in a forest, where over the last 13 years, he has already installed 3,400 funeral urns, sold at €3,000 each. With this income, he intends to buy forests so as to create "untouched" forest reserves dedicated to biodiversity, where he organizes visits. As most Germans, he also likes the warmth of wood and burns every year 10 m^3 of wood, that is, 10 times the national average.

Boutefeu (2008) recalls that "forests are sensitive objects. […] The attention given to the senses generates a feeling of calm, of appeasement synonymous with inner peace. Thanks to memories of personal history and to an emotional state, forests are intimate places and have a mirror effect allowing everyone to resonate with one's interior space."

However, Van den Berg and Konijnendijk (2012) describe how forest can also provoke ambivalence or contradictory feelings (see also Chapter 6). By offering solitary experiences of the forest at night, Terrasson (1998) intends to wake up fears confined in the unconscious, connected with the perspective of the return to the original status of nature, thus sending everyone back to his instinct, immoral, and animal ego.

Cultural memory is often associated with the core of social representation. The cultural memory of forests and civilization goes back a long way, for instance, in Lucrecia's De Natura "day by day they forced the forests to retreat further up the mountain, abandoning the lowlands to the crops…"

BOX 7.2   CEMETERIES IN FORESTS

For a long time, trees have been linked to memory and regeneration in the face of death (Rival 1998; Schama 1995). However, the symbolism associated with it has changed strongly over time, from the utilitarian function and amenity to a symbolic passage, or even the connection to Nature (Pearson 1982). In the middle of the 19th century, the first "garden cemeteries" or "gardens of graves" were created in the United States in response to the saturation of cemeteries close to churches. Some of these early cemeteries are today splendid arboreta offering public spaces of ecological and social interest, close to the centre of cities. On the other hand, in Europe, the cohabitation of trees and cemeteries seems weaker: trees may be planted around the graves and in the paths, but burials are more associated with tombstones and monuments. However, this is not clear-cut: forest cemetery initiatives have existed for over a century in Europe. Thus, the Munich Waldfriedhof ("wooden cemetery") dates from 1907 and occupies 170 hectares, and Skogskyrkogården ("woodland cemetery") in Stockholm was created between 1917 and 1920 by two young architects and is part of World Heritage since 1994.[6]

Now widespread in the world, the dispersion of human ashes is expanding, a practice for which the forest symbolism brings a highly sought-after complementarity. Among possible sites for the dispersal of the ashes in a natural context (air, sea, and rivers), forests are regularly chosen. Trees offer a symbolic passage, and they are enduring loci for meditation. Thanks to this long-lasting "temporal anchorage," the deceased and the families also accomplish a gesture for the environment. The tree then becomes sacred and performs functions that are far more sensitive than simple economic ones.

In Belgium, the Foundation "Les Arbres du Souvenir," created in 2015, is a case in point. The foundation aims to offer an alternative way to mourn, offering places of memory, relaxation, appeasement and serenity in nature. In its forests, it offers to dedicate a tree to the memory of a lost relative and offers the possibility to accommodate their ashes (Figure 7.4).

At the same time, this project, acknowledged by public authorities, also develops environmental, citizen, and art projects on its sites. For families, planting a tree in the forest on the ashes of a deceased person and scattering or burying the ashes in the roots of an old tree are gestures that contribute to the grieving process. The testimonies of families are that trees "bring serenity and patience," "create a link with nature," and are "places of life." Since 2015, a community has been formed around the forest; all the families of the deceased gather near the trees. Together, they participate in the embellishment of the forest, its management, and its communication. Biodegradable urns are offered to families; at times, it contains a space filled with soil and the seed of a tree. Far from silvicultural considerations, the tree roots on the ashes of the deceased. The tree is no longer a setting for the cemetery but an actor in the process of mourning.

FIGURE 7.4   "Trees of remembrance": cemetery in the forest in Belgium. (Photo by Olivier Baudry.)

In an extensive essay on the imaginary dimension of forests, *Forests: The Shadow of Civilization*, Harrison (1992), traveling through history and literature, describes how forests have been sacred, profane, shelter, sanctuary, outlaw, or dangerous in turns. He further intends to demonstrate in a very elegant way that cutting trees, or more generally deforestation, have strong symbolical meanings in Western societies: "how many buried memories, ancestral fears and dreams, popular traditions, myths, and more recent symbols flare into the fires of deforestation. These fires move us for reasons that escape our understanding: they make us react on another level, that of our cultural memory." Forests, the demonstration goes, can be considered as metaphors of remembrance and still lie deep in our cultural memory: despite attempts at rationalizing since the Enlightenment period, they are still connected with human transcendence and our intimate relation to our natural origins (Harrison 1992).

In that sense, deforestation amounts to much more than depleting biodiversity: while cultural memory is deeply embodied in forests, losing forest means indeed losing the access to this collective memory. Citing Vigo, Harrison recalls that Western civilization was built both thanks to and against forests: "This was the order of human institutions: first the forests, after that the huts, then the villages, next the city and finally the academies." In such a context, he emphasizes the particular, high symbolic value of forests for urban dwellers. The city is indeed a clearing, but today, it is felt that there are risks and fears that this might be forgotten because the city has lost its connection to forests, this literal and symbolic boundary "without which the human abode loses its grounding."

Trees in towns can in this sense play the role of symbolic witnesses or relays. Likewise, Brosseau (2011) recalls that in Japanese cities, "the main building of a 'Shinto' sanctuary or a Buddhist temple is surrounded by trees. High and profuse, they remain a strong visual cue, vestiges of the original sacred forest or its evocation. The tutelary deities reside there, hence the name 'Chinju no mori,' a protective forest. Today, these sacred trees or woods are often inventoried or even protected as natural heritage. Even though they are not as old, nor as 'original' as some think, they transmit the sacred symbolism of the forest, and perpetuate in the city the presence of a nature of cultural and environmental value."

7.9   CONCLUSION

This chapter has identified the great potential and relevance of further developing the use of the social representation concept in forestry and implicitly argues in favor of giving due space to a regular follow-up of people's concerns and sensitivity toward forests. In today's world, with political and environmental threats, as well as information technologies contributing to reinforce feelings of threats, forests and trees seem indeed to become increasingly deeply rooted symbols of peace and security, in particular in urbanized contexts, thanks to their unique role played throughout human history.

The social representations of forests are closely connected to the history of humankind, as clearing land made way for civilization to flourish, but at the beginning of the third millennium, there seems to be a clear trend for a sense of human responsibility toward the protection and survival of forests, if only for the services that they provide us with. Forests are clearly not culturally neutral: as an important part of our "biocultural heritage," they bear the hallmark of local and national history.

The discrepancy between this heritage and the economic importance of the forest is obvious. Preservation and "soft" practice such as contemplation are becoming keywords in people's use of forests. While the media often rely on dominant representations and thus reinforce them, the prominence they give to environmental functions and issues only seems to increase the trend.

The main arguments for forest protection today seem to revolve around environmental drivers (carbon storage, protection of wildlife, climate change, and renewable energy) and health and social needs, with their recreational uses and landscape identity.

Distance, virtuality, and immateriality, which characterized the new link between people and the forest under urbanization, probably contribute to awakening humans' cultural memory. In every European country, among adults and young people, the main function of forests, so it seems, is environmental. The environmental concerns could activate symbols partaking of the core of the forest representations: people share an intimate relationship with their natural origins. The success of a romantic vision of forests could be interpreted as a sign of the predominance of the preservation of heritage among such social groups.

Contradiction and paradox reflect dynamics involved in the development of social representations, for example, the fact that people like wood but disapprove of tree cutting or that they consider leisure as a minor function of forests despite their increasing use for recreation. That probably reflects more the peripheral elements of the representations.

With the caution required due to the very preliminary status of the studies analyzed, a difference appears between children and young people: in their drawings of what the forest means for them, children do not represent an endangered forest, whereas young people (and adults, too) are seriously concerned by forest issues. We could argue that children are relatively immune to the media effect around forests, in particular social media. The media could overplay the dangers to forests by giving a partial viewpoint on related issues, which would reinforce the already pessimistic representations felt by people.

What are some of the main consequences for the forestry sector? On the one hand, those trends could have a high impact when urban citizens and authorities make decisions that affect forest resources and rural populations or when people from developed countries set up a global agenda that affects developing countries on forestry issues. The case of the misconception of deforestation in developed countries, both on the scale and the process itself, that generates an inadequate response and eventually creates a controversial reality is an illustrative case in this regard (Fabra-Crespo and Rojas-Briales 2015). On the other hand, ethical divergence on tree cutting constitutes a challenge for a sector that needs social legitimacy and a "license to operate" in a context where the forestry sector is becoming a huge ally of humans in their intents to mitigate and adapt to the effects of expected climate change, in particular through the development of a greener economy, besides many other vital environmental services (water, soil), biodiversity, and other social services (livelihood in rural areas, recreation, and landscape) (IPCC 2011; EC 2012). This ambivalence remains paradoxical and requires particular attention.

In such a context, the forestry community and, in particular, forest professional would do well to renew its communication strategies as developed in Chapter 10. Far for being a marketing tool intending to convince people with simplistic messages, today's communication should focus on the media skills of the recipients rather than on the message itself, while assuming the complexity of forest-related issues.

The forest community should also review or adapt its position to society, as they have already done in the United Kingdom, for instance (Nail 2008). Still often acting as the symbolic owner of forests and adopting to defensive mode, forest professionals should reverse the perspective to clearly serve society, more explicitly offering a resolution to concrete problems and thus overcome their strict role as forest managers in order to become the managers of the relationships between forest and people (Kennedy and Koch 2004).

Lastly, a deep understanding of the structuring and resilience of social representations of forests has become as important as understanding the functioning of forest ecosystems itself. This implies first and foremost, for forest professionals in particular, acknowledging the existence and the power of persuasion of people's social representations and accepting their legitimacy.

Facing these challenges could imply exploring new scientific spheres, in particular in the human and social sciences, and even fetching novice colleagues on forestry issues. The development of the monitoring of social representations, and even a media observatory, should also contribute to that end. Lastly, it might also be useful to consider framing forest-related discussions and decision-making processes in open and inclusive ethical dialogues (see Chapter 4) allowing to softly, but deeply and explicitly, address divergences and paradoxes while boosting creativity.

REFERENCES

Abric, J.-C. 1976. Jeux, conflits et représentations sociales. PhD Dissertation, Marseille: Université de Provence.

Abric, J.-C. 1993. Central system, peripheral system: Their functions and roles in the dynamics of social representation. *Papers on Social Representations* 2:75–78.

Bardin, L. 2007. *L'analyse de Contenu*. Paris: Presses Universitaires de France.

Bauer, M. and G. Gaskell. 1999. Towards a paradigm for research on social representations. *Journal for the Theory of Social Behaviour* 29:163–186.

Bodson, D. 2005. Comprendre les perceptions, les usages et les significations de la forêt en 2005. *Forêt Wallonne* 79:19–28.

Boutefeu, B. 2008. La forêt, théâtre de nos émotions. *RDV Techniques* 19:3–8.

Brosseau, S. 2011. Du territoire rural aux parcs urbains: Métamorphoses du *satoyama* dans la métropole de Tokyo. Agriculture métropolitaine/Métropole agricole 196–210.

Buijs, A. 2009. Public natures: Social representations of nature and local practices. PhD Dissertation, Wageningen: University of Wageningen.

Dassié, V. 2016. Cyber-plantations: Les arbres entre politique et communication. In *Forêt et Communication: Héritages, Représentations et Défis*, eds. Dereix, C., Farcy, C., and F. Lormant, 247–271. Paris: L'Harmattan.

de Bonville, J. 2006. *L'analyse de Contenu des Médias: De la Problématique au Traitement Statistique*. Bruxelles: De Boeck.

Demougin, J. 1995. Forêt, nature et société. Conclusions. In *La Forêt: Les Savoirs et le Citoyen*, eds. Meiller D. and P. Vennier, 63–65. Chalon-sur-Saone: Agence Nationale de Création Rurale.

Deuffic, P., Granet, A. M., and N. Lewis. 2004. Forêt et société: une union durable. 1960–2003: Evolution de la demande sociale face à la forêt. *RDV Techniques* 5:10–14.

Dobré, M. and A.-M. Granet. 2007. La forêt des jeunes. *RDV Techniques* 17:61–67.

Dobré, M., Lewis, N., and A.-M. Granet. 2006. Comment les français voient la forêt et sa gestion. *RDV Techniques* 11:55–63.

Dobré, M., Cordellier, M., and A.-M. Granet. 2016. Enquête nationale Forêt-Société 2015: Premiers résultats. Papier présenté au colloque du Réseau SHS d'ECOFOR Regards croisés sur les valeurs de la forêt.

Dobsinska, Z. and Z. Sarvasova. 2016. Perceptions of forest owners and the general public on the role of forest in Slovakia. *Acta Silvatica and Lignaria Hungarica* 12:23–33.

Durkheim, E. 1898. Représentations individuelles et représentations collectives. *Revue de Métaphysique et de Morale* 6:273–302.

Eizner, N. 1995. La forêt, archétype de la nature. In *La Forêt. Les Savoirs et le Citoyen*, eds. Meiller, D. and P. Vennier, 17–19. Chalon-sur-Saone: ANCR.

EC (European Commission). 2012. *Innovating for Sustainable Growth: A Bioeconomy for Europe*. Brussels: EC. http://ec.europa.eu/research/bioeconomy/pdf/official-strategy _en.pdf (Accessed August 12, 2014).

EU (European Union). 2014. *Special Eurobarometer 416 Attitudes of Europeans towards the Environment*. Report. Brussels: European Union.

EU. 2015. *Special Eurobarometer 436 Attitudes of Europeans towards Biodiversity*. Report. Brussels: European Union.

EU. 2016. *Special Eurobarometer 440 Europeans, Agriculture and the Common Agricultural Policy*. Report. Brussels: European Union.

Fabra-Crespo, M. and E. Rojas-Briales. 2015. Comparative analysis on the communication strategies of the forest owners' associations in Europe. *Forest Policy and Economics* 50:20–30.

Farcy, C., Herrezeel, A., Matagne, J. et al. 2015. Qu'est-ce que la forêt pour moi? Éléments émergents pour une typologie des représentations sociales de la forêt par les enfants de l'Union Européenne. Paper accepted for the World Forestry Congress.

Flament, C. 1989. Structure et dynamique des représentations sociales. In *Les Représentations Sociales*, ed. Jodelet, D., 204–219. Paris: Presses Universitaires de France.

Food and Agriculture Organization of the United Nations. 2003. *Unasylva* 213.

Forestry Commission 2017. *Public Opinion of Forestry 2017, UK and England*. Edinburgh: Forest Research.

Gendron, C. and D. Dumas. 1991. Culture écologique: Etude exploratoire de la participation de médias québécois à la construction de représentations sociales de problèmes écologiques. *Sociologie et Sociétés* 23:163–180.

Glaser, B. G. and A. L. Strauss. 1967. *The Discovery of Grounded Theory: Strategies for Qualitative Research*. Chicago, IL: Aldine.

Granet, A. M. 2011. Au-delà des usages et des pratiques: Les représentations partagées de la forêt idéale. In *Forêt et Paysage*, ed. Corvol, A., 273–287. Paris: L'Harmattan.

Granet, A. M. and R. Amand. 2016. Les ruraux sont-ils des urbains comme les autres? Analyse des pratiques et représentations de la forêt dans les territoires ruraux. In *Forêts, Savoirs et Motivations*, eds. Farcy, C. and N. Huybens, 67–83. Paris: L'Harmattan.

Gubbels, J. S., Kremers, S. P. J., van Kann, D. H. H., Stafleu, A., Candel, M. J., and P. C. Dagnelie. 2011. Interaction between physical environment, social environment, and child characteristics in determining physical activity at child care. *Journal of Health Psychology* 30(1):84–90.

Guimelli, C. 1993. Locating the central core of social representations: Towards a method. *European Journal of Social Psychology* 23:555–559.

Hallé, F. 2011. *Du Bon Usage des Arbres. Un Plaidoyer à L'attention des Elus et des Enarques*. Arles: Actes Sud.

Harrison, R. P. 1992. *Forests: The Shadow of Civilization*. Chicago, IL: Chicago University Press.

Heino, J. and J. Karvonen. 2003. Forests—An integrated part of Finnish life. *Unasylva* 213:3–9.

Hooton, C. 2016. People are buying woods (for less than you'd think) as an escape from hyperconnected society. *The Independent*. November 2016. http://www.independent.co.uk/arts-entertainment/people-are-buying-woods-uk-as-an-escape-from-hyperconnected-society-a7418666.html.

Höijer, B. 2011. Social representation theory: A new theory for media research. *Nordicom Review* 32:3–16.

Hubert, A., Farcy, C., and P. Fastrez. 2016. L'arbre et la forêt dans les récits des jeux de société francophones. In *Forêt et Communication. Héritages, Représentations et Défis*, eds. Dereix, C., Farcy, C., and F. Lormant, 107–131. Paris: L'Harmattan.

Hunziker, M., Frick, J., Bauer, N. et al. 2013. *La Population Suisse et sa Forêt: Rapport Relatif à la Deuxième Enquête Menée dans le Cadre du Monitoring Socioculturel des Forêts (WaMos2)*. Berne: Office Fédéral de l'Environnement, Institut fédéral de recherches sur la forêt, la neige et le paysage

IPCC (Intergovernmental Panel on Climate Change). 2011. *IPCC Special Report on Renewable Energy Sources and Climate Change Mitigation*. Prepared by Working Group III of the IPCC. New York: Cambridge University Press.

Jodelet, D. 1984. Représentations sociales: phénomènes, concept et théorie. In *Psychologie Sociale*, ed. Moscovici, S., 357–378. Paris: Presses Universitaires de France.

Johnston, M. 2012. Trees in towns II: A new survey of urban trees in England and their condition and management. *Arboricultural Journal* 34(2):119.

Juster, F. T., Ono, H., and F. P. Stafford. 2004. *Changing Time of American Youth: 1981–2003*. Research report. Ann Arbor, MI: Institute for Social Research, University of Michigan.

Kalaora, B. 1997. Du musée vert au musée écologique. Illusion ou réalité. In *La Forêt. Perceptions et Représentations*, eds. Corvol, A., Arnould, P., and M. Hotyat, 303–311. Paris: L'Harmattan.

Kennedy, J. J. and N. E. Koch. 2004. Viewing and managing natural resources as human-ecosystem relationships. *Forest Policy and Economics* 6:497–504.

Le Roy, E., Karsenty, A., and A. Bertrand. 1996. *La Sécurisation Foncière en Afrique, Pour une Gestion Viable des Ressources Renouvelables*. Paris: Karthala.

MacCleery, D. W. 2011. *American Forests: A History of Resilience and Recovery*. Durham: Forest History Society.

McComb, B. C. 2016. *Wildlife Habitat Management: Concepts and Applications in Forestry*. Boca Raton, FL: CRC Press.

Matagne, J. 2017. Littératie médiatique et environnement: Évaluation de l'autonomie cognitive des jeunes envers les médias traitant des forêts. PhD dissertation, Louvain-la-Neuve: University of Louvain.

Michel-Guillou, E. 2006. Social representations and social practices: The example of the pro-environmental commitment in agriculture. *Revue Européenne de Psychologie Appliquée* 56:157–165.

Moliner, P. and C. Guimelli. 1995. A two-dimensional model of social representations. *European Journal of Social Psychology* 25:27–40.

Moliner, P. and C. Guimelli. 2015. *Les Représentations Sociales: Fondements Théoriques et Développements Récents*. Grenoble: Presses universitaires de Grenoble.

Moscovici, S. 1963. Attitudes and opinions. *Annual Review of Psychology* 14:231–260.

Moscovici, S. 1969. Préface. In *Santé et Maladie: Analyse D'une Représentation Sociale*, ed. Herzlich, C. Paris: Lahaye Mouton.

Moscovici, S. 1988. Notes towards a description of social representations. *Journal of European Social Psychology* 18:211–250.

Mottet, J. 2017. *La Forêt Sonore: De L'esthétique à L'écologie*. Ceyzérieu: Champ vallon.

Nail, S. 2008. *Forest Policies and Social Change in England*. New York: Springer.

O'Brien, L. and J. Morris. 2013. Well-being for All? The social distribution of benefits gained from woodlands and forests in Britain: Local environment. *International Journal of Justice and Sustainability* 19:356–383.

Paré, I. 2012. L'influence du documentaire militant sur les conceptions de la forêt—Une analyse des représentations sociales de la forêt entourant la diffusion de L'erreur boréale dans la presse écrite. PhD dissertation, Quebec City: Université Laval.

Pergams, O. R. W. and P. A. Zaradic. 2008. Evidence for a fundamental and pervasive shift away from nature-based recreation. *Proceedings of the National Academy of Sciences of the United States of America* 105(7):2295–2300.

Perraud, P.-P. 1997. Perceptions et représentations de la forêt à travers des questions d'enfants. In *La Forêt: Perceptions et Représentations*, eds. Corvol, A., Arnould, P., and M. Hotyat, 341–352. Paris: L'Harmattan.

Pearson, M. P. 1982. *Mortuary Society and Ideology: An Ethnoarchaeological Study*. New York: Cambridge University Press.

Rametsteiner, E. and F. Kraxner. 2003. Europeans and their forests: What do Europeans think about forests and sustainable forest management? In *Ministerial Conference on the Protection of Forests in Europe*. Vienna: Liaison Unit Vienna.

Rametsteiner, E., Oberwimmer, R., and I. Gschwandtl. 2007. Europeans and Wood: What Do Europeans Think about Wood and Its Uses? Ministerial Conference on the Protection of Forests in Europe. Warsaw: Liaison Unit Warsaw.

Rametsteiner, E., Eichler, L., and J. Berg. 2009. *Shaping Forest Communication in the European Union: Public Perceptions of Forests and Forestry*. Brussels: European Commission.

Rival, L. 1998. *The Social life of Trees: Anthropological Perspectives on Tree Symbolism*. Oxford: Berg.

Robert, A., Farcy, C., and D. Bodson. 2012. Strategic framework for tourist valorization of forests in Wallonia (Belgium). Paper presented at Forest for People, Alpbach.

Schama, S. 1995. *Landscape and Memory*. New York: AA Knopf.

Schmithüsen, F. 1999. Perceiving forests and their management. *Annales de Géographie* 108:479–508.

Schoentjes, P. 2015. *Ce qui a lieu: Essai d'écopoétique*. Marseille: Editions Wildproject.

Scott, A., Gilbert, A., and A. Gelan. 2007. *The Urban-Rural Divide: Myth or Reality?* Policy Brief Series 2. Aberdeen: The Macaulay Institute.

Smajda, A.-C., Farcy, C., and T. De Smedt. 2016. Etude de l'évolution de la représentation sociale de la forêt dans la publicité commerciale d'un magazine généraliste. In *Forêt et Communication: Héritages, Représentations et Défis*, eds. Dereix, C., Farcy, C., and F. Lormant, 273–285. Paris: L'Harmattan.

Sussebach, H. and F. Jänicke. 2016. Le choix des arbres. *Courrier International* 1324:42–45.

Terrasson, F. 1998. La Peur de la Nature: Au Plus Profond de Notre Inconscient, Les Vraies Causes de la Destruction de la Nature. Paris: Sang de la Terre.

Van den Berg, A. E., and C. Konijnendijk. 2012. Ambivalence towards nature and natural landscapes. In *Environmental Psychology: An Introduction*, eds. Steg, L., Van den Berg, A., and J. I. M. De Groot, 67–76. Chichester: BPS Blackwell.

Vergès, P. 1992. L'évocation de l'argent: une méthode pour la définition du noyau central d'une représentation. *Bulletin de Psychologie* 45:203–209.

Wagner, W. 1998. Social Representations and Beyond: Brute Facts, Symbolic Coping and Domesticated Worlds. *Culture and Psychology* 4:297–329.

Westoby, J. 1987. *The Purpose of Forests*. Oxford, UK: Basil Blackwell.

Wohlleben, P. 2015. *Das Geheime Leben der Bäume*. Munchen: Ludwig Verlag.

ENDNOTES

1. http://ec.europa.eu/agriculture/survey/index_en.htm.
2. Full title: "Absorbing carbon": absorbing carbon dioxide, contributing to fight climate change and its detrimental effects. "Habitat for animals": providing animals natural habitats, preserving the different types of animals and plants and conserving nature. "Protecting from disasters": protecting people from natural disasters such as floods and avalanches. "Renewable energy": providing renewable energy using wood as fuel. "Wood and paper": providing wood to produce furniture, paper or construction material. "Healthy leisure": providing healthy leisure activities. "Jobs and rural development": contributing to jobs and rural development.
3. https://www.theguardian.com/environment/2017/apr/07/polish-law-change-unleashes -massacre-of-trees (Accessed November 9, 2017).
4. Green Exercise Partnership videos can be viewed at https://www.youtube.com/watch ?v=78JCeh0my08&feature=youtu.be; https://youtu.be/fd1_H73BWkE, https://youtu .be/rPZ04oJUh-8; https://youtu.be/RFzbn9qQ5ko.
5. http://www.newforestnpa.gov.uk/info/20096/unspoilt_landscape.
6. http://whc.unesco.org/fr/list/558.

8 Human Desertification and Disempowerment of Rural Territories

Eduardo Rojas-Briales, Rafael Delgado-Artés, and Miguel Cabrera-Bonet

CONTENTS

8.1 INTRODUCTION: RELEVANCE AND ROLE OF RURAL SOCIETIES IN DIFFERENT DEVELOPMENT STAGES

Rurality has been historically the most usual living environment for millennia. The emergence of urban settlement is a clear sign of specialization and development starting in ancient cultures such as Egypt, Mesopotamia, or China (Mazoyer and Roudard 1997). Civilization drawbacks such as the collapse of the Roman Empire brought back the predominance of rurality where the relevance of cities diminished considerably. Since the eighteenth century though, the proportion of population living in urban environments has grown sustainably until it has reached a certain stabilization around 80% in the most developed countries. In fact, in the developed world a growing part of citizen living in rural areas show clear urban lifestyle.

Societal progress could be characterized by four stages: primitive, agrarian, industrial, and tertiary:

- *Primitive* corresponds to presettlement nomadic societies based on hunting; fishing; and collection of wild vegetables, fruits, and honey. Dense and open forests were crucial for sustaining population and a high interdependence from forest resources was obvious. A significant caesura was linked to the use of fire for cooking, protection from vermin, and controlled burning in order to improve the grazing quality for wild ungulates as well as for preventing huge fires.
- *Agrarian* societies last for a long period from the moment of domestication of cattle and beginning of agriculture until our modern rural societies. Grazing and agriculture including irrigation as well as later processing of agricultural crops for food and clothes as well as forestry as provider of energy and material for nearly everything formed the backbone of those societies. The share of farmers was all the time higher than 1:1, so one farmer was needed to sustain another person (trade, manufactures, army, church, nobles, etc.).
- *Industrial* or secondary societies emerged in the late eighteenth century linked to the philosophical progress of the Enlightenment period and the following Industrial revolution that demanded a huge amount of work force that was drained from the rural areas. Improvement in agriculture—i.e., genetics, irrigation, fertilization, machines, or new crops such as potatoes or corn—eased transition of an important part of rural population to urban areas although the numbers remained quite high for a few generations due to improved health and hygiene as well as high birth rates.
- *Tertiary* or services societies progressively emerged from World War II once the main economies moved from heavy to consume goods industry, a process that was lastly accelerated with the emergence of the information technologies including the Internet and smartphones. A significant part of today's welfare is not material but services based. Most developed countries show at least 2/3 of its economies in the services sector including tourism or financial services (see Chapter 11).

These four different stages do not happen strict successively, but frequently, they overlap. While some indigenous tribes still live in primitive phases, in rural Africa, the agrarian one predominates despite the emergence of some elements from the tertiary one such as cell phones. Finally, a significant part of the Asian urban population lives still in an industrial phase even if considerable elements of tertiary ones are as well observed. In some cases, there is even a direct jump from primary phases to tertiary ones ignoring the industrial one (Nair 2004).

Demands on forests differ a lot in the different phases. While the strong and direct linkage of primitive societies is obvious, agrarian societies need forests in a more instrumental way as source of energy, pools, and wood for their buildings, source of water to irrigation, and areas to graze. Industrial phases request low-cost commodities in considerable dimensions paying little attention to local population or environmental services. Finally, tertiary societies show a dramatic turn toward social (tourism, health, and recreation) and environmental issues (biodiversity and climate change) in which the material demands seem to become negligible, leaving rural population with a marginal profile.

The aim of this chapter is to analyze rural decline processes in their relation with forest resources and correspondent policies as observed in the developed world. Rural decline is a symmetric process following urbanization, which has received much less attention. Spain is identified as an interesting country as it has suffered important rural decline along the past century, while its forest resources expanded considerably. This is a shared process with other Southern European countries, but in other cases, their hinterland is either short (Italy) or their mountain areas are restricted to a smaller part of the country (France). The chapter is divided into four main parts, starting with a review of the past and forecast rural population globally as well as in the different regions, followed by a review of the Spanish situation in terms of rural population decline and forest area enlargement. A detailed research at district level analyzing in detail the shifts between the different land uses and vegetation cover follows. Finally, the challenges for declining rural societies and proposals how to face them are presented including further research priorities.

8.2 DEVELOPMENT AND FORECAST OF RURAL POPULATION GLOBALLY BETWEEN 1950 AND 2050[1]

As presented in Chapter 5, urbanization or urban transition is a global process in which the population moves from dispersed small rural settlements having agriculture as their economic backbone to a more concentrated dense urban settlement where industry and services prevail (Montgomery et al. 2004). In Europe and North America, fast urbanization processes were observed over the nineteen and twentieth centuries which were strongly linked to the Industrial revolution and rapid economic growth. A similar process can be observed in recent times in many parts of Latin America and the Caribbean and Eastern Asia.

In 2014, 54% of the world's population lived in urban areas compared to 30% in 1950 or the expected 66% in 2050. In 2007, for the first time in history, more people lived in urban areas than in rural ones. Urbanization patterns obviously differ from region to region as well as inside regions (Figure 8.1).

North America, Oceania, and Europe patterns show similarities. All started above 50% in 1950, 64% in the first two regions while Europe showed a lower share (52%). The later growth of urban population reported similar levels in all three regions, reaching in 2014, 75% in North America, 71% in Oceania, and 70% in Europe. North America's and Europe's forecast for 2050 predict similar levels of 87% and 82%, respectively, while in Oceania, the share of urban population is expected only to grow marginally. Asia and Africa started very low at 17% and 14%, respectively, in 1950, and sped up during the following 65 years, but Asia accelerated in the 1980s reaching in 48% in 2014, whereas Africa reached 40%. Both are expected to rise until 64% and 56% in 2050, respectively. Latin America and the Caribbean started in between with 41% in 1950 and observed a sped up urbanization process during the following 50 years, but for the past 15 years, the process has significantly slow down reaching 80% in 2014 and expected to reach 86% in 2050 (Figure 8.2).

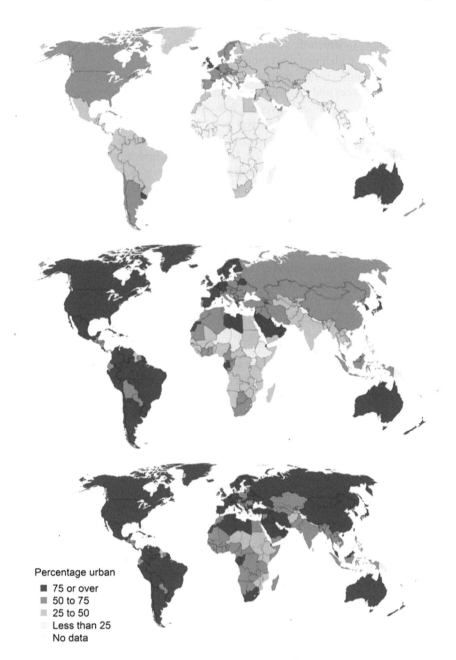

Percentage urban
■ 75 or over
■ 50 to 75
▨ 25 to 50
　 Less than 25
　 No data

FIGURE 8.1 Percentage of urban population as a proportion of total population in 1950, 2014, and 2050. (Extracted from UN, *World Urbanization Prospects: The 2014 Revision* (ST/ESA/SER.A/366), Department of Economic and Social Affairs, Population Division, UN, New York, 2015. Copyright © 2014, UN.)

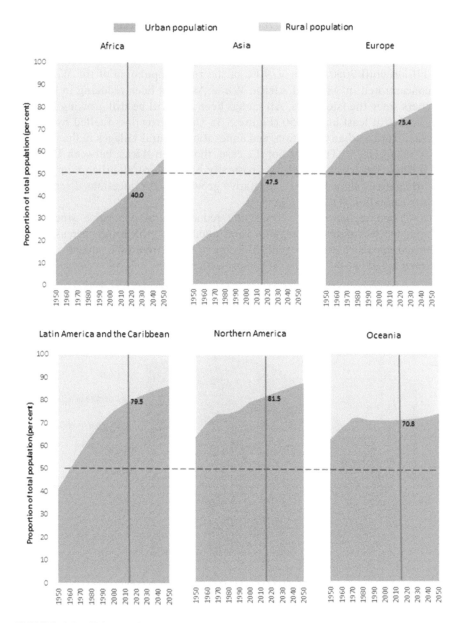

FIGURE 8.2 Urban and rural population as a proportion of total population by regions (1950–2050). (Extracted from UN, *World Urbanization Prospects: The 2014 Revision* (ST/ESA/SER.A/366), Department of Economic and Social Affairs, Population Division, UN, New York, 2015. Copyright © 2015, UN.)

Despite the permanent reduction of the share of rural population, its absolute figures have increased, albeit at a slower path, from 2 billion in 1950 to 3.4 billion in 2014. It is expected to reach its peak in 2020 and then fall slowly to 3.2 billion until 2050. By then, 90% of the rural population of the World will be concentrated in Asia and Africa. While Asia has been reducing in absolute numbers since the late 1990s, Africa has been and will be still growing in absolute figures at least until 2050 (Figure 8.3). Urban growth is fuelled by natural increase, rural–urban migration, and annexation of rural villages in the frame of urban enlargement. This was not the case, though, in Africa between 1970 and 2000 where the urban growth was basically based on rural–urban migration as the urban areas show negative vegetative growth due to infectious diseases due to poor sanitation and health services.

Developed regions have permanently reduced rural population since 1950, whereas the less developed region's rural population growth has increased until 1970, and since then, it is constantly reducing its speed from 1.8% annually, and it is expected to fail to −0.4% in 2050 (Figures 8.4 and 8.5).

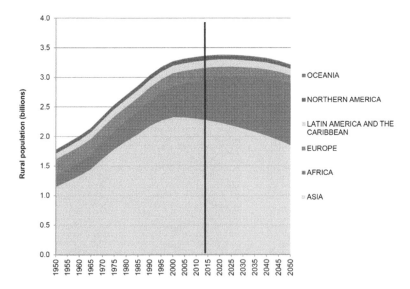

FIGURE 8.3 Rural population by regions (1950–2050). (Extracted from UN, *World Urbanization Prospects: The 2014 Revision* (ST/ESA/SER.A/366), Department of Economic and Social Affairs, Population Division, UN, New York, 2015. Copyright © 2015, UN.)

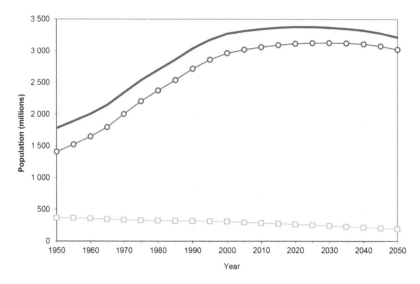

FIGURE 8.4 Estimated and projected rural populations of the world, the more developed and less developed regions (1950–2050). (Extracted from UN, *World Urbanization Prospects: The 2014 Revision* (ST/ESA/SER.A/366), Department of Economic and Social Affairs, Population Division, UN, New York, 2015. Copyright © 2015, UN.)

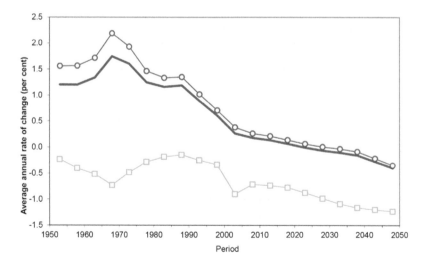

FIGURE 8.5 Average rate of change of the rural populations of the world (*blue*), the more developed (*green*), and the less developed regions (*red*) (1950–2050). (Extracted from UN, *World Urbanization Prospects: The 2014 Revision* (ST/ESA/SER.A/366), Department of Economic and Social Affairs, Population Division, UN, New York, 2015. Copyright © 2015, UN.)

8.3   RURAL POPULATION DECLINE AND FOREST COVER AT NATIONAL LEVEL: THE SPANISH CASE

8.3.1   RURAL POPULATION DEVELOPMENT[2]

The official classification in Spain (Law 45/2007)[3] defines rural areas as municipalities with less than 30,000 inhabitants and with population being below 100 inhabitants/ km^2 (Figures 8.6 and 8.7).

Following that definition, from a total number of 8,124 municipalities in Spain, 97% (7,873) are defined as rural; 1,143 municipalities lie in the range between 5,000 and 30,000 inhabitants, where below 5,000 inhabitants and 6,730 municipalities are found. In total, they add 40.5% to the Spanish population occupying 94.6% of the territory, while the small municipalities below 5,000 inhabitants occupy 84.9% of the land but host just 18.3% of the population with an average density of 18 inhabitants/ km^2. The evolution over time of the distribution of population between the different municipality classes can be found in Figures 8.8 and 8.9.

Whereas the population in Spain increased during the twentieth century by 119.4%, the share of people living in small villages below 2,000 inhabitants dropped from a share of 30% to just 7% during that period (Figures 8.10 and 8.11).

This maps show a dramatic emptying of inner Spain in a rather short period benefiting the coast and some development pools such as Madrid or the Ebro valley and a general tendency to move to the Southeast, the driest part of the country, challenging the correspondent water supply. Until 1960, the rural population was rather stagnant, moving the population growth to the periphery and inner development pools such as Madrid or Ebro valley. From 1960, the rural population dramatically declined with

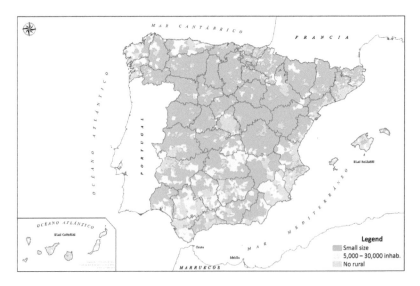

FIGURE 8.6 Municipalities by population size in Spain: dark green: below 5,000 inhabitants; light green: between 5,000 and 30,000; and gray: nonrural municipalities. (Extracted from MARM, *Población y Sociedad Rural: Análisis y prospectiva*, MARM, Madrid, 2009.)

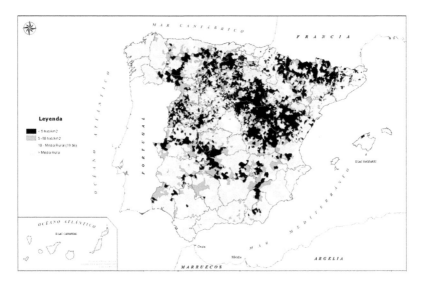

FIGURE 8.7 Distribution of population density at municipal level in Spain: black: below 5 inhabitants/km^2; green: between 5 and 10 inhabitants/km^2; white: between 10 inhabitants/km^2 and rural areas average (19.56); and cream: above 19.56. (Extracted from MARM, *Población y Sociedad Rural: Análisis y prospectiva*, MARM, Madrid, 2009.)

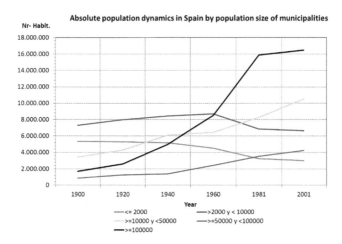

FIGURE 8.8 Population dynamics in Spain breakdown by municipal population size. (Extracted from MARM, *Población y Sociedad Rural: Análisis y prospectiva*, MARM, Madrid, 2009.)

important conglomerations emerging. Since the 1980s, the loss of rural population continued, especially in the northwest and less in the southwest, while some growth of medium cities could be observed as well.

The main reasons for that intense process is linked to massive and concentrated industrial developments such as car factories (Zaragoza, Barcelona, Vigo, or Valladolid), the soaring of the coastal areas due to tourism, and especially permanent

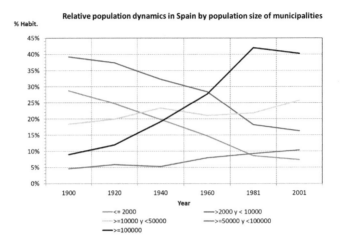

FIGURE 8.9   Relative population dynamics in Spain breakdown by municipal population size. (Extracted from MARM, *Población y Sociedad Rural: Análisis y prospectiva*, MARM, Madrid, 2009.)

FIGURE 8.10   Population density in Spain in 1900. (Extracted from MARM, *Población y Sociedad Rural: Análisis y prospectiva*, MARM, Madrid, 2009.)

foreign residents and related services. The decentralization process of the early 1980s broke to some extent interregional migration, but not the internal one, in favor of the regional capitals and few other development pools. The European Union (EU) membership since 1986 mobilized important resources both for infrastructures in rural areas as well as for some agricultural crops through the Common Agricultural Policy (CAP) slowing down to some extent the potential decline. In recent years, migration both from southern countries such as Maghreb, sub-Saharan Africa and

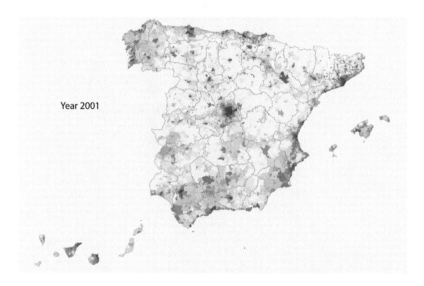

FIGURE 8.11   Population density in Spain in 2001. (Extracted from MARM, *Población y Sociedad Rural: Análisis y prospectiva*, MARM, Madrid, 2009.)

Latin America, or Eastern Europe, most as unskilled labor, and the west, central, and northern of Europe, as pensioners, has strengthened even more the polarization tendency of the Spanish population during the past 60 years.

   The less populated areas are generally those quite distant from urban areas (inaccessible) and with poor natural conditions (mountains). It is someway a paradox that while the most developed part of Spain is in the northeast, the largest low-population area falls in its middle. Additionally, aging, masculinization, and primary sector predominance are qualitative strengthening factors of rural decline as can be observed in Figures 8.12 through 8.14. While the average Spanish aging and masculinization indexes are not very far from the average rural one (respectively 28.5 vs. 31.1 and 110.6 vs. 112.6), the key low populated areas affected by strong population decline (center–north interior) show significantly worse indexes. Whereas agricultural workforce presently occupies 4.3% of the total labor in Spain, in rural areas, this increases to 27%. While labor productivity in urban areas reaches €52,689/employee and even €74,853/industrial employees, it falls to €24,969/employee in agriculture and €39,733/employee in forestry. The poverty and social exclusion index was 35% in rural areas in 2013, 11% higher than in cities or 7% higher than the Spanish average. It can be stated that there is a clear relation between low population densities, aging, masculinization, and strong weight of the primary sector as evidence for the most affected areas of rural decline. If we compare them with natural factors, attitude, high forest cover, and remoteness would complement the socioeconomic parameters. It is quite obvious that the answer by the affected population is youth migration to urban areas, whereas the older population remains resilient due to the difficulties to find new labor opportunities, just postponing the collapse of the population for a few decades. In times of unemployment such as the period of 2009–2015 and before, job searchers preferred urban environments rather than rural areas due to the higher likelihood of finding a suitable job. This approach is replicated as well by southern migrants.

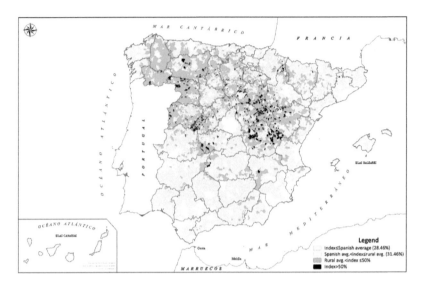

FIGURE 8.12 Aging index in Spain (2007): brown: below Spanish average; light green: between rural and Spanish average; dark green: between rural average and 50%; and black: above 50%. (Extracted from MARM, *Población y Sociedad Rural: Análisis y prospectiva*, MARM, Madrid, 2009.)

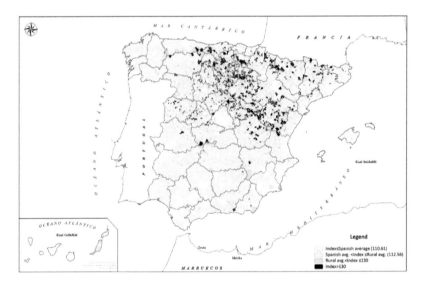

FIGURE 8.13 Masculinization index in Spain (2007): brown: below Spanish average; light green: between Spanish and rural average; dark green: between rural average and 130; and black: higher than 130. (Extracted from MARM, *Población y Sociedad Rural: Análisis y prospectiva*, MARM, Madrid, 2009.)

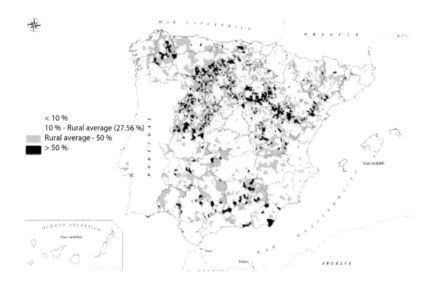

FIGURE 8.14 Share of agricultural employment at municipal level (2007): brown: below 10%; light green: between 10% and the rural average (27.6%); dark green: between rural average and 50%; and black: higher than 50%. (Extracted from MARM, *Población y Sociedad Rural: Análisis y prospectiva*, MARM, Madrid, 2009.)

Rural population in Spain has recovered slightly during the past 10 years (+0.5%); though, the process has strengthened the polarization, with significant growth in bigger rural villages, while a strong reduction is observed in the smaller ones. The northwest-located regions such as Asturias and Galicia are those who lose more population as well as those with higher age indexes. The lowest population densities can be found in Aragón, Castilla y León, and Castilla-La Mancha. The digital divide is weakening despite the availability of reliable and fast Internet connections in small villages is still far from satisfactory. It is also observed that while the core areas from big cities such as Madrid or Barcelona show reduction of population, their second and third rings are growing considerably. A positive sign at the end of the tunnel might be found in some valleys of the core Pyrenees where population has started to rise (Figure 8.15).

8.3.2 FOREST COVER DEVELOPMENT[4]

In parallel to the rural decline, especially in less advantaged areas such as mountains, marginal agriculture and extensive grazing have been massively abandoned since the late nineteenth century, while wood fellings remained stable for the past 60 years (average 18 million m[3]).

During the past 42 years, the forest area[5] in Spain has expanded 7.2 million ha or 62%, whereas other forestlands formed by shrubs, extensive grasslands, rocky areas, or high mountains diminished to 5 million ha or 33%. In total, forest use

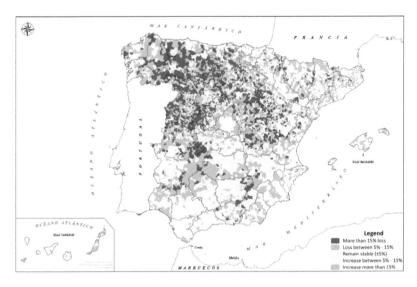

FIGURE 8.15   Recent evolution of population in Spain between 1997 and 2007 in munici-palities below 30,000 inhabitants and 100 inhabitants/km². (Extracted from MARM, *Población y Sociedad Rural: Análisis y prospectiva*, MARM, Madrid, 2009.)

increased to 2.9 million ha or 11%. The forest cover increased from 23% to 38%, while other forestlands reduced from 27% to 17%. Forests use increased from 51% to 55% (Figures 8.16 through 8.22). The reason for these very significant increases are the afforestation program of the Franco Regime (3 million ha), the EU-driven afforestation program of the 1990s that aimed at setting aside agricultural farmland

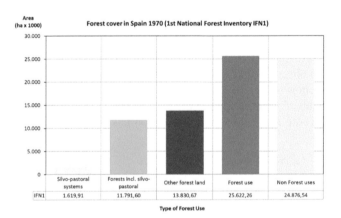

FIGURE 8.16   Forest cover in Spain 1970 (1st National Forest Inventory—IFN1): from left to right: silvopastoral systems, forests including silvopastoral, other forestland, forest use, and nonforest uses. (Extracted from MAPAMA, *Inventario Forestal Nacional*, MAPAMA: Madrid, Spain, 2017.)

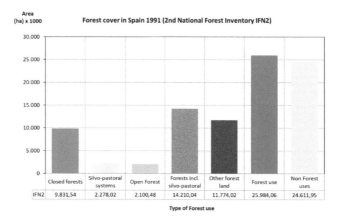

FIGURE 8.17   Forest cover in Spain 1991 (2nd National Forest Inventory—IFN2): from left to right: closed forests, silvopastoral forests, other open forests, forests including silvopastoral and open ones, other forestland, forest, use and nonforest uses. (Extracted from MAPAMA, *Inventario Forestal Nacional*, MAPAMA: Madrid, Spain, 2017.)

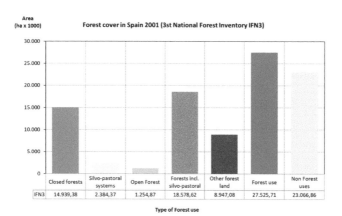

FIGURE 8.18   Forest cover in Spain 2001 (3rd National Forest Inventory—IFN3): from left to right: closed forests, silvopastoral forests, other open forests, forests including silvopastoral and open ones, other forestland, forest use, and nonforest uses. (Extracted from MAPAMA, *Inventario Forestal Nacional*, MAPAMA: Madrid, Spain, 2017.)

(0.6 million ha), abandonment of marginal agricultural land, the collapse of extensive farming, and the rural population decline in the more disadvantaged areas. Most of that increase, especially after 1970, has not been due to afforestation but to the spontaneous recovery of forest vegetation. The process has been more intensive inside the forest use than the overall expansion of forestlands. While the process was very intensive, especially in the 1990s, it seems to be culminating during the past decade.

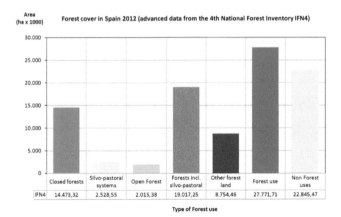

FIGURE 8.19   Forest cover in Spain 2012 (advanced data from the 4th National Forest Inventory—IFN4): from left to right: closed forests, silvopastoral forests, other open forests, forests including silvo-pastoral and open ones, other forestland, forest use, and nonforest uses. (Extracted from MAPAMA, *Inventario Forestal Nacional*, MAPAMA: Madrid Spain, 2017.)

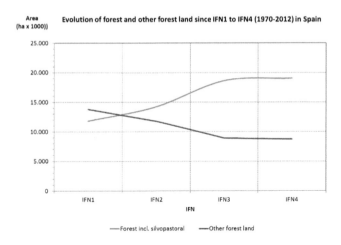

FIGURE 8.20   Evolution of forest and other forestland since INF1 to IFN4 (1970–2012) in Spain. (From MAPAMA, *Inventario Forestal Nacional*, MAPAMA: Madrid, Spain, 2017.)

The distribution of forests shows that the areas more affected by rural decline coincide with those with higher forest area share, especially mountain areas. That is why the forest and rural decline agenda are strongly interlinked and should be tackled in a coordinated manner. Rural decline has been acknowledged for a longer time, but just in the past years, it has been able to both consolidate important research and publications and enter the political agenda. In fact, in early 2017, the Spanish government established a commissioner tasked exclusively with this issue.

The intense forest area recovery process identified in Spain is comparable with the other southern European countries as shown in Figure 8.23 as the main underlying

FIGURE 8.21   Land use cover of Spain 1956–1966 (IFN1 and Forest Map 1966): cream: nonforest uses; dark green: forests; and light green: other forestland. (Extracted from MAPAMA, *Inventario Forestal Nacional*, MAPAMA: Madrid, Spain, 2017.)

FIGURE 8.22   Land use cover of Spain 2012 (IFN4): cream: nonforest uses; dark green: forests; and light green: other forestland. (Extracted from MAPAMA, *Inventario Forestal Nacional*, MAPAMA: Madrid, Spain, 2017.)

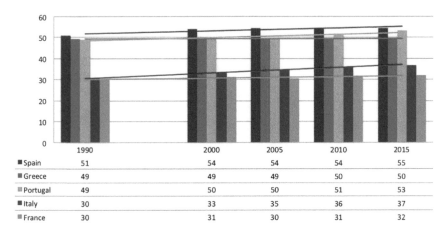

	1990	2000	2005	2010	2015
■ Spain	51	54	54	54	55
■ Greece	49	49	49	50	50
▣ Portugal	49	50	50	51	53
■ Italy	30	33	35	36	37
▨ France	30	31	30	31	32

FIGURE 8.23 Trend of the forestland use in the main Mediterranean countries. (Extracted from FAO, *Forest Resource Assessment 2015*, FAO, Rome, 2015.)

causes for forest area recovery are shared throughout the region. In fact, the northern Mediterranean region is one of the areas with more intense forest recovery in the World (FAO 2015).

8.3.3 ANALYZING LAND USE CHANGES DUE TO RURAL DECLINE
AT OPERATIONAL LEVEL: DISTRICT CASTELLÓ[6] OF THE REGION VALENCIA

The twentieth century has brought a major transformation of rural areas in the European Mediterranean region with massive emigration to urban and coastal areas (Collantes 2007). Whereas in the reception areas, a massive transformation could be observed including a fast transition from agrarian to tertiary societies with a short industrial phase (Chica et al. 2012), the human landscape modulating factor since the Neolithic has practically disappeared in the remaining areas. This has generated a vast break of social and natural structures and uses generating a deep spatial crisis (Domingo 1994; Tello 1999).

While this process was progressive in central and northern Europe, in the southern European countries, it came late, sudden, and unplanned (Mataix-Solera and Cerdà 2009) as shown, for example, in the region of Valencia in Spain as a representative example of a Mediterranean Spanish region. In 1960, 43% of labor was related to the primary sector, whereas in 2013, it had dropped to just 4%. The weight in terms of gross domestic product (GDP) dropped during the same period from 29% to 2%. The strongest fall concentrated in the period o f1960–1975 (CCINV 2014; Soler 2015).

Forest inventories show that the forest cover (forests and shrubs) has significantly expanded at the expense of the agricultural abandoned land (MARM 2004), while since the 1970s, big fires have increasingly taken a de facto key role in landscape management (Vélez 2000). Although despite their important size, forest fires

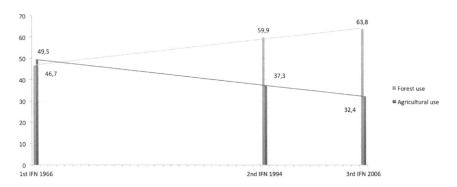

FIGURE 8.24 Trend of forest (including other forestland) and agricultural land use in the district of Castelló (Valencia region) based on the National Forest Inventory. (Extracted from MARM, *Tercer Inventario Forestal Nacional 1997–2007*, MARM, Madrid, 2004.)

have not been able to revert the strong increase of forest area in all the area (FAO 2015; Figures 8.23 and 8.24).

Without doubt, the second half of the twentieth century has brought a new forest paradigm in which the perturbation pattern based on fires has dramatically changed, opening new questions to the future evolution in both quantitative and qualitative terms in an area that covers most of the Iberic Peninsula. This fact differs from France and Italy, where temperate and sub-Mediterranean forests prevail.

A detailed research conducted in Castelló (Figure 8.25), the most northern of the three districts that form the region of Valencia, allows deepening the process called *silvogenesis* or the spontaneous recovery of forest vegetation after land abandonment. The forest cover evolution was compared through aerial photos of 1957 and 2007 applying the same methodology and analyzing the trends by different patterns; 1% of the land area (6632 km²) was monitored in a total of 6632 plots (Tables 8.1 and 8.2).

According to selected literature regarding land use and cover (Bakker and Veldkamp 2008), a two-level categorization was applied allowing multiperspective analysis (Anderson et al. 1976).

The main land uses evolved as presented in Figure 8.26.

The highest relative growth happened in the urban uses following the global tendency (UN 2012), although in absolute terms, the changes are modest and restricted to the coastal areas due to the concentration tendency of population in the coast and the growth of infrastructures. Although the relative share of land use might seem modest at the district level (3%), in the coastal areas, it has already reached a considerable share (in 2007, between 11% and 7% in the most populated counties). The migration to the coast had its origin in its own hinterland but came from other regions of Spain as well as from other countries and continents as work force (Maghreb, sub-Saharan Africa, Latin America, and eastern Europe) or retirees from central and northern Europe.

The vast majority of new urban land is formerly agricultural land, a considerable share of which corresponds to irrigated agriculture (Guinot and Selma 2008).

FIGURE 8.25 Region of Valencia and the district of Castelló (*top*) and counties of the district Castelló (*bottom*). (Extracted from Delgado, R., Análisis de los patrones de evolución de las coberturas forestales en la provincia de Castellón en los últimos 50 años, PhD Tesis, Universitat Politècnica de València, Valencia, 2015.)

This phenomenon coincides with a general trend in Asia where city enlargement is consuming the best agricultural land. The four main uses of the new urban uses are city enlargement, industrial uses, second residences, and infrastructures.

TABLE 8.1

Cover and Land Use Applied to the Castelló District (Aerial Photos 1957 and 2007)

Level I: Cover and General Uses	Level II: Cover and Specific Uses
1. Forest use; [cover]	11. Tree cover (forests) (FF); [cover]
	12. Shrubs (FS); [cover]
	13. Extensive grasslands and rocks (FGR); [cover]
2. Abandoned agriculture; [cover]	21. Abandoned agriculture (AA); [cover]
3. Agriculture; [use]	31. Herbaceous agriculture rain feed (AHR); [use]
	32. Fruit agriculture rain feed (AFR); [use]
	33. Herbaceous agriculture irrigated (AHI); [use]
	34. Fruit agriculture irrigated (AFI); [use]
4. Urban; [use]	41. Urban residential (UR); [use]
	42. Urban industrial (UId); [use]
	43. Urban infrastructures (UIf); [use]
5. Other uses; [use]	51. Other uses and cover incl. water (O); [mixt]

Source: Delgado, R., Análisis de los patrones de evolución de las coberturas forestales en la provincia de Castellón en los últimos 50 años, PhD Tesis, Universitat Politècnica de València, Valencia, 2015.

TABLE 8.2

Trend in the Different Subcategories of Forestland Use in 1957 and 2007

	FF	FS	FGR
Area 1957 (ha)	84,144	74,963	173,884
Area 2007 (ha)	135,740	101,267	148,939
Increase (ha)	51,596	26,303	−24,946
Increase (%)	61	35	−14

Source: Delgado, R., Análisis de los patrones de evolución de las coberturas forestales en la provincia de Castellón en los últimos 50 años, PhD Tesis, Universitat Politècnica de València, Valencia, 2015.
Note: Acronyms in Table 8.1.

Agriculture shows very different tendencies depending of the location: coastal and interior/mountain areas. Whereas the abandonment is massive in the interior/mountain part, in the coastal, the substitution of rainfed agriculture in favor of irrigated fruit trees is undoubtable. Herbaceous crops, generally cereals, have dropped from 17% to just 3%, while rain feed fruit trees (olive, almond, carob, or wine) have considerably reduced from 22% to 11% of the total land area in just 50 years. The single agricultural land use that has increased during this period is irrigated fruit, mainly citrus, from a modest 1.5% to 6% (Calatayud 2012), while irrigated vegetables have dropped from 1.5% to just 0.5% most likely due to city growth. In summary,

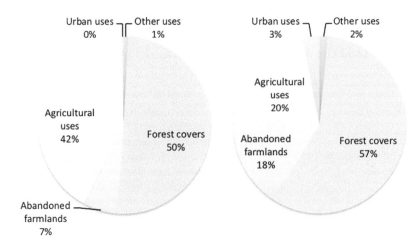

FIGURE 8.26   Land use and cover in the Castelló district between 1957 (*left*) and 2007 (*right*). (Extracted from Delgado, R., Análisis de los patrones de evolución de las coberturas forestales en la provincia de Castellón en los últimos 50 años, PhD Tesis, Universitat Politècnica de València, Valencia, 2015.)

the single remaining vital agriculture is related to export-oriented irrigated fruit crops, although smallholdings are jeopardizing this development and even a significant area planted to citrus is being abandoned (Francés and Romero 2014).

Without a doubt, the most relevant land use change has been the massive abandonment of rain feed agriculture in the interior/mountain areas. Low profitability, smallholdings, emigration to coastal areas, and site restrictions explain most of that process (Domingo 1982). Before croplands show forest cover characteristics, it passes a longer period of grass cover. Already in 1957, 7% of the land was identified as abandoned cropland, moving to 18% in 2007, showing that the process started earlier and will still continue for decades. The more intensively affected regions are located in the intermediate areas between the coast and the mountain areas previously used for cereal-based agriculture.

Forestland use covered 50% of the district Castelló in 1957, whereas in 2007, it grew up to 57%. The fact that after 50 years, just the equivalent of the already abandoned cropland has been colonized by forest vegetation shows a relatively slow process with an increasing land area joining this pool.

Inside the forest use, a clear tendency toward denser forest vegetation such as forests or shrubs at a higher speed than the overall growth of the forestland use is observed while extensive grasslands and rocks is the single forest subcategory losing share and area due to progressive colonization by pioneer trees. This has been as well observed by other authors (MARM 2004) (Figure 8.27).

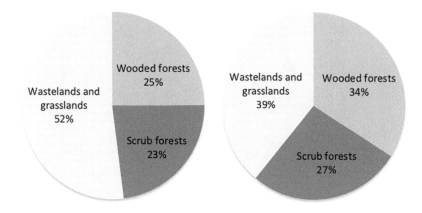

FIGURE 8.27   Share of the different subcategories of forestland use in 1957 (*left*) and 2007 (*right*). (Extracted from Delgado, R., Análisis de los patrones de evolución de las coberturas forestales en la provincia de Castellón en los últimos 50 años, PhD Tesis, Universitat Politècnica de València, Valencia, 2015.)

The Castelló district was subjected to a considerable extensive grazing during centuries that collapsed around the midtwentieth century (Enne et al. 2002; Royo 2011). This could explain the densification of the forestland use in favor of forests and shrubs in the sense of vegetation progression (Margalef 1998) despite important forest fires. The fact that the main tree species is a serotine pine (*Pinus halepensis*) and the lack of any other deforestation factor might explain this apparent contradiction (Figure 8.28).

Despite the expansion of both forests and shrubs, the rate in the case of forests (+61%) is double the rate in the case of shrubs (+35%). This might be explained by the climatic restriction of shrubs as well as by the pioneer nature of some of the tree species dominating in the area, especially *Pinus halepensis* and, to a lesser extent, *Pinus sylvestris* (Figure 8.29).

Regarding the geographical preference, the most intensive expansion of forests happened in the drier part of the district in the first and second mountain rows near the coast, while the colonization by forests is much slower in the northwest of the district despite higher rainfall but more continental climate.

As a conclusion, it might be stated that agriculture is about to lose its structural landscape role and social function while forest use is slowly but constantly occupying the remaining land outside the coastal planes. Due to the relative slowness of the process and considering the highest forest fire risk linked to abandoned cropland, the risk of extended fires in an expected tougher climate will considerably grow in the coming decades and could become the determinant landscape management factor of the near future.

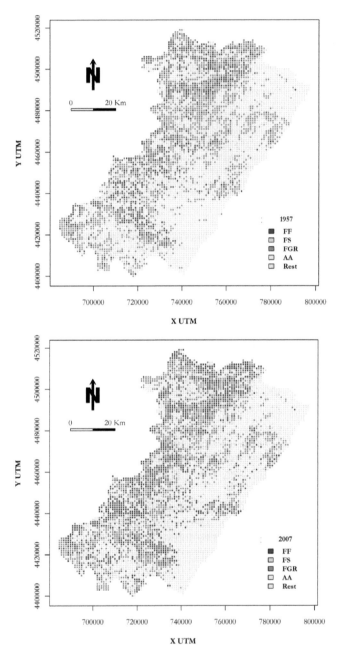

FIGURE 8.28   Forest cover subcategories in 1957 (*top*) and 2007 (*bottom*). (Extracted from Delgado, R., Análisis de los patrones de evolución de las coberturas forestales en la provincia de Castellón en los últimos 50 años, PhD Tesis, Universitat Politècnica de València, Valencia, 2015.)

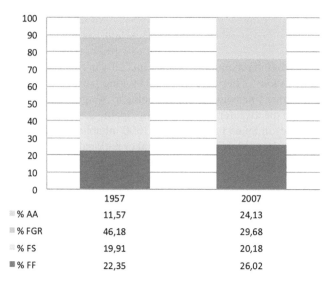

	1957	2007
% AA	11,57	24,13
% FGR	46,18	29,68
% FS	19,91	20,18
% FF	22,35	26,02

FIGURE 8.29   Share of each of the subcategories of forest cover in 1957 and 2007. (Extracted from Delgado, R., Análisis de los patrones de evolución de las coberturas forestales en la provincia de Castellón en los últimos 50 años, PhD Tesis, Universitat Politècnica de València, Valencia, 2015.)

8.4   INCREASED CHALLENGES FOR RURAL COMMUNITIES IN URBANIZED SOCIETIES

As stated in Section 8.1 regarding the societal phases, the remaining population of rural areas face increasing challenges from the time population decline phase enters into speed and even if later it might seem to stabilize at a very low level. The main identified challenges are as follows:

- *Population collapse*: Increasing age of the remaining population accelerates the demographic collapse due to the lack of sufficient reposition by births. Gender imbalance, in the Spanish case, due to masculinization, but in other cases such as in Ecuador or Mexico (Martínez-Iglesias and Alarcón 2013), due to feminization, exacerbates the problem additionally.
- *Public services withdraw*: Public services, especially education and health unit costs, become very expensive and progressively withdraw from demographically collapsing areas, generating negative incentives for the remaining (elder) population or for raising children for the small younger remaining population or eventual newcomers. Regarding the health services, too long distances become challenging both for elderly people, pregnant mothers, or those with small children. Schools cannot be maintained if a minimum number of students are not ensured. In Spain, the fact that for the past

25 years, the age for entering high school was reduced to 12 years has generated vast areas with no available high school creating a major incentive for moving out of the remote areas. Up to a 100 km distance between high schools is usual in mountain regions, especially around regional borders.

- *Local institutions*: These are affected by long periods of population loss facing reducing financial resources, political weight, and most likely the worse: a perspective for the future. In Spain, 3,897 of a total of 8,119 municipalities in Spain have less than 1000 inhabitants (Arroyo and Zoido 2003). Their electoral weight for regional or national parliament diminishes election after election. A similar situation can be observed in France with 30,000 municipalities, most of them of small and shrinking size.

- *Heritage*: A huge material (houses, churches, and other buildings) and immaterial (cultural legacy, vocabulary, and know-how) heritage is being threatened or lost without a strategy on how to preserve it. Rural tourism is only compensating a part of this loss in the material part but not for the immaterial one.

- *Tenure arrangements*: The simultaneity of the different phases and some legal inertia explains why still today many norms affecting rurality and forests in particular are rooted in the agrarian phase due to legislative inertia despite the weight of industrial and, very especially, tertiary, services being very obvious. This fact challenges the advisable law adjustments. Tenure issues such as access, hunting, land rent, heritage, or even the right to pick the most volatile nonwood forest products are still strongly anchored in past agrarian practices and do not adapt to today's general economic frame. In Spain, the number of private forest owners, who own two-thirds of the country's forest land, is estimated to be 2 million, with an average size of 10 ha each (MMA 1999). The very fact of the lack of precise information regarding this is an indicator of the underlying problem shared with most of Europe's southwest (France, Portugal, Italy, and Belgium). In agriculture, the challenge has been easier faced due to the generalization of land rent despite some rigidities in land rent regulations.

- *Communal forests*: The most favorable forest ownership structures counting nearly 30% of Spain's forests with an average size of 500–1000 ha are communal forests. Although the old-fashioned management conducted by the regional forest services has been successful for preserving the forests and its estate, it has detached municipalities from a more proactive role. Neither the forest service nor the municipalities at the end have shown the desirable ownership. The predominance of implementing public forest management through short-term micro-contracts, be it harvesting, investments, or tending, has not been able to generate secure reliable labor offer in rural areas even if its potential would be moderate. Also, the abuse of public companies hoarding most of the forest budget challenges the emergence of a vital private sector. Most of the needed labor is transported over long distances considerably reducing its efficiency.

- *Megaforest fires*: The collapse of subsistence agriculture and extensive grazing as already described has allowed an impressive recovery of forests. This fact coincides with all northern Mediterranean forests including Turkey or Israel

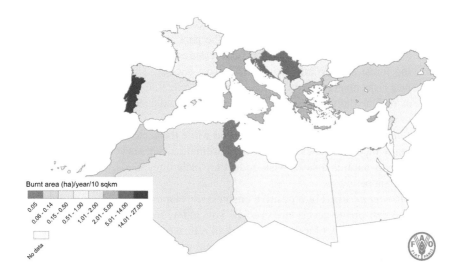

FIGURE 8.30 Average burnt area (ha/km²) per year in the Mediterranean region, 2006–2010. (Extracted from FAO, *State of Mediterranean Forests 2013*, FAO, Rome, 2013.)

and aligns with the forest transition theory (Mather 1992). Despite the evident advantages of a certain forest recovery like climate change mitigation, soil protection, biodiversity, landscape values or forest production, it opens an emerging risk of huge fires that on the long run may generate considerable damages and become the key landscape-managing factor in the absence of a human-driven one as happened over the past millennia (Figure 8.30).

- *Low level of public funding, especially for forests and biodiversity*: For the past 30 years due to the high dependency of the EU CAP and its lack of forest competence. At EU level, 20% of the CAP funds are devoted to rural development, from which slightly less than 5% is allotted to forest measures, arriving to an overall share of 1% at EU and 3% at Spanish level (Marongiu et al. 2017; Szedlak 2017). Additionally, most of the domestic resources in the Mediterranean countries are spent in firefighting related issues both for means, especially expensive aerial and survey staff, only exceptionally linked to fire prevention and forest management.

- *Biodiversity, water, or climate change policies* have not been able to actively incorporate forest management, wasting considerable potential synergies in line with COP13 CBD (2016)[7] main theme: mainstreaming biodiversity in agriculture, forestry, fisheries, and tourism. On the contrary, the prevailing biodiversity doctrine has generated considerable restrictions to forest management and failed to integrate the local stakeholders.

- *Lack of private funding*: A decline scenario is not the most palatable for mobilizing private funding needed as prerequisite for enterprise development able to create valuable employment. In the frame of aging, population decline, and Internet-based banking, bank offices withdraw from the less populated areas increasing the access threshold for investment projects.

- *Failed priorities* ignore the real challenges due to a lethal cocktail of inertial legal prescriptions anchored in the past and cliché-based approaches, e.g., prioritizing passive protection measures in a time of massive expanding forest area. While spatial risks have declined in developed countries, new risks emerge, such as forest fires, and punctual or linear demands such as residue dumps, electricity lines, or motorized access are either overseen or symptomatically approached.
- *Integrated spatial planning* has failed while the sectorial one prevails with many inconsistencies and lack of coordination. This is very critical in relation to natural disaster risk reduction including floods, forest fires, and avalanches.
- *Internet access*: For a positive business development, high-quality Internet and cell phone access are crucial. Still today, remote areas, especially in mountains, lack of it due to the lack of return for the service providers.
- *Lack of GDP visibility*: Many traditional primary phase arrangements are cash free (barter), e.g., hunters might not pay a rent on return but maintain roads or help in case of fires. Additionally, the lack of consideration of environmental and social services provided by natural resources, namely, forests, causes a structural underallocation of resources called macrodecoupling (Mendes 1999) as well as a political underestimation of the relevance of forests, apparently just 1% of GDP globally as well as in many countries such as the United States, EU, or Russia (FAO 2015).
- *Security*: While rural areas have been traditionally very safe, the collapse of its population opens new risks due to the unlikelihood of maintaining cost-free surveillance by local rural population as until present.

8.5 CONCLUSIONS: POSSIBLE ANSWERS AND FURTHER RESEARCH

8.5.1 Comprehensive Midterm Transition Strategies for Rural Areas

Substantive population shifts as humankind is living today from rural to urban areas need adequate planning in both cases. In order to unlock the opportunities of urbanization, adequate planning and governance are required (UN 2015). In the opposite case, downsizing processes affecting extended rural areas need adequate comprehensive midterm planning able to guide this complex process taking advantage of the opportunities opened and tackling the associated risks. Designing in due time the rural areas of the future is crucial in order to avoid a chaotic and unplanned depopulation process leaving many of the already commented challenges behind as described and analyzed previously. A consistent national debate may ensure the broad ownership of such plans and permanent political support needed during its undoubtedly long implementation phase (20–30 years). The following key elements shall be included: vision, opportunities and risks, priority actions and monitoring, and adjustment mechanisms.

The vision should be guided by the *right to equal living conditions* for citizens regardless if they chose to live in urban or rural areas as well as the concretization of adequate *minority rights* in order to avoid overrunning in political decision processes that might affect their core living conditions due to their small demographic and, due to it, political weight. E.g., establishing even small industries might become an impossible endeavor in forest-dominated areas due to environmental constraints in comparison with highly urbanized areas.

Further elements to be considered in the response to rural depopulation processes could be as follows:

- Overcoming *institutional minifundium*: Municipalities are frequently too small to be able to ensure adequate public services. Concentrating the most important ones at county level, establishing or strengthening that governance level would be crucial. They shall act as the first dam against depopulation. *School facilities* need to be located in reasonable distance from any settled area until the end of high school.
- Identifying *development strategies* for each area including all the endogenous resources overcoming sectoral divides. Forests are called to be crucial in many of them. *Bioeconomy* opens excellent opportunities regarding this, especially related with energy supply, small and medium enterprises (SME), and nonwood forest products.
- Establishment of *tax and social insurance exemptions* for labor and SME effectively working in the most affected areas, well known regarding disadvantaged groups in society (young, elderly, woman, and long-term unemployed).
- Developing entrepreneurship oriented funding mechanisms for rural decline affected areas is crucial.
- Improving the *full cover access to Internet* regardless of the location in order to ensure SME fair conditions overcoming any spatial bias.
- Ensuring *certification of meat from extensive grazing* and *forest food* such as honey or mushrooms.
- *Strengthening civil society* in the different areas as well second-level structures in order to ensure adequate representation at regional, national, and EU levels in the area of advocacy but as well sharing experiences, partnering, and marketing.
- *Overcoming land minifundium* is an urgent requisite in order to unlock key rural potential be in agriculture or forestry. Both substantive legal reforms and social architecture are needed.
- Proactive *communication strategies* are needed in order to overcome broad perception that rural culture might be outdated. Young people will hardly bet to stay or even move in if the link it to overcome times is not broken. Technological development and attractive living conditions are crucial.
- It would be advisable in *all related disciplines* (agriculture, forestry, geography, political sciences, law, infrastructures, education, health, communication, energy, etc.) to develop special attention both in research and in teaching to *rural decline areas* in order to adapt correspondent science-based responses to these challenges and ensure the availability of sufficient skilled staff.

Obviously, *further research* is still needed in order to improve the efficiency of the policies applied such as the following:

- Analyzing the main causes for the success or failure under similar conditions in rural decline processes including policies, funding, and institutions
- Cost-efficient ways in which forest, agriculture, extensive grazing, and hunting could contribute to preserving biodiversity, preventing disasters, optimizing carbon sequestration, improving water yield and landscape values, or reducing forest fire risks in order to shift the prevailing stick approach to the carrot approach as well as to objectivize public interventions too frequently based tacitly or explicitly in nonaction approaches
- Adaptation of pedagogical methodologies and school structures to low populated areas
- Analyzing success and failure examples of policies tackling *minifundium*
- Consolidating broadly accepted legal doctrine regarding the tenure of environmental services including reference levels, ownership, threshold, additionality, and responsibilities

Despite the huge challenges in areas where chaotic rural depopulation processes have prevailed, interesting examples showing successful strategies can be found. In the past two decades, thousands of EU immigrants, especially from the United Kingdom, have moved to the hinterland of southeast Spain or other rural areas of southern EU searching for favorable climate and living conditions. The areas affected have dramatically changed positively, the main drivers being determination, critical mass, cultural level, and financial resources. Most likely, the worst factor in rural depopulation processes is the lack of future perspective as a complex sociological problem that should be recognized and faced. The EU CAP has spent huge resources in rural areas but has not been able to significantly stop rural decline, eventually due to its relative short time scale (7 years) and sectoral biases and inertia. Identifying the problem is the first step, but if no further action follows, it might even be counterproductive deepening the sociological brake that the lack of future perspective causes.

REFERENCES

Anderson, J. R., Hardy, E. E., Roach, J. T., and R. E. Witmer. 1976. A Land Use and Land Cover Classification System for Use with Remote Sensor Data. Reston, VA: US Geological Survey.

Arroyo, A. and F. Zoido. 2003. Capítulo 1. La población en España. In *Tendencias demográficas durante el siglo XX en España*, eds. Zoido Naranjo, F. and A. Arroyo Pérez. Instituto Nacional de Estadística (INE), Ministerio de Industria, Energía y Competitividad: 19–75. https://dialnet.unirioja.es/servlet/articulo?codigo=4731350 (Accessed August 3, 2018).

Bakker, M. and A. Veldkamp. 2008. Modelling land change: The issue of use and cover in wide-scale applications. *Journal of Land Use Science* 3(4):203–213.

Bernabé, J. and J. M. Albertos. 1986. Migraciones interiores en España. *Quaderns de Geografía: Universitat de València* 39–40:175–202.

Collantes, F. 2007. La desagrarización de la sociedad rural española, 1950–1991. *Historia Agraria: Revista de Agricultura e Historia Rural* 42:251–276.

Calatayud, S. 2012. Desarrollo agrario e industrialización: Crecimiento y crisis en la economía valenciana del siglo XX. *Historia Contemporánea* 42:105–147.

CCINV (Cámara de Comercio, Industria y Navegación de Valencia). 2014. *La Comunidad Valenciana en cifras.* Valencia: Cámara de Comercio, Industria y Navegación de Valencia.

Chica, J. A., Pérez, M. L., and J. M. Barragán. 2012. La evaluación de los ecosistemas del milenio en el litoral español y andaluz. *Ambienta: La revista del Ministerio de Medio Ambiente* 98:92–104.

Delgado, R. 2015. Análisis de los patrones de evolución de las coberturas forestales en la provincia de Castellón en los últimos 50 años. PhD Tesis, Valencia: Universitat Politècnica de València.

Domingo, T. 1982. Algunos aspectos de la estructura agraria en el País Valenciano y su explicación en base a las tesis clásicas. Revista *de Economía Política* 91:179–207.

Domingo, C. 1994. La profundización de los desequilibrios territoriales en Castellón. *Quaderns de Geografia: Universitat de València* 56:135–154.

Enne, G., Pulina, G., M. d'Angelo et al. 2002. Agro-pastoral activities and land degradation. In Mediterranean Areas: Case Study of Sardinia: Mediterranean Desertification: A Mosaic of Processes and Responses, eds. Geeson, N. A., Brandt, C. J., and J. B. Thornes, 71–82. Hoboken, NJ: John Wiley & Sons.

FAO (Food and Agriculture Organization of the United Nations). 2013. *State of Mediterranean Forests 2013.* Rome: FAO. http://www.fao.org/docrep/017/i3226e/i3226e.pdf (Accessed April 16, 2018).

FAO. 2015. *Forest Resource Assessment 2015.* Rome: FAO. http://www.fao.org/forest -resources-assessment/en/ (Accessed April 16, 2018).

Francés, M. and J. Romero. 2014. *La Huerta de Valencia: Un paisaje cultural con futuro incierto.* Valencia: Universitat de València.

Guinot, E. and S. Selma. 2008. L'estudi del paisatge històric de les hortes mediterrànies: una proposta metodològica. *Revista valenciana d'etnologia* 3:100–124.

MAPAMA (Ministerio de Agricultura, Pesca, Alimentación y Medio Ambiente). 2017. *Inventario Forestal Nacional.* MAPAMA: Madrid, Spain. http://www.mapama.gob .es/es/desarrollo-rural/temas/politica-forestal/inventario-cartografia/inventario -forestal-nacional/ (Accessed August 3, 2018).

Margalef, R. 1998. *Ecología.* Barcelona: Omega.

MARM (Ministerio de Medio Ambiente, Medio Rural y Marino). 2004. *Tercer Inventario Forestal Nacional 1997–2007.* Madrid: MARM.

MARM. 2009. *Población y Sociedad Rural: Análisis y prospectiva.* Agrinfo12. Madrid: MARM.

Marongiu, S., Chiozzotto, F., and L. Cesaro. 2017. Forestry measures in the European Rural Development programs 2014–2020: Planning expenditure and priorities in the EU Member States. *Austrian Journal of Forest Sciences* 134(1a):81–100.

Martínez-Iglesias, M. and A. Alarcón. 2013. Sociedades rurales, migración masculina y posición de resguardo femenina. Paper presented at the 11th Spanish Congress of Sociology.

Mataix-Solera, J. and A. Cerdà. 2009. Incendios forestales en España: Ecosistemas terrestres y suelos. In *Efectos de los incendios forestales sobre los suelos en España. El estado de la cuestión visto por los científicos españoles.* Valencia: Càtedra de Divulgació de la Ciència, Universitat de València. València: p. 529.

Mather, A. S. 1992. The forest transition. *Area* 24(4):367–379.

Mazoyer, M. and L. Roudart. 1997. *Histoire des agricultures du monde: Du néolithique à la crise contemporaine.* Paris: Seuil.

Mendes, A. M. S. C. 1999. Portugal. Forestry in changing societies. *In Information for Teaching module II: Country Reports*, 295–322. Silva-Network, ICA, and University of Joensuu. http://www.wzw.tum.de/silva-network/pdf/publications/Silva_Publications_2000 _FCSE_PartII_Country-Reports.pdf (Accessed August 3, 2018).

MMA (Ministerio de Medio Ambiente). 1999. *Estrategia Forestal Española*. Madrid: MMA.

Montgomery, R. M., Stren, R., Cohen, B., and H. E. Reed. 2004. *Cities Transformed: Demographic Change and Its Implications in the Developing World*. London: Earthscan.

Nair, C. T. S. 2004. Reinventing forest education. *Unasylva* 216(55):3–9.

Nicolau, R. 1991. Trayectorias regionales en la transición demográfica española. In *Modelos regionales de la transición demográfica en España y Portugal*, Livi Bacci, M. (coordinador), Instituto de Cultura Juan Gil-Albert, Alicante: 49–65.

Pérez, P. 1971. Natalidad, mortalidad y crecimiento demográfico en las comarcas del País Valenciano. *Quaderns de Geografia: Universitat de València* 8:15–34.

Royo, V. 2011. L'influence de l'élevage dans l'organisation du paysage du village de Culla (XVe-XVIIe siecles): Domitia. *Revue du Centre de Recherches Historiques sur les sociétés méditerranéennes* 12:123–142.

Sempere, J. D. and E. Cutillas. 2017. *La población en España*. Alicante: Universidad de Alicante, Departamento de Geografía Humana.

Soler, V. 2011. *Economia espanyola i del País Valencià*. València: Universitat de València.

Szedlak, T. 2017. *EU Forest Strategy and forestry measures in rural development programmes for 2014–2020*. Brussels: DG Agriculture, EU Commission. https://www .forestry.gov.uk/pdf/FR_PESFOR_PRESENTATION_Tamas_Szedlak_EU_Forest _Strategy_and_forestry_measures_in_rural_development.pdf/$FILE/FR_PESFOR _PRESENTATION_Tamas_Szedlak_EU_Forest_Strategy_and_forestry_measures_in _rural_development.pdf (Accessed April 16, 2018).

Tello, E. 1999. La formación histórica de los paisajes agrarios mediterráneos: una aproximación coevolutiva. *Historia Agraria: Revista de agricultura e historia rural* 19:195–214.

UN (United Nations). 2012. *Sustainable Land Use for the 21st century*. New York: United Nations. https://esa.un.org/unpd/wup/Publications/Files/WUP2014-Report.pdf (Accessed April 16, 2018).

UN. 2015. *World Urbanization Prospects: The 2014 Revision* (ST/ESA/SER.A/366). New York: Department of Economic and Social Affairs, Population Division. https://esa.un.org/unpd/wup/Publications/Files/WUP2014-Report.pdf (Accessed April 16, 2018).

Vélez, R. 2000. *La defensa contra incendios forestales: fundamentos y experiencias*. Madrid: Mc Graw Hill.

ENDNOTES

1. Except otherwise stated, the information of this section is based on the study by the UN (2015).
2. Except otherwise stated, the information of this section is based on the study by the MARM (2009). Further information can be found on the birth rate evolution in Pérez (1971), internal migrations (Bernabé et al. 1986), demographic transition by Nicolau (1991), or demographic linked to spatial development (Sempere and Cutillas 2017).
3. Law 45/2007, 13.122007, on sustainable rural development (BOE 299 de 14.12.2007).
4. Except otherwise stated, the information of this section is based on the study by MAPAMA (2017).
5. Forestland is classified in Spain residually as any nonurban and nonagriculture land including extensive rangelands. Inside the forest use, two categories are defined; forests including both dense and open ones on one hand and other forestlands (*montes desarbolados*) are differentiated. The last includes high mountains, wetlands, shrubs, rocky areas, or rangelands.
6. Except otherwise stated, the information of this section is based on the study by Delgado (2015).
7. Convention on Biological Diversity (CBD). 2016. Strategic actions to enhance the implementation of the Strategic Plan for Biodiversity 2011-2020 and the achievement of the Aichi Biodiversity Targets, including with respect to mainstreaming and the integration of biodiversity within and across sectors, Decision XIII.3.

9 Dendroculture
Social Imaginations, Art, and Culture in Forests

Martí Boada, Roser Maneja, and Maribel Lozano

CONTENTS

9.1   INTRODUCTION

The structure of the chapter includes the description of the phenomenon of dendroculture, or tree culture, in the context of the environmental crisis; the perspective of trees and forests in art, folklore, culture, and religion; and the history of valorization of trees and forests in the last centuries. Finally, we propose an innovative taxonomic classification of trees of natural–cultural base. Emphasis is rather on the Mediterranean region, although examples coming from different parts of the World are also used.

The twenty-first century is determined by a socioenvironmental crisis of a civilizing nature, given that it affects not only a certain social culture but also, transversally, all contemporary models of society, although with different intensities. For instance, we are facing a scene marked by global change, characterized in particular by global warming and land use changes, and by the processes of bioinvasion. We now live an existential moment with growing and acute socioeconomic and environmental transformations that are generating new challenges for humanity. Phenomena such as globalization and population growth are exerting pressure on natural resources and are generating a situation of vulnerability and uncertainty in relation to planetary ecosystems.

The predominant economic systems now prevailing in the world have encouraged the excessive exploitation of natural resources including those of energy, forestry, agriculture, and fisheries. According to the Food and Agriculture Organization of the United Nations' Global Forest Resources Assessment 2015 (FAO 2016), between 1990 and 2015, about 129 million ha of forests equivalent to the area of South Africa, were destroyed, with the major losses taking place in the tropics, particularly in South America and Africa.

In fact, in temperate climate latitudes, in countries of medium and high incomes, the proportion of forests is higher. That is the case in Europe, where the stable growth of the forest cover can be explained by the migration from the countryside to the city, the processes of urbanization, and the consequent social desertification of rural areas (see Chapter 8).

In the case of the north Mediterranean, as a consequence of its own distinct patterns of energy use with the hydrocarbons replacing coal and wood in the 1960s, together with minimal forest management, the forests have shown a constant increase in terms of area and growing stock. This process has produced an effect of accumulation in biomass, with increasingly bigger trees, and an increase in dry material. All that, added to an expansion in forest areas close to urban settlements, has really generated a new forest, without documented precedents, with an ecological predisposition to ignite and a fuel charge availability (volatile hydrocarbons, dry matter, etc.) propitiating great forest fires, being without any doubt, one of the most notorious socioenvironmental problems of the current Mediterranean society.

9.2   FORESTS AND SOCIETY

In the context of global change, there is a broad consensus acknowledging that if something in particular characterizes today's society, it is its dynamicity and its constant capacity to change and that everything goes faster. Some authors suggest that this acceleration of processes responds more to the second law of thermodynamics than to strict social factors. It is only a couple of decades ago that we are in a generalized way in the digital era, under the dominion of technologies that we massively use and that often we do not control. Our technological artifacts are modified in a short time, and our scientific knowledge is constantly being transformed, and the dynamics of change affect the field of our thinking and our social perception.

Conceptually, the forest does not escape from this accelerated dynamic of change. Despite the importance and usefulness that trees and forests have had throughout history, the changes in perception of today's urban society with regard to how we use trees are becoming ever more and more evident. Indeed, we move away from seeing the forest and its trees from a productivist and extractive perspective, but rather from the perspective of its aesthetic and environmental properties, including those of offering a place of play and leisure (see Chapter 6). How we perceive forests may denote a "change of era," a hybridization between natural and social history.

The social perception of the forest is dynamic, as the forest itself is changing over time in various ways according to different civilizations. Trees and the forest in general have given rise to different types of perceptions. Authors such as Urteaga (1993) note that during the course of history, there have been stages considered as tree-phobia, periods in which forests were considered miasmas, unhealthy, ungovernable places, and refuge for evil doers and outlaws. In Spain, during the reign of King Felipe II, the systematic burning and elimination of forests were decreed in some regions, to end the social banditry of the Baroque period.

Tree-philia would have the opposite meaning, representing human enchantment and respect for forest communities and trees, especially those having a symbolic charge.

Some classical authors, such as Pliny the Elder, Columella, Marcus Varro, and Cato the Elder, prescientifically pointed out the respect for the forest. Pliny, in his far-reaching tome on "Natural History," refers to the importance of conserving the forests and affirms that the shade of tree was the first temple used by humanity.

Bauer (1991) cites Capitullare de Villis (800 AD), a relevant document that makes explicit the need for respecting and planting trees of special social interest in the forest, being for food, for therapeutic purposes, for construction, and for their energy.

One thousand years had to pass for science to be applied to the generation of sustainable forest management. During the eighteenth century, central European forests were seriously reduced by logging and agriculture. First, von Carlowitz (1713), in his work "Sylvicultura oeconomica," introduced the concept of forest management, and then later Johann Heinrich Cotta suggested to restore them by applying scientific principles. He founded the first forest academic institution in Tharandt in 1811 and applied the principle of measuring the productive capacity of forest while taking into account social needs. This project would be influenced by a small group of intellectuals who shared his vision with Goethe and Alexander von Humboldt (see Chapter 3).

With all the civilizational variables that can be nuanced, the forest has been and is still considered a "legacy." Even for some cultures, it represents the home of the human soul. More generally, the forest has been essential as a resource of firewood, of timber, and of fruits and seeds. It has also been seen as a mythological grove, a religious shrine, with the tree as a temple.

The forest has also provided literary inspiration, a place of poetry for romantic authors, and a space for pictorial and artistic recreation. Forests are eternal carriers of culture; they have become referents of the landscape and of social and artistic imaginary (Boada and Boada 2011). The tree, specifically, has represented the bridge between heaven and earth, between the gods and humans.

In the context of the current environmental crisis, the significance of forest is clear in the provision of environmental services; in the social, public, and human health function; and in the fields of global change research and environmental pedagogy.

9.3   DENDROCULTURE

In general, the forest represents the expression of maturity of a terrestrial ecosystem being its main component trees. In fact, trees are the most complex vegetables and the single living being that can be substantively modified its environment. To understand the complexity of a forest including at the scale of an organism, and of a tree, as the eminent ecologist Ramon Margalef would indicate, beyond scientific knowledge, a poet is needed.

James George Frazer (1951), in his reference work *The Golden Bough, Magic and Religion*, speaks of the presence of the tree in all cultures, stating in particular that no culture has been indifferent to this exceptional living organism in all its varieties and diversity. Indeed, according to the *State of the World Forest Genetic Resources* report (FAO 2014), more than 80,000 wild species of tree are spread across the planet.

In modern times, in interpreting and giving value to the tree, we have tended to distil out the biological aspects when diagnosing its values in terms of conservation strategies, limiting almost exclusively to biological aspects. Without denying the importance of conservation and the biology behind it, we should consider other functional aspects and include the aesthetic, chemical, physical, silvicultural, and productive dimensions, as well as the social and cultural roles of trees in cosmology.

Considering the cultural dimension of trees in the broad sense, this chapter focuses on the perspective of "dendroculture," referring to those cultural forms of human relationship with specific trees, being therefore protected and observed over time, giving away monumental or singular forms for various reasons of social nature.

Through dendroculture, the sociocultural dimension of remarkable trees, forever noticeable for their function as symbol and for those variables associated with their form, use, size, and longevity, can, hopefully, recover its due value.

9.3.1   FOREST ART

Modern art has given a new creative perspective in which the forest and tree have become essentially calligraphic. Both generate their own description, in arboreal morphology, thereby creating their own descriptive literature: the forest landscape, as a land-based chronicle of natural and social history. "The forest becomes like a piece of literature in itself and the trees, its words," a perspective formulated by the artist Perejaume Borrell. Referring to the geographical antipodes, Albeda and Saborit (1997) describes how in Kimberley (Australia), the local aboriginal culture understands the forest as a writing that the people know how to read and of which they know the alphabet.

We do not discover anything new in recording the role that forests and trees have played throughout the different manifestations of art, from the perspective of naturalistic painting and the different variables of landscaping. In Spain, the research group Arte y Naturaleza of the Polytechnic University of Valencia stands out in this respect. One of its members, J. Albeda, argues that art never ceases to reflect the cultural desire to impose on the landscape a "natural order," which can find itself symbolically subjected by means of garden or uneven-aged forms, to some symbolic victory over the natural chaos that the forest could represent. By way of contrast, dendroculture describes the forest and tree as an interbreeding between nature and culture, like a "continuum" without frontiers.

In modern times, there is a current of art explicitly forest related, the so-called Forest Art that emerged in 1968 under the auspices of England's Public Forest Estate. It was a type of art that was deployed under the auspices of England's Art Council with the complicity of the Forestry Commission and resulted, in 2012, in the first Forest Art Program to encourage the representation of the forest in art, and of art in the forest, also meaning located within the forest.

Currently, the dimension reached by Forest Art is both broad and complex and manifests itself in diverse forms: painting, sculpture, literary work, dance, etc. It displays a diversity that makes systematization difficult, although two main currents can be distinguished.

- Forest Art is produced, shown, and integrated in situ in the forest landscape itself. Well known in Spain are the painted forests by the Basque Agustín Ibarrola, one in Urdai Bai Biosphere Reserve (Basque country) and the other in Allariz (Galicia). In the first photo, the artist develops his work in a pine forest, while in the second, he uses an alder plantation, playing with the organismicity and verticality of the trees. His way of pictorially intervening is to color the bark, thereby creating sculptured/painted manifestations of renowned aesthetic impact (Figure 9.1).

FIGURE 9.1   Allariz Forest by Agustin Ibarrola, Spain (Courtesy of Foto A. Ibarrola).

FIGURE 9.2   Forest Art by Martí Boada (Photo by M. Boada).

- The other current departs from a self-sculpturing of those elements that form the forest system, focusing instead on surplus elements, such as those necrotic parts that proceed from the overall structure of the tree. Some authors see in such creation not only that it is within the scope of art but also that it goes beyond, reaching what they have formulated as "post-art." Namely, art as a way of approaching and learning about the functioning of the forest in the broad sense. Forest Art is a proposal for contributing to overcoming the context of the environmental crisis of the current civilization moment (Figure 9.2).

9.3.2   Folklore, Culture, and Religion

Since their origin, trees and forests have played a vital role in the different cultures of the world, in all likelihood, thanks to their undeniable practical utility as a source of energy, food, wood, and timber. This is thanks to their physical peculiarities such as their longevity and perenniality as well as their nutritive and medicinal properties, or even their mystical dimension that has designated them as the dwelling of the gods and, for some, gods in themselves. These attributes have given rise to infinity of beliefs, myths, and legends that, far from being forgotten, remain alive in the animist cults, whereby the trees are venerated as supernatural beings, the home of spirits of the vegetation and of fertility.

Since ancient times, trees and forest form part of culture, language, folklore, and religion of the peoples of the world. For primitive peoples, they signify a source of light, food, refuge, and heat, as well as primary material for the development of subsistence tools such as bows, lances, and harpoons. With the evolution of the times, their importance transcended to incorporate itself in the popular wisdom as protagonists of myths and legends.

Like divine symbols, trees and forests represent in diverse cosmovisions a channel of communication between the gods and the humans. Their magic and splendor have reached a religious meaning, secularly associated with prophets such as Buddha or Jesus who found, respectively, in their figure the illumination and the crucifixion.

The tree of life is an example of the intention to associate the human condition with divinity, philosophy, and religion. As a universal myth, it is represented in the symbolism of Judaism, Islam, Buddhism, Hinduism, and Christianity. For example, the Bible, in the Book of Genesis, makes reference to the tree of life in the Garden of Eden, as a representation of humanity in a state free of sin. It is an important symbol in the cosmology of all peoples of the world, being present in their myths, legends, and folklore (Crews 2003).

Other rituals in which trees act as protagonists correspond to the relationship between female fecundity and nature. Such rituals are valued for their role in propitiating fertility, conception, and life. In some Greek and Roman myths, named in the reference work of Ovidio's (2008) *Metamorphosis*, the maidens or nymphs persecuted by gods were turned into trees. For example, Daphne, the goddess of the Earth's fertility, was turned into a laurel to escape the god Apollo.

Another example of an intimate relationship between humans and trees can be found in the ancient Celtic civilization in which the name of certain trees are attributed to the characters of the alphabet and even to some months of the calendar. There is a Celtic calendar based on seasonal links with significant forest species, even links between the date of birth and what would be the specific personal tree.

In the multisecular cultures of Mediterranean Europe, the presence of a centennial tree in front of the peasant houses represented the place where the deals and consultations of all signs were better closed, under the testimony of the imposing canopy of the temple tree, which conferred beneficial protection. At the same time, a summer livestock function, the canopies of such trees in the countryside offered shade for sheep in summer periods and times of maximum thermal intensity.

In relation to human traditions tied to trees, I. Abella (2012), in his book *The Great Tree of Humanity*, relates how, for 40,000 years, prehistoric humans represented trees, forests, and landscapes in their rock art. In his analysis, he makes clear that all cultures hold the tree as a common denominator, as the center axis of life, with a religious significance combined with a tradition of cult and folklore. Certain species of tree play a role in those myths and legends in which trees are associated with the origin of the world: "All those belong to a same universal culture in which the tree was the place of parliament and meeting—a living symbol of the tribe and its territory. From that point of view, all human beings share a same culture. Whatever may be our origins, race or system of beliefs, the tree appears to have been a point of agreement and natural encounter, apart from all the differences" (Abella 2012).

The primitive tree is also very popular in the myths of American cultures like that which gives rise to rivers, mountains, plants, and animals and a point of connection between the astral people and the people of the Earth. It is also considered as the tree of life, representing links and alliances and the injection of forces, powers, and influences that circulated in the world and that were in the care of priests, mohans, and shamans, who revered, fed, and, at the same time, controlled these (Rozo 2006).

For Francis Hallé (2016), renowned French botanist, the ancient mythological figures of the "green man," conceived as hybrids between humans and plants, with faces of leaves, as guardians of nature, had their origin in ancient European art and in the pagan religions and that with the passage of time, managed to integrate themselves and survive to the artistic manifestations of Christianity since the fourth century. However, it was not until the Classic Gothic period (twelfth to fourteenth century), characterized by a growing interest in naturalism art, that its peak was observed as an image of the spirit hidden in nature together with real or fantastic animals and vegetable forms with symbolic meanings being represented in the decoration, structure, and sculptures of important European churches and cathedrals.

The "green man," as a mythological personage persists up to the present time in the paintings of the Italian artist, Arcimboldo, and in popular folklore related to the forest in Europe as "Wilde Mann" in Germany and Switzerland, as "Basajaun" in the Basque Country, as "Ojáncanu" in the Spanish region of Cantabria, and as "Jack in the Green" in England (Abella 2012). He is also present in other latitudes, and for the Amerindian Tikuna of the Amazon rainforest, he is known as "Kurupira," and in the Andean forests of Colombia, as "Mohán."

Meanwhile, the "Xanas" are feminine characters very popular in the mythology of Asturias and León in the Iberian Peninsula. Popularly known as fairies in distinct European folklore traditions, they can be related to the figure of the goddess Diana or Jana, the latter being connected to the sorceresses in the Middle Ages. Those mechanisms of construction of the cultures around the tree are an example of the way in which the origin and identity of the tribe and territory were united to the myths of the tree. They illustrate the way in which the consciousness of belonging to the land and to the forest as a home was built.

In a recent script, Professor Manuel Frochoso narrates a story that alludes to the mythological characters appearing in the Iberian forests: "The Xanas laughed and the Ojáncanu looked surprised without understanding anything at all. They did not know the concept. Better I did not explain it and we walked together towards the Bayas River" (Frochoso and Castañon 1990).

The sacred tradition of forests extends to this day. The United Nations Organization for Education, Science and Culture includes in its Intangible Heritage a list of forests that are recognized as being sacred thanks to their nonmaterialistic cultural patrimony, their therapeutic properties, their spiritual values, and environmental services. This is the case of the Central East Rainforest Reserves of Queensland, Australia; the sacred forest of Ochún–Oshogbo, Nigeria; the Kaya Woods, Kenya; the Cedar

FIGURE 9.3   Cedar trees in the Shouf Biosphere Reserve, Lebanon (Photo by R. Maneja).

Woods of the Shouf Biosphere Reserve, Lebanon (Figure 9.3); and the sacred wood in Luzón, the Philippines, for the rice ceremonies (Crews 2003).

Modern society has been losing links with those traditional cultures in which trees and forests play a role; those cultures, which still today persist in folklore, are becoming ever less common. In today's world, their protection depends on legal instruments and protective status, mainly through the application of the notions of Biosphere Reserve or Protected Natural Area.

9.3.3 Singular Trees

In the case of the so-called singular trees, they are usually because of their monumentality, although they can also be for other different reasons, not strictly of size or longevity. For example, the tree of the Sad Night in Mexico and the oak of Guernica in Euskadi are used for swearing the bask rights by the kings. On the other hand, there can be ritual connotations such as those tied to the tradition of the May tree in the Pinares County of Castile, where popular ancestral festivals are organized around May 1, closely linked to similar traditions in central Europe. Other examples are the General Sherman, a sequoia (*Sequoiadendron geganteum*) of the Sequoia

National Park of California (United States) whose total biomass (including roots) makes it the tree of greatest volume in the world. Also in California, the forest of the pine, *Pinus aristata* subsp. *longaeva*, in Inyo National Forest, in which trees as old as 5000 years can be found. Or the Tule's tree, Oaxaca, Mexico, the tree with largest perimeter on the planet. In those cases, their monumentality and longevity make them living beings objects of outstanding cultural and scientific curiosity.

In front of the beauty and spectacularity provoked by the contemplation of a singular tree, the observer does not remain indifferent, giving himself/herself exceptional opportunities to generate information about basic processes of natural systems. From an exercise of recognition of the species itself, starting from the thread of the specific description, one can work the socioecology of the species and its functionalism, taking advantage of this scenario of receptive goodness: photosynthesis, water cycling, nutrient cycling, biodiversity, and ecology should be complemented with cultural and social uses as well as, if there are, highlight the legendary or magical dimension of the tree.

The singular tree can represent an opportunity to promote sustainable tourism of a region (Boada and Gómez 2011) and help with training and motivation to understand, face, and help humanity to overcome the environmental crisis.

The present chapter offers a proposal to classify trees taxonomically that in accounting for their particular biological characteristics and their significance from a cultural, religious, political, and social point of view, will highlight their remarkable features. The suggested model would bring about a strategy of protection that could be incorporated into conservation policies and would function as an innovative proposal to revalue biological and cultural diversitiesas well as the promotion of relevant knowledge that, in turn, would motivate activities of environmental education and sustainable tourism.

9.3.4 Phenomenon of Dendroculture in the Context of Environmental Thinking: A New Approach to Trees

The object of this work, rather than simply displaying an inventory of characteristics associated with the tree, aims to propose a system for innovative classification of singular trees, based not on the criteria of classical botany, even without abdicating the Linnaean perspective of the current taxonomy, but on cultural and civilizational criteria with the aim of revealing the significance of the cultural dimension.

Now, well into the twenty-first century, the persistence of renowned forest forms with significant monumental trees represents an accumulated multibenefit in the form of a patrimonial value for a region and, at the same time, become a starting point for programs of environmental education as well as training to provide the basis of essential knowledge and awareness.

The ways for the formal knowledge of the environment, in a classical way, have been through scientific methodology. Nevertheless, some authors such as S. Funtowicz and J. Ravetz (2003) push for the necessary incorporation of other forms of knowledge. Along the same lines, the Mexican ethnoecologist, Victor Toledo (2008) proposes the "dialogue of knowledge (wise people)" as a formulation with a basis on the scientific method, but that must incorporate empirical, popular knowledge, especially indigenous knowledge, knowledge of women (for its existential role with the environment), and peasant knowledge. The processes associated with dendroculture that, even though complex, display a universality in their contemporary expression will help know and understand the environment. Despite their biogeographical and morphological differences, the various examples of tree culture have a singularity in common which makes them socially remarkable.

As pointed out, the persistence to date of trees, often millennial or at least several hundred years old, beyond their biology, has a component of environmental history, sometimes evident, sometimes not so much. Their persistence until now expresses a cultural need for protection, especially when we take into account that such singular trees are likely to be the most remote example of conservation, far from those contemporary protectionist models. Indeed, their persistence until today is for cultural reasons rather than for biological ones; or at least, it would be the result of a combination between the long-lived capacity of the species and the social cause of its safeguard.

9.3.5 BACKGROUND IN THE ENHANCEMENT OF REMARKABLE TREES

The first inventories of remarkable trees in Europe occurred in England, with the intention of establishing the legal measures to protect them. The main reason that led to make this decision was surely the fact that the forests of the British Islands had been drastically deforested due to agriculture; grazing; and the massive exploitation of timber, wood, and coal during centuries. The most important actor in this act of protection was the founding of the "English Nature," an important government organization dedicated to the conservation of nature in England.

Today, many European Union (EU) countries have carried out a national inventory of monumental trees, elaborated with more or less standard protocols. By means of declaring that a tree has monumental or historic interest, the inventories give weight to the legal instruments of protection and conservation. Table 9.1 presents a set of representative bibliographic references related to the issue.

9.3.6 DOES AN ECONOMY OF DENDROCULTURE EXIST?

Singular trees, with distinct levels of public appreciation and varying from case to case, constitute an important element of attraction in sustainable tourism. The environmental crisis produces a perceptive shift in the current tourism model, particularly in the most sustainable forms of this activity. The new "green" tourists are more interested in visiting and observing well-conserved landscapes, particularly the protected natural areas.

In reality, in the policies of sustainable tourism, the remarkable trees have an interesting attractor effect toward the potential visitor and, in some instances, become a driving element of the local and/or regional economy.

In the United States, as mentioned, one of the many examples is found in the Sequoia National Park, where the central motivating element is the sequoia General Sherman, considered to be the tree with the largest biomass on the planet (Figure 9.4). The total number of visitors to the park exceeds three million annually, all of which represent a direct income of more than US$40 million.

TABLE 9.1

Studies on Monumental Trees in Europe (Set of Representative Bibliographic References)

Year	Title. Author(s). Country
1965	*Dictionnaire du bois, ses dérivés.* Moirant, R. France/Belgium
1970	*Know Your Conifers.* Edlin, H. United Kingdom
1973	*The International Book of Trees.* Johnson, H. United Kingdom
1978	*The International Book of Wood.* Johnson, H. United Kingdom
1978	*A Field Guide to the Trees of Britain and Northern Europe.* Mitchell, A. United Kingdom
1985	*Trees of Britain and Northern Europe.* Mitchell, A. United Kingdom
1987	*The Illustrated Encyclopaedia of Trees.* Edlin, H., Nimmo, M. United Kingdom
1989	*Mythologie des arbres.* Brosse, J. France
1995	Árvores isoladas, maciços e alamedas de interesse público. Instituto Florestal. Portugal
1995	*La route vers les arbres millénaires.* Monnier-Berhidai, J. France
1996	*La magia de los árboles: Simbolismo, mitos, tradiciones, plantación y cuidados.* Abella, I. Spain
1998	*Estimating the Age of Large and Veteran Trees in Britain.* White, J. United Kingdom
2000	Árboles monumentales y singulares de la región de Murcia y territorios limítrofes. Carrillo, F. et al. Spain
2002	*Remarkable Trees of the World.* Pakenham, T. United Kingdom
2003	*Forest and Tree Symbolism in Folklore.* Crew, J. Italy
2003	*Irish Trees: Myth, Legend and Folklore.* Mac Coitir, N. Ireland
2003	*Árboles monumentales de España.* Moya, B. et al. Spain
2004	*Olivos Monumentales de España.* Muñoz, C. et al. Spain
2004	*Árboles Singulares de España.* Mundi Prensa. Spain
2004	*The Heritage Trees of Britain and Northern Ireland.* Stokes, J., Rodger, D. United Kingdom
2005	*Guía de los árboles singulares de España.* Palacios, J., Redondo, J. Spain
2005	*Bosques Monumentales de España.* Rigueiro, A. Spain
2005	*Los árboles que forman el mundo.* Petherick, T. United Kingdom
2005	*Olivos de Castellón: paisaje y cultura.* Gimeno, J. Spain
2006	*Arbres i arbredes singulars del Solsonès: catàleg dels arbres monumentals i singulars, arbredes i formacions vegetals d'interès natural.* Guixé, D. et al. Spain
2006	Árboles singulares de Zaragoza. Conde, O. Spain
2006	*Arbres i arbredes singulars del Montseny.* Broncano, M. et al. Spain
2006	Woodlands. Rackham, O. United Kingdom
2006	*Heritage Trees of Scotland.* Ogilvie, J. et al. Scotland
2006	*The Secret Life of Trees: How They Live and Why They Matter.* Tudge, C. United Kingdom
2006	*Findings.* Jamie, K. United Kingdom
2006	*The Trees that Made Great Britain.* Miles, A. United Kingdom
2007	*Arbres monumentals de Catalunya: 18 anys des de la primera protecció.* Parés, E. Spain
2008	*The Economics and Social Values Forests for Recreation and Nature Tourism.* Simpsom, M. et al. EU
2009	*Guide des arbres remarquables de la France.* Cousseran, F., Feterman, G. France
2011	*Arbres remarcables de Catalunya.* Boada, M., Boada, A. Figueres: Brau, 243 pg.
2012	*Arbres d'exception: Les 500 plus beaux arbres de France.* Feterman, G. France.

(Continued)

TABLE 9.1 (CONTINUED)
Studies on Monumental Trees in Europe (Set of Representative Bibliographic References)

Year	Title. Author(s). Country
2013	*Study: Old Trees: Cultural Value*. Blicharska, M. and Mikusiński, G. Sweden
2014	*Essay Incorporating Social and Cultural Significance of Large Old Trees in Conservation Policy*. Blicharska, M. and Mikusiński, G. Sweden
2014	*Arbres extraordinaires de France*. Feterman, G. France
2014	*Ancient Trees, Portraits of time*. Moon, B. United States
2015	*Bigtrees4life: Proyecto Europeo para mejorar la conservación de árboles monumentales en España*. Fundación Felix Rodriguez de la Fuente. Spain
2015	*Arbres remarquables en Bretagne: Un patrimoine à découvrir*. Jézégou, M. France
2016	*Arbres remarquables de Haute-Savoie: Du Mont-Blanc au Léman*. Rougier, H. France
2016	*Arbre et patrimoine de France*. Mezinski, Z., Feterman, G. France

FIGURE 9.4   Sequoia National Park, United States (Photo by M. Boada).

For some economists, the singular tree constitutes an accumulated capital gain. From the point of view of ecological economics, it would be considered a type of environmental good of symbolic value, sustained in the sphere of value of human preferences, associated with the value of inheritance and existence and of cumulative and intrinsic value. Nevertheless, some ecological economists are reluctant to talk about an economy of singular trees. Even recognizing the attractor character they raise, the market value they represent could not be applied given their incommensurability, due to the complex variable of space and time, almost remote.

Recovering the new dimensions and significance of traditional knowledge will allow opening a space so that from the dendroculture, trees and forests can be considered as referents of the understanding and care of the environment. This can contribute to gestating a change in the ways in which the society interacts with trees and forests, to achieve a positive behavioral change toward them, valuing their strategic role for life, climate, water, land, landscape, and true sustainability in all the plethora of variables entailed in the relationship of humans with the planet.

9.3.7 DENDROCULTURE: A TAXONOMIC CLASSIFICATION SYSTEM BASED ON CULTURE AND NATURE

The proposed taxonomic system based on dendroculture focuses on the recognition of both culture and nature using examples taken from the interaction of human societies with trees and forests over time (Figure 9.5). The idea is for a categorization based on the relationship between humans and remarkable trees to act as a simple and easy guide that encompasses the association of sociocultural values with the natural world.

It is a practical classification integrating scientific and traditional knowledge with the main objective of giving value to socioenvironmental diversity. It has a broad spectrum of usefulness with a base supported by popular knowledge that recovers the understanding of the environment that society has historically built around trees and forests.

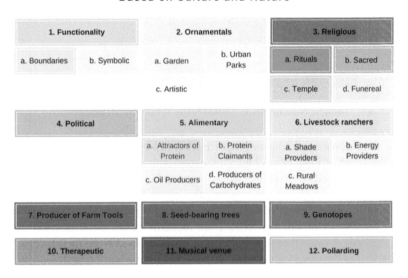

FIGURE 9.5   Taxonomic classification based on culture and nature.

FIGURE 9.6 Cork oak of Tiétar. Arenas de San Pedro, Avila, Spain (Photo by M. Boada).

9.3.7.1 Functionality

Trees with a function related to a use or sociocultural point of reference: They represent an explicit language of territorial use in toponymical or topographical name. They may even contain a code in the language of passersby.

1. *Boundaries*: Trees that take on the role of milestones or indicate the limits of municipalities or between properties. They define boundaries and internal and/or external demarcations; they have value in the territorial topographical representation and in the cadastral definition and a referential value as points of location and/or encounter (Figure 9.6).
2. *Symbolic*: Trees that symbolize a tradition or an accepted social convention that has been historically recognized. It is either they may make explicit an idea or they may represent a service. An ancient cypress can be taken as an example: it indicates hospitality and greeting in the tradition of pilgrims and wayfarers. A cypress in front of a farmhouse or convent tells the passerby that he or she has right to bread and water; two trees, right to a plate of food; and three trees, the right to stay overnight (Figure 9.7).

9.3.7.2 Ornamentals

Trees characterized by their aesthetic or decorative purpose: They are located in public spaces of social encounter. Outstanding examples may pertain to historic and private gardens.

1. *Garden*: Trees used in gardens having relevant specimens that have been preserved over the years by means of strategies to provide green spaces in the urban environment, including public gardens or that have been part of

FIGURE 9.7   Cypress of Sant Climent, Les Gavarres, Catalonia, Spain (Photo by M. Boada).

the family history or tradition. The purely ornamental sense explains the exotic nature or the distant origin of species exhibited. Often, their form has been shaped to fit into a topiary or to create a living sculpture. In sustainable gardening, a criterion applied may be the autochthony of the selected species. Florescence and especially fruiting as well as the architecture of the tree are determinants in the type of biodiversity that is likely to be attracted, especially birds (Figure 9.8).

2. *Urban parks*: Monoculture plantations arranged as if in a meadow, thereby forming groups of trees that are especially apt for social and leisure events in urban spaces. The shading structure is especially relevant for social events during the summer. The greening of the urban space by adding trees adds aesthetics, form, coloring, and functionality in ways that are clearly beneficial (Figure 9.9).

FIGURE 9.8   Sauce, Quebec, Canada (Photo by M. Boada).

FIGURE 9.9   An urban park in Girona, Catalonia, Spain (Photo by M. Boada).

3. *Artistic*: Trees that appear in different formats given that their morphology has been worked ornamentally. On some occasions, it has proved possible to mold the tree structure into a creative form. One example is the basket tree (*Ficus* sp.) or some urban trees that have been worked with ornamental pictorial forms in the structure of the trunk itself. Finally, another group consists of sculptural works in structures of senescent centennial trees (Figure 9.10).

FIGURE 9.10   *Pinus silvestris* transformed into a sculpture, Burgos, Spain (Photo by M. Boada).

9.3.7.3   Religious

Trees that are characterized by their ritual and mystical function: Taking into account all the recognizable cultural variables, they have one component in common: the temple tree possesses a large space beneath the canopy that provides protection and lends itself to the meeting taking place below (Figure 9.11).

1. *Rituals*: Trees identified especially for their ritual gifts and kindnesses. Traditional meetings and nuptial ceremonies were held under the canopy of these trees; in these meeting areas, the uses of agricultural land were decided and agreements were reached regarding the distribution of irrigation water and grazing rights (Figure 9.12).
2. *Sacred*: Trees considered sacred, such as the African baobab that welcomes the spirits of the ancestors that live in them, taking care of the living (Figure 9.13). European examples include the thorn tree of Glastonbury, a place of pilgrimage, on account of its being a cutting of Christ's cross. The tree of Tule, in the Mexican state of Oaxaca, was planted by Aztec priests and is still alive and vigorous (Figure 9.14).

FIGURE 9.11   Cedar of Lebanon, Cloister of the Abbey of Moissac, France (Photo by M. Boada).

FIGURE 9.12   Marriage tree, Ibagué, Colombia (Photo by M. Lozano).

FIGURE 9.13   African Baobab, Tanzania (Photo by R. Maneja).

FIGURE 9.14   Tree of Tule, Mexico (Photo by M. Boada).

3. *Temple*: Trees associated with a spiritual leader, prophet, or saint. Such trees generate veneration and pilgrimages for their religious significance, such as the Olive of the Rocio Virgin (*Olea europaea*) in Spain (Figure 9.15). Another well-known example is the tree of Bodhi in Thailand, in which the branches ascent to the heavens and the roots descend into the underworld.
4. *Funeral*: Trees of sacred character as a tomb or ossuary for holding the corpses or the remains of the defunct or simply planted in the holy grounds. In Spain, the yew has a significant meaning in the cemeteries of the mountainous hamlets of Asturias-León and of Galicia. The Yew of the genus *Taxus* is also common in the cemeteries of the Celtic region of the British Islands (Figure 9.16).

FIGURE 9.15   Olive tree of the Rocío Virgin, Spain (Photo by M. Boada).

FIGURE 9.16   Yew of Bermiego, Asturias, Spain (Photo by M. Boada).

9.3.7.4   Political

Trees considered as the symbol of a country: They represent principles, interests, values, and ideals of a society or social groups inhabiting in a given territory. The tree of Guernica (Euskadi) is one example (Figure 9.17), placed in the renamed city bombed by the German air force in support of Franco during the Spanish Civil War and retained by a landmark painting of Picasso. In Mexico, the Tree of the Sad Night (Figure 9.18), a decimated ahuehuete (*Taxodium mucronatum*), symbolizes Hispanicness, and is defended by the supporters of the symbolism of colonization and attacked by local ideologies. Another politics-based example can be found in some places in Nicaragua, where painted trees show the colors of the flag of the Sandinista Front of National Liberation and are the symbol of the country's freedom (Figure 9.19).

FIGURE 9.17   Tree of Guernica, Euskadi, Spain (Photo by M. Boada).

FIGURE 9.18   Ceiba tree of Hernán Cortes (1520), La Antigua, Veracruz, Mexico (Photo by M. Boada).

FIGURE 9.19   Tree with the colors of the Sandanista flag, Nicaragua (Photo by M. Boada).

9.3.7.5   Feeding

Trees that have been a source of food for the human diet through animal trophic attraction:

1. *Attractors of protein*: Trees that carry out the function of providing protein for a peasant community. Their fruits have a massive attractor effect for some species of sedentary and migrating birds, which are then captured in large quantities by means of a mistletoe or natural glue (Figure 9.20).
2. *Protein claimants*: Trees that act as a claim, through hunting gear, generally using a claim, to massively capture birds in migratory passage (Figure 9.21).
3. *Oil producers*: Centuries-old olive trees that, despite their longevity, produce high-quality oil which has proven to be an important alimentary resource for tens of generations. The slow growth of the trees generates outstanding natural and beautiful sculptural forms. A variety of different vertebrates lives in its hollows (Figure 9.22).
4. *Producers of carbohydrates*: Centuries-old trees, especially chestnuts, oaks, and beeches, which have historically been a source of carbohydrates

FIGURE 9.20   Frozen Hackberry, Conca d'lsona, Catalonia, Spain (Photo by M. Boada).

FIGURE 9.21   Holm Oak for attracting migrating birds, Puig sa Guàrdia, Pla de l'Estany, Catalonia, Spain (Photo by M. Boada).

FIGURE 9.22   A century-old olive tree, L'Arión, Ulldecona, Montsià, Catalonia, Spain (Photo by M. Boada).

FIGURE 9.23   Chestnut tree, El Bierzo, León, Spain (Photo by M. Boada).

and starch for human diet and feed for animals. Their particular biology produces a cortical growth accompanied by a necrosis in the center of the trunk, which has come to be used as a refuge for lumberjacks and charcoal burners (Figure 9.23).

9.3.7.6   Livestock Ranchers

Age-old trees with ample canopies traditionally used to provide shelter for flocks and herds:

1. *Shade providers*: Trees with a crown canopy, shaped by the cattle themselves at their base, thereby serving as a living shelter. Thanks to the shade provided during the summer, the herd of animals will come together under

its branches. Furthermore, such trees provide cover during bad weather and cold winds, therefore helping to reduce the stress to the flock and herd of animals caused by high temperatures and the hot winds of summer. Shepherds and ranchers place salt lick rocks at the base of such trees for the health of the animals (Figure 9.24).

2. *Energy providers*: The main species has classically been the carob tree (*Ceratonia siliqua*), the fruit of which, carob bean, has traditionally been used as the staple food for draught animals, prior to the use of hydrocarbons in mechanical traction engines. The carob is highly appreciated for its abundant production and its energizing of working animals. The architecture of the carob, with its singular beauty, is host to a remarkable animal biodiversity (Figure 9.25).

FIGURE 9.24   Shade trees for camels, Desert Erg Chebbi, Morocco (Photo by M. Boada).

FIGURE 9.25   Carob, Sierra de Las Nieves, Andalusia, Spain (Photo by M. Boada).

FIGURE 9.26   Holm oaks, Cáceres, Spain (Photo by M. Boada).

3. *Rural meadows*: These are naturally diverse. In the Iberian Peninsula, one can find outstanding formations of Holm Oak (*Quercus ilex*) and Cork Oak (*Quercus suber*). As examples of natural forest, these formations of trees are renowned for their high productivity. In addition, they hold an amazing biodiversity of fauna among the most notable in all Europe. In this respect, we should highlight the alignment of cypresses, lime, and planes trees which embellish the landscape of carriageways and paths (Figure 9.26).

9.3.7.7   Producers of Farm Tools

Throughout the centuries, farming communities have been obliged to be self-sufficient including by creating their own tools. Historically, they have developed special techniques using those trees that naturally have vegetable fibers of the length and strength necessary for farm tools. The shape of the pruning serves to obtain the wood used to make utensils. This is the case of the hackberry (*Celtis australis*) from which the gallows and hoes were made. These practices have produced large trees near the country houses (Figure 9.27).

9.3.7.8   Seed-Bearing Trees

Ancient trees kept for their production of seeds so as to enable the regeneration of the trees of the forest: This practice has resulted in the survival of magnificent trees (Figure 9.28).

9.3.7.9   Genotopes

Trees of appropriate structure and size for the location of bird colonies of gregarious reproductive behavior (herons, spoonbills, storks, etc.): This is the case of the

FIGURE 9.27   Basket willow of La Comella, Catalonia, Spain (Photo by M. Boada).

FIGURE 9.28   Seed-bearing beech L'Espinal, RB Montseny, Catalonia, Spain (Photo by M. Boada).

FIGURE 9.29   Communal nesting, Doñana Natural Park, Spain (Photo by M. Boada).

famous aviaries of the Doñana Natural Park, Spain. On another scale, some mature holly specimens (*Ilex aquifolium*), due to the so-called igloo effect, constitute a very well-known breeding area for forest birds such as the common bullfinch (*Pyrrhula pyrrhula*) (Figure 9.29).

9.3.7.10   Therapeutic

Some trees are known for their therapeutic properties. The lime tree (*Tilia* sp.), for example, when the tree is in full bloom, emits a volatile substance which has a tranquilizing effect (Figure 9.30).

9.3.7.11   Musical Venue

Trees beneath their canopies can house concerts and folkloric events. They are specimens of wide glass of wingspan that, by their morphology, have acoustic virtues and a strong symbolic load and of welcome (Figure 9.31). Some authors claim that trees make their own noise at a particular frequency, attracting those species with an affinity for the same sound. The wood sculptures created by the French forest artist Jose Le Piez can be used to generate an unique sound, achieved simply by passing one's hand over the surface of the cut wood, thereby allowing the voices of the trees to express themselves (old oak, biloba ginkgo, and cedar of Lebanon) (Sirven 2016). Through the research of Le Piez, it has been proven possible to rediscover the origins of one of the ancestral musical instruments that the French anthropologist and ethnomusicologist André Schaeffner pronounced in his book *The Origins of Musical Instruments* to be one of the strangest human-made inventions. It is the Livika, found in the Province of New Ireland in Papua New Guinea, where it is used during the funereal rituals of the natives to imitate the voice of the wood spirit and the soul of the dead.

9.3.7.12   Pollarding

An intervention consisting of a careful cut just below the tree canopy such as to bring about a branching that allows wood to be extracted for charcoal or timber as well as

FIGURE 9.30   Lime tree of Freiburg, Germany (Photo by M. Boada).

FIGURE 9.31   Grevolosa Beech (Courtesy of Jordi Baucells).

FIGURE 9.32   Pollard Beech in Zegama, Oria, Biscaia (Photo by M. Boada).

fodder for cattle without having to cut down the entire tree. The resulting, aesthetically pleasing form of the tree favors the refuge and breeding of different forest animals (Figure 9.32).

9.4   CONCLUSION

Forests and the trees have been predecessors of all the civilizations of the world and symbols of reference for culture, religion, and art since times long past. As a source of inspiration, the landscape has integrated itself, like a piece of painting, sculpture, literature, poetry, and folklore, into the social and collective imagination, where it reads like a chronicle in which is narrated the history of a people and its relation with the land.

Dendroculture seeks to recover the form in which trees and forests can be appreciated, by incorporating a holistic vision, in which the environmental, social, and cultural aspects can be highlighted and that unites natural history with social history, recovering the symbolic, cultural, and artistic relationship that humanity has built around remarkable trees and thereby encouraging the safeguarding that allowed the survival of century-old, even millennial-old trees.

Monumental trees, by being associated with dendroculture, can add patrimonial value to a territory, a capital gain, and thereby rescuing popular recognition of their empirical importance. They can dynamize sustainable tourism and environmental education.

In Europe, particularly, the opportunity exists to foment a dendroculture in the forests that have become more extensive due to the processes of urbanization and the progressive decrease of rural population. In contrast, in tropical zones, it can play a role in encouraging the conservation of trees and forests threatened by the processes of deforestation by identifying remarkable trees and thereby strengthening their protection and safeguarding.

In essence, in the face of today's human-caused environmental crisis, dendroculture offers an opportunity to recover the sociocultural component linked to forests and create awareness of the importance of the relationship between human beings and the natural world.

REFERENCES

Abella, I. 2012. *El gran árbol de la humanidad*. Barcelona: Integral.

Albelda, J. and J. Saborit. 1997. *La construcción de la naturaleza*. Valencia: Conselleria de Cultura, Educació y Ciència, Generalitat Valenciana.

Bauer, E. 1991. *Los Montes de España en la Historia*. Madrid: Fundación Conde del Valle de Salazar.

Boada, M. and A. Boada. 2011. *Arbres remarquables de Catalunya*. Figueres: Brau.

Boada, M. and F. J. Gomez. 2011. *Boscos de Catalunya*. Barcelona: Lunwerg.

Crews, J. 2003. Forest and tree symbolism in folklore. *Unasylva* 54(213):37–43.

FAO (Food and Agriculture Organization of the United Nations). 2014. *The State of the World's Forest Genetic Resources*. Rome: FAO.

FAO. 2016. *Global Forest Resources Assessment 2015. How Are the World's Forests Changing?* Rome: FAO.

Frazer, G. 1951. *La Rama Dorada*. México: Editorial Fondo de Cultura Económica, S.A. de C.V. México.

Frochoso, M. and Castañon, J. C. 1990. El Medio Natural (I): Los Rasgos Generales. In *Historia de Asturias (Tomo I)*. Meres: La Nueva España.

Funtowicz, S. and J. Ravetz. 2003. *Instituto para la protección y seguridad del ciudadano*. (IPSC), (EC-JRC), TP 268, 21020 Ispra (VA). London.

Hallé, F. 2016. Atlas de Botanique Poétique. Paris: Arthaud.

Ovidio, P. 2008. *Metamorfosis: Libros I–V*. Madrid: Gredos.

Rozo, J. 2006. *Bachue: Relación mito-arte rupestre*. http://www.rupestreweb.info/bachue .html (Accessed January 23, 2018).

Sirven, B. 2016. *Le Génie de l'arbre*. Paris: Actes Sud.

Toledo, V. M. and N. Barrera-Bassols. 2008. *La Memoria Biocultural: la importancia ecológica de las sabidurías tradicionales*. Barcelona: ICARIA.

Urteaga, L. 1993. *La teoría de los climas y los orígenes del ambientalismo*. Barcelona: Universitat de Barcelona, Facultat de Geografia i Història.

von Carlowitz, H. C. 1713. *Sylvicultura oeconomica, oder haußwirthliche Nachricht und Naturgemäße Anweisung zur Wilden Baum-Zucht*. Reprint of 2nd edition, 2009. Remagen-Oberwinter, Germany: Verlag Kessel.

10 Communicating to Support the Comprehension of Forest-Related Issues by Nonexpert Audiences

Julie Matagne and Pierre Fastrez

CONTENTS

10.1 INTRODUCTION

Media[1] are a central element in the relationship between citizens and forests. On the one hand, forests are ever present in the media. They have long held a significant place among the settings of a variety of works of fiction, from classic fairy tales (*Little Red Riding Hood* and *Hansel and Gretel*) to contemporary films (*Avatar*, *The Blair Witch Project*, and *The Cabin in the Woods*) or now video games (Bonvoisin 2017). They are also part of societal issues reported by the news media, such as climate change or sustainable development. On the other hand, in today's increasingly urbanized societies, it is fair to say that what most individuals know or believe about forests heavily depends on the media they consume.

Despite this constant presence in the media sphere, there is a complexity to contemporary forest-related issues that is seldom addressed by the media, making them difficult to grasp by nonexpert audiences. Even though anyone identifies what a forest is (Chapter 2), it may actually mean many different things to different people among the general population. The concept of forest is a complex one. It encompasses various dimensions, roles, and functions: the forest is an ecosystem participating in biodiversity, a green lung playing an essential role in the absorption of CO_2, a resource that can be exploited and provides jobs, a home for hosting animals, a hunting ground, and a place of leisure or where recharging, wandering, and pondering on the meaning of existence. There is not only *one* forest, but many forests, depending on the perspectives. The relationship to forests is be shaped by the multiple forms of individual, collective or societal experiences that tap into our history, our ethics, our myths, and our expertise (Paré 2012).

The complexity of forest issues is further reinforced by the evolution of the forestry sector, which, with the emergence of new policies related to sustainable management, counts multiple players: forest managers, industry operators, private and public owners, lobbyists, international organizations and institutions, environmental associations, or citizen movements. Each of this community seeks to gain visibility in the public sphere, to inform and to influence the opinion of the general public with strategies and objectives that match their respective interests and, as a result, are often divergent. The messages are numerous, sometimes contradictory.

Several studies have shown the high affective value that the public gives to the forest: citizens associate the forest with calm, renewal, life, freedom, continuity, and immutability (Barthod 1995; Bodson 2005), as well as with negative emotions, such as anxiety or fear (Boutefeu 2008; see Chapter 6). Our relationship to forests builds on the sensory (moisture, freshness, color, etc.) and emotional experiences (childhood memories and feeling of freedom) they provide us with, rather than on rational reasoning (Boutefeu 2008).

Humans have an intense and special affective connection with forests and trees, which dwelled on Earth long before them, and which will live long after them (Harrison 1992). As a consequence, media messages on trees and forests yield very intense reactions from the public. The emotional connection to forests could very well be further enhanced by the growing urbanization that feeds the fantasies of city dwellers. The recent success of Peter Wohlleben's book *The Secret Life of Trees*, which sold more than 1 million copies in Germany and was translated into 32 languages, is probably a good example of this current emotional connection between people and trees. In particular, Wohlleben compares forests to human communities and often uses anthropomorphism wording. He argues, in particular, that trees are sensitive beings able to communicate with each other, of mutual help and friendship.

The emotional connection to forests is also related to its strong symbolic value, which is based on the consensus, among the population, on the positive image of the tree, which seems to have only virtues (De Smedt et al. 2016). Tree symbolizes values such as nature, future, steadiness, generational lineage, past, heritage, rootedness, quality of life, grandeur, sustainability, and life. All these virtues make a tree seems untouchable, and its standing in the collective imaginary lends them a strategic status in the context of media communication. Media producers tend to use the symbol of the tree as a generic means to generate affective reactions to manipulate the public as part of strategies of persuasion (De Smedt et al. 2016).

The context we have just described—the omnipresence of forests in the media, the multiplicity of stakeholders voicing their perspective on forests, the emotional connection to forests and the symbolic value of trees, and their exploitation by strategic communication—highlights a societal stake for forest communicators: making the public able to understand the complexity of forest-related issues, such as their sustainable management or the coexistence of their ecological, economical, and social functions, so as to support the democratic debate these issues call for.

This chapter will include the following. First, we will start by attempting to explain existing evidence of the relative incompleteness of the representations the public holds on forests. Next, we will introduce and contrast two models of media communication that forest communicators may choose to adopt and highlight their respective potential in supporting audiences in constructing more complex and more accurate representations of forest-related issues. We will then present the recent results of two studies completed by the authors on how teenagers and young adults are able to understand, contextualize, compare, and evaluate forest-related media. Finally, based on the previous sections, we will formulate recommendations on how communication on forests should be practiced to support the appropriation of forest-related societal issues by the public.

10.2   THE PUBLIC FACED WITH FOREST-RELATED ISSUES: A LACK OF KNOWLEDGE

Given the complexity of forest issues, developing a critical understanding of forest-related media messages seems like a daunting task, for which the public is ill equipped, both in terms of prior knowledge and attitudes. As Farcy et al. mentioned in Chapter 7, several studies have examined how the no-expert public perceives forests, and it is often far from the reality of the sector.

In particular, studies observed a gap between public perceptions and the reality of the state of the forests (Rametsteiner and Kraxner 2003; Matagne 2017): the public would think, for example, that the forest cover in Europe is declining, whereas, in fact, it keeps increasing. Forest stakeholders also face a paradox: the forest industry suffers from a negative image while the forest and its products, on the contrary, are generally associated with a positive image, which Suda and Schaffner (2004) designate as the "slaughterhouse paradox": the public loves forest, and wood as a material, but considers cutting down trees as a crime, rather than a levy on a resource that can be managed from sustainable way (Gadant 1994; Suda and Schaffner 2004; cited by Janse 2006).

When thinking about forests, most people feel concerned about the environmental benefits of forests, long before their social or economic role. Moreover, the public seems to generally consider that the protection of nature (involving the preservation of forest and biodiversity, as well as halting deforestation) is difficult to reconcile with its exploitation as a resource. Faced with this apparent incompatibility, people seem to be in favor of protecting trees. The difficulty of choosing between "protecting or exploiting" forests lies in the definition of "sustainable management."

For Lakoff (2010), the public fails to develop a complex vision of the environment because of what he calls a problem of hypocognition: people lack cognitive frames that capture all the dimensions of the complex reality of the environment. Frames, or schemas, are cognitive data structures that represent stereotyped situations (Minsky 1974). Frames are "packets of information that contain variables" (Rumelhart and Norman 1988). They include "semantic roles, relations between roles, and relations to other frames" (Lakoff 2010). For example, a restaurant frame includes roles for clients, cooks, waiters, a kitchen, a dining room, a menu, and checks. Each of these roles includes a fixed part (e.g., the skills of the cook) and a variable part (e.g., their identity). A central feature of frames is their interconnectedness: information packets are connected together through roles, and frames can embed one into another. For example, a frame for the order is embedded into the restaurant frame. Frames can also exists at different levels of abstraction: a frame for fastfood restaurants and a frame for gastronomic restaurants is subsumed by a more general frame for restaurants. Similarly, our frames for forests, or for oceans, are subsumed by a more general frame for the environment.

Unlike other areas of everyday knowledge (like how to interact in commercial settings such as restaurants), the environment represent a knowledge domain covering a variety of areas and factors for which the majority of people lack proper frames.

We are suffering from massive hypocognition in the case of the environment. The reason is that the environment is not just about the environment. It is intimately tied up with other issue areas: economics, energy, food, health, trade, and security. In these overlap areas, our citizens as well as our leaders, policymakers, and journalists simply lack frames that capture the reality of the situation (Lakoff 2010).

The lack of frames regarding the environment results in an inability to "connect the dots," i.e. to hold a coherent, structured view of environmental issues, leading to inadequate or missing associations of ideas. Lakoff's thesis on hypocognition that affects the way the environment is perceived and understood echoes Meunier's (1998) argument about different forms of knowledge.

> [...] what is happening in areas of knowledge that are far removed from everyday life and have not been systematically learned? These areas have developed considerably thanks to the media, which disseminate a great deal of information on all kinds of domains that are often far removed from ordinary life [...] but which nevertheless can have a considerable impact on it [...]. It can be hypothesized that knowledge about these domains is based on schemas whose inferential power is very limited. They are not generally based on more basic levels from which they would have been extracted, and as a result they are poor in sub-schemas that can be specified in various ways depending on the circumstances. [...] Another characteristic of such schemas is often to highlight only certain partial and sometimes marginal aspects of the phenomena. (Meunier 1998; authors' translation)

Both Lakoff's and Meunier's arguments seem particularly applicable to forests. Like other environmental issues, forest-related issues call upon knowledge that is relatively remote from the everyday lives of today's citizens. Hence, these citizens may be limited in their ability to develop a comprehensive view of forests, specifying the roles of forests, the roles of humans with respect to forests, and the multiple dimensions of the relationship between forests and humans (pertaining the economy, energy, food, leisure, habitat, etc.). Their forest-related knowledge may be ill structured and include disconnected, unidimensional (and largely incompatible) views, all due to a lack of frames to organize them.

The question that arises is the following: how are these limited, incorrect, or even lacking representations problematic? We believe that they are because they prevent a nuanced, correct, and complex understanding of economic and sustainable forest issues. The quote from Jim Bowyer, a professor of forestry at the University of Minnesota, perfectly illuminates the problem. In his article "Fact vs. Perception," Bowyer (1995) questions "is it possible to discuss forest management with someone who believes trees will live forever?" How can we have a discussion about plantation issues or sustainable management if people think that cutting a tree is killing the forest? How can we approach the issue of reasoned tree cuts if people are against cutting trees by principle, because to them it represents a threat to the forest?

The challenge is therefore to make more complex such limited representations and to correct false associations of ideas. It represents a long and slow work, especially as such representations are connected to emotions that play a vital role in the building and strengthening of frames.

10.3   FOREST COMMUNICATION MODELED ON THE DOMINANT COMMUNICATION IN THE MEDIA

In today's urbanized societies, the relationship between the public and forests has gradually changed. Most of the world's population is now living in cities. Urban dwellers have a relationship to the virtual, immaterial forest, which is mainly based on mediated experiences. The way in which forests are represented in the media has a potentially decisive influence on the public's knowledge of forests.

Currently, most forest stakeholders tend to adopt the type of communication that dominates the media sphere: strategic messages that mainly aim at influencing their audience through short repetitive messages, which rely on simple slogans, on images rather than on text, and on the elicitation of emotions rather than of reasoning. In 2006, one of the reports of the European Forest Institute (EFI), entitled "European Cooperation and Networking in Forest Communication," explicitly advocates for this type of advertising communication, stating that "it appears that more attractive media, especially visual methods, are needed to get to the public. Evoking emotion with 'warm, friendly images' and using simple slogans are very useful in this respect—something Environmental NGOs (ENGOs) have understood already for a long time" (Janse 2006).

Such communication strategies are built upon the view that audiences are inherently "passive" and "massive." Indeed, a simplified and universal message can only be considered effective if the audience is conceived as a "mass," all members of which are considered to be identical and expected to react in similar ways to a given message. In this approach, the audience is also considered as passive, since it is assumed to be the inert receptacle of the various messages they receive, unable to criticize the representations they are submitted to. The opinion of a "massive and passive" audience can then only correspond to the reproduction of the messages they receive, since all its members would react identically, without critical distance.

The adequacy of this particular type of communication for the respective goals of the different forest stakeholders does not seem to be questioned. The actors in charge of communication do not question their practices in any way and do not plan to evaluate the effectiveness of their actions (Courbet and Fourquet-Courbet 2010).

More than anything, this way of considering the public belongs to a particular model of communication among others. To highlight this, we will refer to Meunier's (1994) description of two alternate theoretical models of communication, corresponding to two different views of the role of communication and the nature of the social relationship between the sender and the recipient of a message.

The first model, the code model, considers communication as a process of persuasion, emphasizing the influence of the sender on the recipient. It frames communication in terms of impact on a target. The recipients are conceived as a passive mass, oblivious to the persuasive intentions of the senders. As a result, the recipients' representations are expected to be the simple reproduction of the messages they were previously exposed to.

This model is grounded in a technical understanding of communication, which equates communication with data transmission. Its roots can be traced back to information theory (Shannon 1948), which was proposed to model and quantify the necessary redundancy in the coding and decoding of information in any data transmission for the transmission to tolerate a given signal/noise ratio.

> This model—characterized as the telegraph model, the conduit model, or the code model—unfurls communication on a line from the sender to an individual described as a receiver, and emphasizes channel quality and the encoding and decoding operations that ensure the correct transmission of the message, that is to say its restitution at the end of the chain. (Meunier 1994; authors' translation)

In this linear model, the recipient's (cognitive) activity is limited to mechanically decoding a message, using the code previously used by the transmitter. This model neglects the fact that the recipient is a unique person characterized by a unique experience. It underestimates the richness of the processes of signification. Reciprocity is excluded. The social relationship between the transmitter and the receiver consists of the influence the former has on the latter (Meunier 1994). In the context of media communication, this model emphasizes the design of media messages over their interpretation, as it assumes that the way all members of their audience will understand (i.e., decode) them will correspond to their producer's intentions.

Meunier's second model, the cooperative model, views communication as a process of negotiation and cooperation between the sender and the recipient. The centre of the relationship is no longer the individual, but the interaction between the communicative agents, and the world they share. The receiver leaves its role of passive decoder to become a key player in the construction of meaning. Their cognitive activity involves mechanisms of inference and decentration.

[…] The inferential activity of a recipient whose goal is to answer the question "what did he want to tell me about our mutually shared world?" obviously entails decentration and the interiorization of the other's point of view; and the answer of the same subject, now a speaker in the interaction, supposes that he returns the intra-individual to the interindividual after having integrated his own point of view, and so on. (Meunier 1994; authors' translation)

This second model assumes that communication relies on a common conceptual ground (Tomasello 2008) between the communicative partners and on their ability to adapt their perspective to the perspective of the other and their respective contributions to this conceptual common ground. From the perspective of the recipient, this means being able to ask oneself both "what is one trying to tell me?" and "what do I already know?" This ability to articulate multiple perspectives, resulting from successive communicative interactions with different partners, opens the possibility of grasping the complexity of the object of communication.

From a historical point of view, these two models of communication correspond to an evolution in the scientific study of communication, from theories considering communication linearly to models that conceive of communication in a circular way. The latter introduced a place for the recipient, his/her relation with the addressee and the context. This evolution has been described as the transition from telegraphic models to orchestral models (Winkin 1981).

Patriarche (2008) explained the history of communication and the evolution of the role assigned to audiences in communication theories. "Linear models of mass communication (Shannon and Weaver's model, Lasswell's model, etc.) confined the audience to the role of recipient of messages. The notion of feedback had to be imported from cybernetics (Wiener), from general systems theory (Von Bertalanffy) and from systemic therapy (Watzlawick) so that communication was reconceptualized as a circular process. The notion of feedback, by drawing attention to the reactions (spontaneous or provoked) of the audience that are likely to influence the media supply, therefore allows "to consider the 'recipient' pole from a different standpoint than that of a simple ultimate agent in a linear process (Antoine 2004). The audience is no longer excluded from production; to some extent, they become co-authors of the messages."

Since the 1960s, a multitude of research works has questioned "the degree of passivity or activity of media reception by the public" (Patriarche 2008) and opposed the traditional view according to which the meaning conveyed by the media is exclusively located in the message itself. According to Piette (2000), audience studies have long recognized that "the audience is not composed of passive individuals, but of individuals who contribute an important part to the process of constructing the meaning of messages. It is believed that the media can never be completely certain of how the public decodes messages. […] The notion of reception of messages is thus problematic: the meaning of the messages is never definitively fixed, because the public is engaged in a permanent process of negotiation of meaning."

In the world of social science research, therefore, a conception of communication in which the public would be passive has been dated for decades. However, it is precisely this conception that continues to reign in today's communication practices: this model is the one favored by the players in the forestry world (Matagne et al. 2014), and it is the one that predominates in the media sphere in general. This model remains a major implicit reference for many communication actors.

Why did the practices of the communicators not follow the evolution of the theories of the communication? How can one explain the gap between the theoretical advances of the information and communication sciences and the practices of the actors of the field? Two hypotheses may help find an answer to these questions.

The first is that in a context of overabundance of media messages in our societies, the media are caught in a competition for their audiences' attention, which pushes the producers of media messages to bet on the quantity of short and decontextualized messages rather than on the quality of elaborate messages which would leave a place for the recipient in the construction of meaning. Such short and simplistic messages implicitly adopt the code model of communication, where each message sent requires the recipient to simply decode and accept its perspective. The multiplication of messages creates an addition of independent, disconnected perspectives, where each sender hopes that theirs will be the one the audience will remember and accept.

This is the argument of Potter (2004): faced with a media world in which the pace of information is constantly accelerating, the public is forced to develop avoidance strategies. Media designers are therefore working to reduce message costs (in terms of time, attention, etc.) by developing messages that condition the audience in traditional and automated reception patterns. To maximize their chances of being heard, communicators produce shorter messages that have lost the detail needed to form a context of interpretation. By multiplying short and fast messages, which can be easily repeated, media designers create superficial messages that make it difficult for the public to construct meaning. For Potter, the lack of context concerns both mediated information and the identification of producers.

The second hypothesis, which is complementary to the first, is part of a political vision of the media sphere characterized by power relationships and the predominance of economic interests. According to this view, the vast majority of media messages that circulate are intended to change the behavior; attitudes; and representations of the public toward a particular behavior, an attitude, and a representation in the interest of its author. Most media are strategic, in the sense that they aim to get a particular point of view through to the public.

We then hypothesize that the difficulties of the public to grasp the complexity of forestry issues could be rooted in this type of communication. This hypothesis would be in agreement with Potter (2004) when he asserts that the multitude of messages circulating in the media cause erroneous constructions of meaning for three reasons: (1) the superficial nature of the messages—short and decontextualized messages— in circulation; (2) the economic logic of the media, which aims to maximize the public's exposure to messages and to prevent the critical capacity of the public from being expressed; and (3) the state of automatism in which the public encounters the messages because of informational fatigue.

It is from this hypothesis that we have sought, on the one hand, to evaluate the effect of this type of persuasive communication on the ability of young people to demonstrate cognitive autonomy and, on the other hand, to test whether an alternative model of communication was likely to provide a better understanding of the complexity of forest issues by the public.

10.4   ASSESSMENT OF COGNITIVE AUTONOMY OF YOUNG PEOPLE TOWARD FOREST MEDIATIZATIONS

We conducted two studies to assess the degree of cognitive autonomy of teenagers and young adults toward forest-related media and how it is influenced either by individual differences, or by the communication strategy adopted by the media. We defined the cognitive autonomy toward the media as "the abilities required to understand, contextualize, compare and make critical judgments about media messages" (Matagne 2017).

We evaluated the cognitive autonomy of the public by asking a total of 303 young people to participate in a survey requiring them to read and react to posters from different forest stakeholders, representing a variety of communication strategies.

In the first study, 183 Belgian- and French-speaking high school students (aged from 16 to 19) were asked to react to four posters that corresponded to the combination of two criteria: simple vs. complex and emotional vs. informational. We consider media messages to be simple when they represent a limited number of referents, connected by a single relation. Conversely, complex media messages represent numerous elements connected by various types of relations. We distinguished between emotional and informational messages based on their relative reliance on the generation of an emotional response on the part of the viewer or on the communication of factual content. Of course, both criteria are relative. The four posters selected were, respectively, simpler and more emotional, simpler and more informational, more complex and more emotional, and more complex and more informational.

In the second study, 120 master's degree students (aged from 20 to 23) were asked to react to two posters: one strategic poster and one educational poster. We defined strategic media messages when their authors attempt to influence their audience and establish a relationship of authority with it. Conversely, in educational media messages, the audience is considered as a partner of the construction of the meaning of the poster, which is left open for interpretation.

In both studies, the participants were asked to answer open-ended questions regarding posters. The goals were to diagnose the ability of respondents to think by themselves when faced with a variety of forest-related media and to identify what factors could influence the cognitive autonomy of young people. The main findings are presented below while results are detailed in Matagne (2017).

Respondents to our questionnaires showed a limited, but effective, degree of cognitive autonomy when asked to understand, contextualize, compare, and position themselves toward media messages dealing with forest issues. They synthesized and partially understood the posters and their messages. They identified their authors as long as they were already known or presented in a transparent way. They were partially able to detect the intentions of the authors and evaluate the media. Identifying the institutional quality of the authors and the limits of the poster's audience was difficult for them, as was comparing posters. Their capacity for taking a stand and formulating critical judgments was mainly dependent on their knowledge of forestry issues.

Regarding the effects of the types of communication used by the posters, the results of the first study clearly showed that respondents were more successful in answering questions for simpler poster than for more complex posters. A message that represents numerous elements connected by various types of relations required more cognitive autonomy than a representation message of the simple referent. Differences between the emotional and informational nature of posters did not appear to be as important. Rather, we suggest that the observed differences were mostly dependent on the specificities of the poster itself.

In the second study, students generally demonstrated less cognitive autonomy in the face of the educational poster compared to the strategic poster. Strategic posters are far more common in the media sphere. Students appeared more comfortable with this type of poster than with the educational poster that left them puzzled. The effort required to understand the educational poster was more important since its message is not given as such, but must be mentally constructed by the recipient. The respondent must pose interpretative hypotheses based on the connections they can establish between the different elements shown in the poster, connections that are less explicit than in the strategic poster. These results tend to show that a media that explicitly requires the recipient to coconstruct its meaning calls for more cognitive autonomy than a media that simply explicitly states its message.

10.5 RECOMMENDATIONS ON HOW TO COMMUNICATE ON FORESTS

Based on our two studies and previous interdisciplinary research (Dereix et al. 2016), the second model of communication presented earlier—the cooperative model—appears as a double-edged sword. By leaving more room for the recipient in the construction of the meaning of media messages, it is not only likely to promote the development of the cognitive autonomy of media audiences, but it also requires more cognitive autonomy and stronger frames to start with. This could lead the authors of media messages to embrace the telegraphic model of communication, which they could perceive as the easy road to getting their audience to understand their message: making it simple and avoiding to leave room for interpretation, since part of the public may not have the necessary skills to coconstruct its meaning. In this model, people are expected to simply "decode" messages and adopt the perspective of the author.

We argue that this view is a dead end for two reasons. On the one hand, simple unequivocal messages are the ones that are the easiest to critically evaluate (as our studies have found). The notion of a passive, compliant audience is thus especially questionable for such messages. On the other hand, in a context in which each author adopts this model, the multiplication of messages in our media-saturated environment will inevitably condemn each of them to produce more and more simplistic messages than their competitors to maximize their chances of being heard, only reinforcing said saturation. This is not a sustainable communication strategy in the long run. Additionally, whereas this option may make sense for communication designed to sell products in a competitive market, it seems at odds with messages associated with issues that should be discussed democratically (such as forests or the environment in general). Such messages call for a complex understanding of their stakes by their audiences: citizens. Hence, we believe there is no other choice than taking the hard road: betting on the intelligence of the audience and supporting them in appropriating these complex issues.

In our opinion, this argues both for an educational approach to communication, in line with the cooperative model, and for the development of media literacy and knowledge for audiences to acquire the competences and frames required to participate in the coconstruction of the meaning of media messages, and their critical evaluation. This leads to share five recommendations for developing a communication about forests in line with this perspective.

10.5.1 CONSIDERING ALTERNATIVES TO PERSUASIVE FORESTRY COMMUNICATION

Communication sciences suffer from a lack of legitimacy that is felt in the practices of communicators. On the one hand, even though models of communication based on the cooperation and coconstruction have gained prominence in communication research decades ago, most communication professionals still behave as if their audiences were a passive mass. On the other hand, when it comes to communicating about forests, foresters themselves are often responsible for communication, implicitly considering that communicating is a skill that everyone has.

When using the media to communicate on any issue, but in particular on complex issues, a number of basic questions need to be asked, such as why do I communicate? (what am I trying to change or generate by communicating? what do I wish to say?) what is the audience to whom I address my message? (who are they? what do they know? how could different audiences interpret my message differently?) or what are the most relevant means to convey this message and my goals to this audience? Asking themselves all these questions is part of the routine work of people trained in communication, but many forest communicators (like other actors not specifically trained in communication) seem not to ask such questions and reflect on their communicative practices. Rather, they tend to mimic what they see and hear in their everyday life: commercial strategic and persuasive communication, which is the dominant form of media communication in our society.

However, this form of communication is implicitly shaped by a telegraphic model of communication, which matches the effects it intends to produce: making one's brand as visible, memorable, and likable as possible for their supposedly passive and obedient audience, with simplistic associations of ideas. By the dint of repetition, the goal is to create automatisms of thoughts in the public. When they adopt this model, professional communicators consciously work to produce these mechanisms. But not all actors who want to communicate through the media are necessarily trying to produce the same effects.

Forest communication seems to call for an explicit reflection on this topic, with the support of communication experts. A necessary step in this process is to consider alternatives to persuasive communication that are not fully represented in dominant media practices, but that have been studied and argued for in communication research for decades (i.e., communication that recognizes the active role of audiences in the coconstruction of the meaning of media messages). Such a reflection would allow forest stakeholders to avoid the simple mimicry of dominant media practices, to make informed choices about the type of communication they wish to engage in and to develop practices that correspond to their objectives, their audiences, and the larger context in which their messages are produced.

10.5.2 Documenting the Public's Representations to Change Them

While being in charge of their own communication, foresters do not always realize the gap that can exist between their views, interests, and knowledge and those of the general nonexpert public. If they communicate about forests without considering this gap, they are likely not to be heard and understood.

Several scientific studies now provide us with information on the nonexpert public's general knowledge on forests, habits in forests, and attitudes toward (sustainable) forest management. In our opinion, these studies need to be furthered to explore the richness of people's social representations of forests and how they integrate dimensions that emerge today as central to their experience of forests: the affective attachment to forests, their symbolic significance, their sacred role, etc. Additionally, further studies are needed to document the diversity of representations among different social groups.

Starting from documented representations, the challenge of communication is to support them in the development of an understanding that captures all the dimensions of the complex reality of the forests. This includes starting from their unstable or limited frames and strengthening and complementing them to gradually change the representations likely to hinder a good understanding of forest issues.

Considering how different audiences experience forests can be seen as an expensive preliminary step. However, we believe that it is essential in order to guide the authors in the design of messages that will maximize their chances of being heard and understood.

Furthermore, it would be particularly interesting to evaluate any communication in terms of effects on the public in order to know whether the expected objectives were actually achieved by this communication. This would allow communicators to adjust their future communication in order to develop ever more appropriate tools.

10.5.3 Daring the Complexity

We believe that if they wish to contribute to a better understanding of forest issues by the public, forest stakeholders must refuse to apply the techniques of advertising—short and repetitive messages—that oppose a critical construction of the public's knowledge (De Smedt et al. 2016). Instead, we think they would be better off inventing a more explicit communication (about who, what, when, where, how, and why) that is appropriate to their subject of communication: forests.

The short format of the posters and their design in a persuasive mode deprives the public from the many implicit and contextualizing elements essential for inferential work, potentially leading them to simplistic or fallacious interpretations.

We could hypothesize that designing long-format media would have more lasting effects on understanding the forest and its universe. Indeed, if they want to shape the representations on the forest of a large nonexpert audience, media designers may be interested in producing elaborate messages allowing the public to grasp all the nuances of its complex reality, such as messages articulating two points of view that may appear totally opposed: exploitation and preservation of the environment (De Smedt et al. 2016).

Without pretending that long formats are automatically complex, the argument is that longer formats are likely to allow time for the public to think. The long format also leaves more opportunities for media designers to accompany the public in its construction of meaning. However, if complex messages are most likely to support a complex and nuanced representation of reality, they require more prior knowledge and a sufficient degree of cognitive autonomy to be interpreted correctly. The more complex a message is, the more it will require skills of understanding, contextualization, comparison, and critical judgment from its audience. An audience that does not have these skills will experience difficulties when interpreting it.

Beyond the objectives that media designers pursue, this may also be an ethical issue. Since more elaborate messages require more elaborate interpretative work by the public, media designers may choose between two options. They can either consider that they cannot run the risk that the public does not have the skills to properly interpret their message and therefore choose to continue to disseminate lacunar and imprecise messages that require less mental effort on the part of the public. They can also take the risk of developing more complex messages and thus, in a collective movement, gradually participate in developing the skills of the public who, as a result of encountering messages that require higher skills more frequently, would be more inclined to adapt to this type of more complex messages. This second alternative would then need to be accompanied by a more sustained effort in the realm of media education on the part of the educational world (De Smedt et al. 2016), since we know that media competences do not naturally develop in individuals. If this is not the simplest solution, it is the solution that opens the way to more democracy and social equality.

The predominance of short and decontextualized messages is probably indicative of the main objectives that motivate media designers: most current media messages are driven by political and economic motives. Their designers are therefore probably not interested in helping elevate critical competences and public knowledge.

10.5.4   STATING WHO IS SPEAKING EXPLICITLY

Another important point for forest communication is to make the institutional quality and the specific skills of the authors explicit in the media they produce, in order to promote a better understanding of all the forest stakeholders and their respective expertise by the public.

The results of our studies showed that young people struggled to understand the institutional quality of media authors and their roles. This is even more the case with posters from the wood industry, which did not name the institution that produced them at all. It would therefore be necessary for authors to make their institutional quality more explicit, in accessible language. This would allow the public to better situate the origin and the intentions of the media messages and, consequently, to better interpret their meaning and scope. This way, the audience would be better equipped to connect the different forest or wood industry stakeholders with respect to their work, their different roles, and their respective interests (De Smedt et al. 2016). It would also be useful to better explain the many facets of the skills and techniques that forest professionals hold, as the public has little knowledge in this area (De Smedt et al. 2016).

In the current media sphere, the effacement of the author seems to be the norm. This makes the ability of young people to understand and identify the institutional nature of the authors all the more necessary, since they often have to guess it. Hence, in parallel with this effort of explanation by forestry communicators, the educational world has an interest in reinforcing young people's knowledge of different types of social, political, cultural, and economic actors. The stronger this knowledge is, the more they can develop their capacity for understanding, contextualization, comparison, and critical judgment of the media these actors produce.

10.5.5 STRENGTHENING MEDIA EDUCATION IN SOCIETY

The relationship between the media practices of the young people we interviewed and their ability to understand, contextualize, compare, and evaluate the forest media, as it appears in our results, confirms an existing suspicion: the simple exposure to contemporary media, whatever the degree of complexity of their associated practices, does not develop these abilities. In our studies, it even seemed to hinder them, especially in more complex tasks such as summarizing the posters, formulating a critical judgment on them, or taking a personal position regarding them.

These results confirm that although media practices can create spontaneous learning, the development of media competences (especially critical, comparative, and evaluative competences) requires training by qualified educators (Potter 2004; Jacquinot 2009). If society wants to see citizens blossom in understanding complex societal issues in order to participate in public debates, it should integrate media literacy into long and structured educational programs. Indeed, our study found that short media education activities did not significantly strengthen the media competences of the youth.

10.6   CONCLUSION: THE CHALLENGES OF FOREST COMMUNICATION

If foresters wish to succeed in communicating on forests with the nonexpert public by helping them understand the complex issues of forests, they must take into account the fact that the public does not think about and perceive forests like them. They need to consider the public's knowledge, beliefs, and attitudes toward forests to design messages that are adapted to their preconceptions and limited knowledge of the sector. Let us take the issue of tree cuts as an example. Before talking about responsible or sustainable forest management, it is necessary to understand what, in the public, creates such a strong reaction and a firm opposition to foresters cutting trees. Understanding and addressing this resistance is a first step in the work of reframing the tree cuts in a wider, more comprehensive depiction of forests and forest management, in which exploitation and preservation are not mutually exclusive.

At the same time, foresters also need to realize that communicating about forests requires thoughtful and careful design work. It is, for example, imperative to precisely know what message one wants to convey and to which audience in particular. Designing vague messages that seek to reach the greatest number is often a mistake. The results of our studies have shown that the public is not familiar with the various forest stakeholders. As a result, foresters have an interest in explaining who they are in their messages and what their specific skills are so that the public can situate them in the forest landscape and understand their intentions. They also have an interest in focusing on more complex and longer messages that allow the public to seize forest-related issues by articulating different perspectives. Everyone is concerned about forests as an inhabitant of the planet. Everyone should be able to be informed and understand these issues in order to participate in the public debate and cocreation of the world of tomorrow. In this sense, foresters should free themselves of communication that seeks to impose their vision of forests to the public. They need to accept that their vision of forests is not the same as the nonexpert public and that this gap is not incompatible with public debate and a shift toward collective decisions.

REFERENCES

Anderson, J. A. and T. P. Meyer. 1988. *Mediated Communication: A Social Action Perspective.* Newbury Park, CA: Sage Publications.

Barthod, C. 1995. Le débat international sur la gestion des forêts. *Forêt méditerranéenne* 16(2):145–52.

Bodson, D. 2005. Comprendre les perceptions, les usages et les significations de la forêt en 2005. *Forêt Wallonne* 79:19–28.

Bonvoisin, D. 2017. Les jeux vidéo sont-ils plus verts que nature? Seminar presented at Point Culture, Louvain La Neuve, November 22, 2017.

Boutefeu, B. 2008. La forêt, théâtre de nos émotions. *RDV Techniques* 19:3–8.

Bowyer, J. 1995. Fact vs. perception. *Forest Products Journal* 45: 17–24.

Courbet, D. and M.-P. Fourquet-Courbet. 2010. *Communication environnementale: Enquête qualitative sur le processus de production des communications pour la prévention des incendies en forêt.* Dijon: SFSIC.

Dereix, C., Farcy, C., and F. Lormant. 2016. *Forêt et communication: Héritages, représentations et défis.* Paris: L'Harmattan.

De Smedt, T., Fastrez, P., Matagne, J., and C. Farcy. 2016. Les recommandations du programme en matière de communication. In *Forêt et communication: Héritages, représentations et défis*, eds. Dereix, C., Farcy, C., and F. Lormant, 381–391. Paris: L'Harmattan.

Gadant, J. 1994. La forêt, les savoirs, le citoyen. *Revue Forestière Française* 46(2):112–115.

Harrison, R. P. 1992. *Forests: The Shadow of Civilization.* Chicago, IL: Chicago University Press.

Jacquinot, G. 2009. De quelques repérages pour la recherche en éducation aux médias. In *EuroMeduc: L'éducation aux médias en Europe: Controverses, défis et perspectives*, 143–51. Brussels: Média Animation

Janse, G. 2006. *European Cooperation and Networking in Forest Communication.* Helsinki: Finnish Forest Association.

Lakoff, G. 2010. Why it matters how we frame the environment. *Environmental Communication: A Journal of Nature and Culture* 4(1):70–81.

Matagne, J., Farcy, C., De Smedt, T., and P. Fastrez. 2014. Assessing the cognitive autonomy of audiences with respect to tree- and forest-related media messages. In *Communicating Forest Science*. IUFRO Communications and Public Relations Working Party and International Union of Forest Research Organizations SPDC, 85–94. Vienna: IUFRO.

Matagne, J. 2017. Littératie médiatique et environnement: Evaluation de l'autonomie cognitive des jeunes envers les médias traitant des forêts. PhD thesis, Ottignies-Louvain-la-Neuve: University of Louvain.

Meunier, J.-P. 1994. Deux modèles de la communication des savoirs. In *Communication des savoirs et publicité sociale (supplément à Recherches en Communication 4, Département de Communication).* Louvain-la-Neuve.

Meunier, J.-P. 1998. Connaitre par l'image. *Recherches en Communication* 10:35–75.

Minsky, M. 1974. *A Framework for Representing Knowledge.* MIT-AI Laboratory Memo 306. Cambridge, MA: MIT-AI Laboratory. https://web.media.mit.edu/~minsky/papers/Frames/frames.html.

Paré, I. 2012. L'influence du documentaire militant sur les conceptions de la forêt: Une analyse des représentations sociales de la forêt entourant le film L'erreur boréale dans la presse écrite. PhD thesis, Quebec City: Université Laval.

Patriarche, G. 2008. Publics et usagers, convergences et articulations. *Réseaux* 147:179–216.

Piette, J. 1996. *Education aux médias et fonction critique.* Paris: L'Harmattan.

Potter, W. J. 2004. *Theory of Media Literacy: A Cognitive Approach.* Thousand Oaks, CA: Sage Publications.

Rametsteiner, E. and F. Kraxner. 2003. *Europeans and Their Forests: What Do Europeans Think about Forests and Sustainable Forest Management?* Vienna: Forest Europe.

Rumelhart, D. and D. A. Norman. 1988. Representation in memory. In *Steven's Handbook of Experimental Psychology: Learning and Cognition*, eds. Atkinson, R. C. and R. J. Herrnstein, 511–587. New York: Wiley.

Shannon, C. E. 1948. A mathematical theory of communication. *Bell System Technical Journal* 27:379–423.

Suda, M. and S. Schaffner. 2004. *Wahrnehmung und Image der Waldbewirtschaftung in der Bundesrepublik Deutschland: Expertise im Auftrag des Holzabsatzfonds.* Munich: Fakultät der Technischen Universität München.

Tomasello, M. 2008. *Origins of Human Communication.* Cambridge, MA: MIT Press.

Winkin, Y. 1981. *La Nouvelle Communication.* Paris: Seuil.

ENDNOTE

1. Following Anderson and Meyer (1988, p. 316), we define a medium as "a recognizable human activity that organizes reality into readable texts for engagement." Media include the traditional mass media (television, cinema, radio, the press, etc.) as well as contemporary networked digital media. The words *readable texts* need to be understood in their most generic sense (i.e., all messages that are in some way interpretive, involving any combination of sound, written text, still or moving images, or any other symbolic means of expression).

Section III

Tertiarization of the Economy

11 Main Findings and Trends of Tertiarization

Maria J. Murcia, Meike Siegner,
and Jennifer DeBoer

CONTENTS

11.1   INTRODUCTION

The academic discussion on tertiarization can be traced back to Allan G. B. Fisher's (1939) seminal article on tertiary production in the late 1930s. Indeed, Fisher is considered one of the founding fathers, together with Colin Clark and Jean Fourastié, of what we now know of as three-sector theory (TST). TST depicts an evolutionary development where the focus of economic activity shifts from the primary sector, through the secondary sector, and finally to the tertiary sector (Fourastié 1949). When a country is dependent on the primary sector, the largest proportion of national income is achieved through the "cultivation or acquisition of food products and in obtaining other raw materials from natural sources" (Fisher 1939), such as forestry, agriculture, or mining, which often coincides with low per capita income and, hence, an early stage of development. When a country's secondary sector expands, the economic development includes "treatment of raw materials, all manufactures, building and construction of all kinds, and gas, water, and electricity supply" (Fisher 1939); a situation that typically corresponds to an intermediate level of national income. Finally, in highly developed countries, services comprise the dominant share of the economy's total output and a higher level of per capita income is achieved. A main tenet of TST is that services' growth would be driven by domestic demand. While services could not by themselves drive higher income levels; instead, they would follow growth (Ghani 2010).

The primary objective of TST research has been to shed light on the pressing question of how to deliver prosperity to all across the challenging interwar and postwar periods of the mid-twentieth century. In Fisher's (1939) own words, "the original purpose in mind in inventing the term 'tertiary production' was to suggest some kind of framework the consideration of which might give a lead in answering the highly important question: 'in what direction is it desirable at this stage of our history to accelerate the rate of economic development?'" (Fisher 1939).

A key milestone of postwar years was the formation of the Organisation for European Economic Co-operation (OEEC) in 1948 to administer American and Canadian aid in the framework of the Marshall Plan for the reconstruction of Europe. Following the 1957 Rome Treaties to launch the European Economic Community, the Convention on the Organisation for Economic Co-operation and Development (OECD) (hereinafter called the "Convention") was drawn up to reform the OEEC. The Convention was signed in December 1960, and the OEEC was officially superseded by the OECD in September 1961. The OECD consisted of the European founding countries of the OEEC, plus the United States and Canada, with Japan joining 3 years later.[1] Within this forum of countries, the previous Great Depression of 1929 had spurred a change in attitudes on the role of the state in the economy, which boosted both the supply and demand of social and producer services. By the 1960s, the growth in the service sector employment was largely due to the expansion of social services, as many countries in Europe moved from a partial or selective provision of social services to a more comprehensive cradle to grave approach, including expanding women and child care, public education, healthcare, social housing, income supplements, employment insurance, and old-age security and pension plans (O'Hara 1999).

In the 1970s and 1980s, the growth in social services declined as the expansion of producer services (i.e., services whose output is primarily purchased by enterprises, such as business, professional, real estate, financial, or insurance services) became more prominent. In addition, employment growth in personal services (i.e., entailing intellectual or manual work in serving a customer, for example, laundry and cleaning services) accelerated in the 1970s (Elfring 1989). The increase of producer services occurred as large corporations shifted away from fully integrated organizations to decentralized structures with multidivisional organizations (Williamson 1964). Each division had greater freedom to choose between internal and external provisions of production processes. Organizational changes opened up new markets, often concentrating on the supply of services to firms (Williamson 1964). In the 1980s, large corporations underwent a long period of downscaling by focusing on a smaller number of businesses along the value chain in order to better cope with regulatory changes and accelerated technological change (Rumelt 1974; Bethel and Liebeskind 1993). These processes fostered firm engagement in global outsourcing arrangements, stimulated by increased competitive pressures (Holcomb and Hitt 2007), which further invigorated the trade of services in international markets.

Finally, the consumption of personal services flourished in the middle of the 1970s. Key factors explaining the sector's growth include increased female labor force participation, which diverted women from household activities, and increased unemployment in the aftermath of the 1973 recession, contributing to the downward pressure on wages and favoring labor-intensive activities.

By 1960, the United States was the world's first society to have the majority of its labor force employed in the service sector. Over the next few decades, nearly all OECD countries followed suit, becoming postindustrial countries (Inglehart and Baker 2000), that is, societies characterized by income growth and, as a result, by increased consumption of services, including healthcare, leisure, education, and cultural activities (Bell 1976).

11.2   CRITIQUES OF THE EARLY THREE-SECTOR THEORY (TST) PROGRAM

Among the notable turning points following World War II was the realization that income and development gaps between most and least developed countries were growing at precipitous rates (Arrighi 2001). These events sparked the establishment of the United Nations (including its subsequent Universal Declaration of Human Rights), the creation of organizations such as the International Monetary Fund and the World Bank (key to designing an open and stable global monetary system and to establishing development and investment programs), and the formation of various world commissions to address issues of global development (e.g., the Brandt Report, Pearson Report, and World Commission on Environment and Development). In addition, in the early 1960s, with the task of rebuilding Europe effectively achieved, a resolution was reached within the OECD to create a body that would not only deal with European and Atlantic economic issues but also devise policies to assist the industrialization of less developed countries and expand global trade (OECD 2017).

These activities were, however, largely initiated by scholars and practitioners based in the global north. A major critique such efforts have faced is whether development issues faced by the global south have been considered in a representative and comprehensive manner within these programs (Borgatta and Borgatta 2017). Such critique was embodied in many theoretical contributions in the 30-year period between 1950 and 1980. Among such key developments, we find the onset of structuralism in Latin America in the 1950s and the dependence theory in the 1970s.

One of the key pillars of structuralism was the so-called Prebisch–Singer hypothesis, which stated that over time, the terms of trade for commodities deteriorate compared to those for manufactured goods, because the income elasticity of demand of manufactured goods is greater than that of primary products. This was a major issue in developing economies, heavily reliant on exports of primary goods. Hence, from this perspective, the policy should focus on the promotion of import substitution industrialization to overcome the structural barriers for economic development in these countries (Prebisch 1962). A follow-up to structuralist thinking, yet also overlapping neo-Marxism and postcolonialism, was the development of dependency theory, which posited that poor nations provide natural resources and cheap labor for developed nations, without which the developed nations could not have the standard of living which they enjoy (e.g., Wallerstein 1974; Wallerstein 2004). From this perspective, the capitalist system unvaryingly creates the conditions for "central" (i.e., developed) countries to pose a major hindrance for the development of "peripherical" (i.e., developing) regions.

In addition, during this same period, China started its own "development miracle," giving rise to a peculiar pattern of simultaneous industrialization and tertiarization efforts, differing from the prescriptions of TST, that is, the idea that a traditional rural sector would be gradually replaced by a growing manufacturing and service economy (Lin 2004). In the 1970s, with the decollectivization of agricultural production and land use rights, Chinese agricultural production and rural incomes experienced a dramatic increase. In the late 1980s, with the reform of

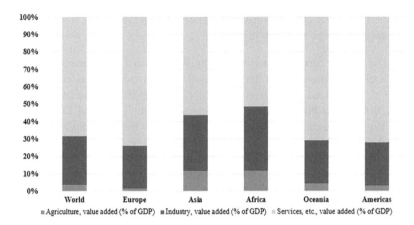

FIGURE 11.1   Value added in agriculture, industry, and services as share of GDP, 2014. (Data from World Development Indicators, World Bank, Washington, DC.)

industrial sectors in urban areas, however, the private sector expanded unprecedentedly. Entry barriers in most sectors were substantially reduced so that many de novo private firms were born, although often associated with government ownership. As the coastal industrial provinces became a growth pole, workers from interior regions migrated there to seek better-paid jobs. A corresponding structural change took place as a more urban population became increasingly employed in services and industry (Fan et al. 2013).

Almost 80 years after the original contributions of Clark, Fisher, and Fourastié, the average share of the service sector in the world output in 2014 was of circa 70% (Figure 11.1). Furthermore, subsequent empirical research on tertiarization has observed a relatively small dispersion between the shares of services in developed and undeveloped countries, this way, challenging one of the pillars of TST, namely, that the pattern of development of services would be a significant function of changes in income per capita (Katouzian 1970). In the end, the tertiarization of the world economy has unfolded as a multifaceted process, reflective of a set of historical, socioeconomic, and institutional factors in different parts of the world, showing substantial differences in productivity, level of labor skill, growth level of value added, and value of gross output (de Souza et al. 2016).

In the forthcoming section, we shall explore in more detail the multiple trends of tertiarization among different groups of countries around the globe.

11.3   TERTIARIZATION AROUND THE GLOBE

Figure 11.2 shows the evolution of the service sector as a percentage of the global gross domestic product (GDP) over the last 20 years, from 1997 to 2014—the last year for which data are available.

While the share of services on the world's GDP presents a smooth upward trend, ranging between roughly 63% at the beginning of the period and 68% by the end of 2014, different world regions exhibit different growth patterns.

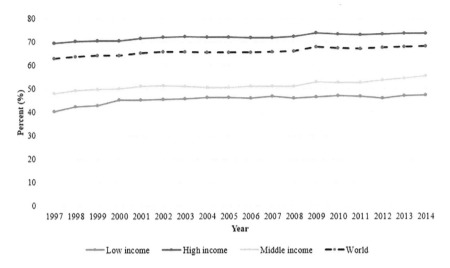

FIGURE 11.2   Services value added (% of GDP), 1997–2014, world regions. (Data from World Development Indicators, World Bank, Washington, DC.)

In developed countries where income per capita is high, GDP composition comprised of the largest share of services, increased consumption of services, such as healthcare, leisure, education, and cultural activities. In these countries, there is a reduced proportion of manual and unskilled labor, and most of the labor force is dedicated to the production of high-value added intangible goods. On the other hand, developing countries have experienced population growth and migration of rural areas to urban areas. As a result, the increased labor supply that could not be absorbed by manufacturing in these countries, complemented by growth in the service sector, for the most part, has been characterized by low-productivity labor in traditional and even informal activities. As illustrated in Figure 11.2, countries with medium to low income show more dynamic rates of services growth vis-à-vis high-income countries. Despite well-documented evidence for developed countries of services positively contributing to economic development, similar statements may not uniformly hold among developing world countries (de Souza et al. 2016).

Figure 11.3 illustrates the evolution of the service sector as a percentage of GDP over the last 20 years for medium–low-income countries. Among this group of countries, the region showing the highest share of services is Latin America and the Caribbean. Propelled by domestic demand and appreciating real exchange rates during this period, economic activity shifted from secondary manufacturing to services, which increased to more than 65% in 2014. In addition, tourism continued to be a major service export for the region (Yeyati and Pienknagura 2017).

Similarly, the services sector has become a major platform for new enterprise start-ups in the former Soviet or transition economies, where emerging entrepreneurs within the countries of central Asia and eastern Europe to deliver previously

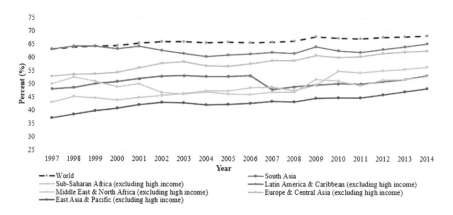

FIGURE 11.3   Services value added (% of GDP), 1997–2014, world vis-a-vis medium–low-income regions. (Data from World Development Indicators, World Bank, Washington, DC.)

neglected personal services to consumers in the 1990s. These activities have been deemed to have offered a safety net in coping with unemployment problems in relation to industrial restructuring (United Nations 2001), thus presenting the second most dynamic growth in the medium–low-income country group.

The service sector has become particularly important in South Asia as well, growing from less than 40% of GDP back in the 1980s to more than 53% of GDP in 2014 (Figure 11.3). According to the accounts on the South Asian service revolution by Ejaz Ghani (2010), this process was not limited to a specific country: the share of services in GDP was more than 50% in India, Pakistan, Bangladesh, and Sri Lanka and 49% in Nepal. This trend has not responded simply to growing domestic demand, but more so to export opportunities of new information technology services in the inception of the Internet era in the early 2000s.

By the mid-2000s, the share of services spiked for the sub-Saharan African countries, taking over south Asian countries, with services comprising around 55% of the GDP, compared to 50% in South Asia (Figure 11.3). Economists have expressed both concern and enthusiasm in this regard.

On the one hand, rising services participation in this context has been seen as a signal of premature deindustrialization. Such an argument, rooted in classical TST, warns against shrinking manufacturing shares in both employment and real value added, turning low- and medium-income economies into service economies without going through a "typical" experience of industrialization (Rodrik 2016). From this vantage point, sub-Saharan African countries would be potentially running out of industrialization opportunities sooner and at much lower levels of income compared to the experience of early industrializers. Similar concerns have been expressed regarding the development of the service sector in the East Asia and Pacific (with the exception of high-income countries in the region, namely, China, Japan, Hong Kong, South Korea, Singapore, and Taiwan) and the Middle East and North Africa regions (World Bank 2017).

On the other hand, the service sector's dynamism has been seen as an auspicious diversification strategy for the commodity-based economies in the region and as a means to alleviate high unemployment rates. Many argue that as services produced and traded across the world expand with globalization, low-income countries will gain enhanced development possibilities based on their comparative advantages. Low-income countries could benefit from a demographic dividend—that is, when the share of the working-age population (15–64) is larger than the nonworking-age share of the population (14 and younger and 65 and older)—which is often associated with a large and increasing supply of labor, increased savings and investments, technological progress, and global growth. Thus, comparative advantage can be gained just as easily through the services sector as it can through the secondary sector (Ghani and O'Connell 2014). Nevertheless, these benefits are not automatic, and latecomers may have less growth prospects.

Henceforth, we turn to providing an overview of the most current state of the art tertiarization research in the midst of global change. We consider how the scholarly conversation has recently shifted toward conceptualizing tertiarization as a possible strategy for dematerialization of the global economy so as to lower its environmental burden. Finally, we close with a discussion and conclusion.

11.4 TERTIARIZATION IN TIMES OF THE "THIRD INDUSTRIAL REVOLUTION"

Thus far, the path to economic growth and development for the global economy has created a series of grand challenges, including a lack of education and development opportunities, growing disparities in income distribution, and the rise of radical political movements and governance failures across a large proportion of the world's population (Bower et al. 2011). In particular, the rise of industrialized production around the world, while having generated an enormous amount of wealth, has simultaneously brought a fundamental change to our climate system, largely due to unprecedented atmospheric concentration of major greenhouse gases, such as carbon dioxide and methane, producing extra warming (Howard-Grenville et al. 2014). Climate change is incomparable to any known turbulence and has been characterized as a substantial disturbance in the broader ecological or social systems of which organizations and economic systems are a part (Winn et al. 2011).

Changes in the way we organize economic life are expected to tackle such major global challenge. The idea of "dematerializing" the economy has been gaining momentum, as several authors proposed the concept of product–services (i.e., providing utility to consumers through the use of services rather than products) as a possible example of a strategy for reducing the global environmental footprint (Mont 2002). According to these scholars, fundamental economic change may occur when new communication technologies converge with new energy regimes, in particular, renewable energy leading to a "Third Industrial Revolution," involving the digitalization of communication, energy, and transportation. Such shifts could not only accelerate efficiency, increase productivity, and dramatically lower the marginal cost of producing and distributing goods and services, but also potentially create the infrastructure for a sustainable postcarbon future (Rifkin 2012). In such a scenario, the dividing line between manufacturing and services would become increasingly blurred (Ghani and O'Connell 2014).

Recent research, however, suggests that while resource productivity metrics would indicate that some developed countries have increased the use of natural resources at a slower rate vis-à-vis economic growth (relative decoupling) or have even managed to use fewer resources over time (absolute decoupling), as income grows, countries tend to reduce their domestic portion of materials extraction through international trade, whereas the overall resource consumption generally increases (Wiedmann et al. 2015). In addition, the landscape of the effects of technology on growth and employment patterns across different economies is still at an early stage but rapidly evolving.

11.5 CONCLUSION

Over the course of the twentieth century, the world has seen rapid industrial development, which has caused entire world regions to transition from primary sector activities and manufacturing, to mass consumption and service driven economies. However, this shift toward the expansion of the tertiary sector, despite presenting a stable upward trend, does not uniformly manifest across different regions of the world. Global statistics show that the services' share of GDP presents different growth patterns in high- and medium- to low-income regions, with the latter group having experienced faster growth in services than the former over the past two decades. In addition, the impact of the global "service revolution" in developed economies has thus far primarily manifested in the form of increased production of high value-added services, whereas the picture in developing economies is mixed. While a few developing countries show growth in high-productivity activities, others have seen growth primarily in low-productivity services.

Tertiarization is thus linked with multiple trends and patterns of global economic development. A main challenge in the twenty-first century will be to determine how to fare with the rapidly evolving digitalization and growth opportunities associated with the transition toward a low-carbon economy. The long-term effects in developed and developing regions have yet to be seen.

REFERENCES

Arrighi, G. 2001. Global capitalism and the persistence of the north–south divide. *Science & Society* 65(4):469–76.

Bell, D. 1976. *The Coming of Post-Industrial Society: A Venture in Social Forecasting.* New York: Basic Books.

Bethel, J. E. and J. Liebeskind. 1993. The effects of ownership structure on corporate restructuring. *Strategic Management Journal* 14(S1):15–31.

Borgatta, E. F. and M. L. Borgatta. 2017. Industrialization in less developed countries. In *Encyclopedia of Sociology.* http://www.encyclopedia.com/social-sciences/encyclopedias-almanacs-transcripts-and-maps/industrialization-less-developed-countries (Acccesed August 17, 2017).

Bower, J. L., Leonard, H., and L. S. Paine. 2011. *Capitalism at Risk. Rethinking the Role of Business.* Boston, MA: Harvard Business Review Press.

de Souza, K. B., de Andrade Bastos, S. Q., and F. S. Perobelli. 2016. Multiple trends of tertiarization: A comparative input–output analysis of the service sector expansion between Brazil and United States. *Economia* 17(2):141–58.

Elfring, T. 1989. New evidence on the expansion of service employment in advanced economies. *Review of Income and Wealth* 35(4):409–440.

Fan, S., Kanbur, R., Wei, S. J., and X. Zhang. 2013. *The Economics of China: Successes and Challenges* (No. w19648). National Bureau of Economic Research.

Fisher, A. G. B. 1939. Production, primary, secondary and tertiary. *Economic Record* 15(1):24–38.

Fourastié, P. 1949. *Le grand espoir du XXème siecle: Progrès technique, progrès économique, progrès social*. Paris: Presses Universitaires de France.

Ghani, E. 2010. Is service-led growth a miracle for south Asia? In *The Service Revolution in South Asia*, ed. Ghani, E., 35–102. New Delhi: Oxford University Press.

Ghani, E. and S. D. O'Connell. 2014. *Can Service Be a Growth Escalator in Low Income Countries?* (July 1, 2014). World Bank Policy Research Paper No. 6971. Available at SSRN: https://ssrn.com/abstract=2470231.

Holcomb, T. R. and M. A. Hitt. 2007. Toward a model of strategic outsourcing. *Journal of Operations Management* 25(2):464–481.

Howard-Grenville, J., Buckle, S. J., Hoskins, B. J., and G. George. 2014. From the editors: Climate change and management. *Academy of Management Journal* 57(3):615–623.

Inglehart, R. and W. E. Baker. 2000. Modernization, cultural change, and the persistence of traditional values. *American Sociological Review* 65(1):19–51.

Katouzian, M. A. 1970. The development of the service sector: A new approach. *Oxford Economic Papers* 22(3):362–382.

Lin, G. C. 2004. Toward a post-socialist city? Economic tertiarization and urban reformation in the Guangzhou metropolis, China. *Eurasian Geography and Economics* 45(1):18–44.

Mont, O. K. 2002. Clarifying the Concept of Product—Service System. *Journal of Cleaner Production* 10:237–245.

O'Hara, P. H. 1999. Welfare state. In *Encyclopedia of Political Economy*, ed. O'Hara, P. H. New York: Routledge.

OECD (Organisation for Economic Co-operation and Development). 2017. *About OECD*. Paris: OECD. http://www.oecd.org/about/ (Accessed June 26, 2017).

Prebisch, R. 1962. The economic development of Latin America and its principal problems. *Economic Bulletin for Latin America* 7:1–22.

Rifkin, J. 2012. The third industrial revolution: How the Internet, green electricity, and 3-D printing are ushering in a sustainable era of distributed capitalism. *World Financial Review* 1:4052–4057.

Rodrik, D. 2016. Premature deindustrialization. *Journal of Economic Growth* 21(1):1–33.

Rumelt, R. 1974. *Strategy, Structure and Economic Performance*. Cambridge, MA: Harvard University Press.

United Nations. 2001. *Services in transition economies: Trade and Investment Guides 6*. Geneva: United Nations Publications.

Wallerstein, I. 1974. The rise and future demise of the world capitalist system: Concepts for comparative analysis. *Comparative Studies in Society and History* 16(4):387–415.

Wallerstein, I. M. 2004. *World-Systems Analysis: An Introduction*. Durham: Duke University Press.

Wiedmann, T. O., Schandl, H., and M. Lenzen. 2015. The material footprint of nations. *Proceedings of the National Academy of Sciences* 112(20):6271–6276.

Williamson, O. E. 1964. *The Economics of Discretionary Behavior: Managerial Objectives in a Theory of the Firm*. New York: Prentice Hall.

Winn, M., Kirchgeorg, M., Griffiths, A., Linnenluecke, M. K., and E. Gunther. 2011. Impacts from climate change on organizations: A conceptual foundation. *Business Strategy and the Environment* 20(3):157–173.

World Bank. 2017. *Global Economic Prospects: Weak Investment in Uncertain Times.* Washington, DC: World Bank http://www.worldbank.org/en/publication/global-economic-prospects (Accessed April 30, 2017).

Yeyati, E. L. and S. Pienknagura. 2017. *Who's Afraid of Tertiarization? VOX CEPR's Policy Portal 10.* United Kingdom: Center for Economic Policy Research, http://voxeu.org/article/services-rise-latin-america (Accessed April 30, 2017).

ENDNOTE

1. http://www.oecd.org/about/.

12 Increasing Role of Services

Trends, Drivers, and Search for New Perspectives

Päivi Pelli, Annukka Näyhä, and Lauri Hetemäki

CONTENTS

12.1 INTRODUCTION

The increasing role of services is recognized as one of the major trends in both the developed and developing economies (de Backer et al. 2015). Often, this trend is described as an increase in the number and volume of services. Today, the service sector contributes 70–80% of employment and value added of the developed economies, and service trade represents 21% of the global trade flows. Manufacturing and service sectors are closely interconnected: the input acquired from external service providers is estimated to contribute 40% of the manufacturing output, and vice versa, services cannot be provided without the necessary infrastructures and other material provisions supplied by the manufacturing industries. Furthermore, 40% of all occupations in manufacturing are in fact service-type occupations, such as research and development (R&D), management, accounting, legal services, marketing, distribution, or after-sales services (Pilat et al. 2006), and 30% of the manufacturing companies themselves also sell services (Neely et al. 2011). Overall services are embedded in all economic sectors. Existing statistical metrics do not fully reveal these developments, and technological advances make the assessment ever more challenging due to new types of services and service concepts emerging in the markets.

During the past decade, the increasing role of services has also gained interest in the forest-based sector research (Hetemäki 2011; Toppinen et al. 2013; Näyhä et al. 2014). The role of the forest-based services—or bioeconomy services—is expected to increase as part of the economic and productive processes of the future circular bioeconomy (Hetemäki et al. 2017). Yet data and information are lagging to assess services and their potential. For example, there is a lack of outlook for the forest-based tourism; health and recreation services; or the exports potential of consultancy, education, and training (Hetemäki and Hänninen 2013; Näyhä et al. 2015). Furthermore, forestry and sustainable forest management provide an important service to society by maintaining and enhancing the ecosystem services [for example, in the EU Forest Strategy, see the study by Pelli et al. (2017)]. These are the services that nature, or forests as such, provide: wood and other raw materials and products, air and soil quality regulation, carbon sequestration, habitat, and biodiversity, as well as contribution to human health, spiritual, and aesthetic experiences (Millennium Ecosystem Assessment 2005).

This chapter provides a broad overview on the increasing role of services: how this phenomenon has been addressed in research on services and what it implies for the forest-based sector. By the use of the term *forest-based sector* instead of *forestry sector* we highlight that in addition to forestry and traditional forest industries of pulp and paper and wood products, other industries and service sectors also utilize forests and forest-based resources for their activities. In addition, new products and services are developed based on forest biomass, for example, for bioenergy and biofuels, textiles or chemicals industries, and nanopulp and microfibrillated cellulose for uses in a variety of sectors. The increasing role of services is embedded in these processes. Overall, the growth of the service sector in general macroeconomic terms has been recognized since the late 1950s, and there are research traditions in several disciplines investigating services, what the increasing role of services implies, and what kind of trends and drivers are detected for future research. Today the multidisciplinary field of services research includes, among others, economics, business sciences, engineering sciences, and socials sciences. The forest-based sector already applies outcomes from these research fields in its own investigations as well as in the company operations. However, the research remains scarce and scattered.

This chapter first summarizes the trends and drivers perceivable in research on services and then gives an overview on the forest-based sector research and its interest in services. The service economy of today is not the same phenomenon than the service economy discussed over the past decades. Taking the observation from the business and engineering sciences, how they have evolved to research services, and the related changes in business and production systems, three forest-based sector developments are described. The examples from biorefineries, building systems, and forest big data illustrate the challenges we face when trying to make an outlook for the future services. Instead of seeking to assess intangible services separately from the tangible production, a synthesis view is taken on service systems. The approach stems from the business and engineering sciences, and today, its concepts are applied, for example, in what is called as 'service science' engineering human and technological systems together. The chosen method does not provide an answer to the question what the future services are or their share in the economy. The chapter seeks to improve understanding about the increasing role of services, how it has been addressed in different fields, and what kind of opportunities and challenges unfold addressing this phenomenon in the forest-based sector context.

12.2 TRENDS, DRIVERS, AND MULTIPLE RESEARCH INTERESTS ON SERVICES

12.2.1 Service Sector, Economic Growth, and International Trade

The service sector contributes 70–80% of national value added and employment in developed countries, and its share is increasing in the developing countries. The traditional division of economic sectors, especially the borderline between manufacturing and services has become blurred: business services contribute an important input to the manufacturing operations, while manufacturing companies themselves also produce services along their value chains and offer services to their customers. Well-developed knowledge-intensive business services (KIBS) are recognized as important for the success of the manufacturing industry in internationally distributed production (de Backer et al. 2015). Business services refer to, for example, financing, legal services, accounting, human resources management, and marketing.

Technological development has speeded up the international trade, and the most competitive combinations of resource use are sought by distributing production internationally. Instead of trade in goods and services produced in one geographic location, interim goods are traded between different locations. For example, in the electronics or automotive industries, interim goods are imported not for producing goods for the domestic markets, but for producing goods for exports. These so-called imports for exports play a significant role for the regions that have specialized in specific production tasks in the supply chain. In such global value chains, the developed countries tend to focus in services tasks, such as headquarter functions, R&D, intellectual property rights (IPRs), design, and marketing, while more routine production and assembly tasks have been located to the low-cost countries. However, also the higher value-added services tasks have become distributed internationally. Information technology (IT) has radically reduced the cost of moving ideas across borders, and the tasks such as management, marketing, and R&D have become offshored (Baldwin 2016). Intangible assets are often the most valuable ones: today, for the value of a manufacturing company, the patent portfolio may be more important than its physical capital assets.

The interdependencies of the services and manufacturing sectors as well as the dynamically changing configurations of the global value networks call for new means to grasp the developments. The work toward this goal is already ongoing, for example, to develop international input–output analyses (Stehrer et al. 2014; Taglioni and Winkler 2016) and trade-in-tasks analyses [summarized by Kenney (2013), Sturgeon et al. (2013), and Ali-Yrkkö and Rouvinen (2015)]. The bioeconomy developments and increasing pressure on renewable resources require more investigations on the interconnections between services and the primary production, too.

12.2.2 SERVICES AND SUSTAINABILITY

In the sustainability studies, for example, product–service systems investigate different forms of sharing, reuse, repair, and other processes that are developed around the tangible products [summarized by Mont (2002) and Tukker (2004, 2015)]. However, the connection between services and sustainability is not unambiguous; although replacing the material goods with increasing use of services is perceived to decrease pressure on natural resources, services use material resources too. In fact, the easier the use of services is made, such as the digital applications in consumers' everyday lives, the more people use them, and the more resources are needed for the necessary infrastructures. Services can affect transformations in the economy, for example, in finding solutions to resource efficiency, environmental concerns, or other societal challenges—but these outcomes are not to be taken for granted (Gallouj et al. 2015).

At the same time, resources are becoming scarce for satisfying the needs of the increasing global population. Access to the raw material sources becomes critical, and resource efficiency needs to be improved. The efforts toward more circular economies and processes of material flows will require new types of services. That is, the tangible production of extraction, processing, production, and assembly can be expected to become ever more amalgamated with the intangible services, such as transportation and logistics; standards and testing; and all data and knowledge flows necessary to design and manage production, distribution, and use [as an example of the reviews across different research fields, see the study by Boehm and Thomas (2013)].

12.2.3 FROM SERVICES TO SYSTEMS IN BUSINESS AND ENGINEERING STUDIES

Company-level investigations on the role of services have been carried out in several fields. Overall, the emergence of services research in the business studies field stems back to the 1970s–1980s, when the growing service companies, such as financing and hospitality businesses, called for the business schools to address the specificities of services as a separate lane of investigation in marketing (Fisk and Grove 2010). Typologies were developed how services were different from tangible goods and, later, how the intangible and intangible value creation could be integrated in an optimal way. One of the popular typologies has been to define services as intangible, deeds, or performance; heterogeneous, different at each time of delivery; inseparable, produced and consumed simultaneously; and perishable, it is thus not possible to store them—these typologies have several challenges [see, for example, the study by Zeithaml et al. (1985)].

Today the business analyses are composed of studies on the service sector companies specifically and on business development in more general terms. Research questions are, for example, how new services are developed, what affects success in services, how value becomes coproduced in interaction between the service provider and the customer, or how markets emerge through resource integration. For the past 10–15 years, service research has gained more and more emphasis in parallel to services research (Bitner and Brown 2008; Ostrom et al. 2010). Highlighting the difference between these two: the research on services is interested on services companies or services business, thus emphasizing the specificities of services, while the service research elaborates a synthesis view and is interested in value creation in any producer–customer or provider–user interaction [for an example of one such approach, see service-dominant logic by Vargo and Lusch (2017)].

In the engineering research, services have been analyzed in industrial marketing and management, supply chain, and operations management, as well as IT and information systems, among others. For example, servitization of manufacturing gained interest from the late 1980s onward. There was a trend recognized in manufacturing companies adding their products with pre- and aftersales services and, later, moving downstream in the value-added chain to sell services mainly (Vandermerwe and Rada 1988; Wise and Baumgartner 1999; Oliva and Kallenberg 2003). Even if services increased the company turnover, they do not necessarily improve profitability of the manufacturing companies (Gebauer et al. 2005). Consequently, the manufacturing companies are not only moving downstream in their value chains, but several strategic choices are made both upstream and downstream [summarized by Baines et al. (2009) and Kowalkowski et al. (2017)]. The engineering and machinery sectors provide several examples of these developments, such as the Rolls Royce changing its business concept from selling engines for aircrafts to selling its customers "power by the hour," i.e., the product and its maintenance invoiced as the time that the aircrafts operate, or the IBM shifting from production of computers to engineering information systems and, further, to the engineering and management of service systems. In other words, today, servitization in manufacturing is understood to also include considerable business model changes regarding how value is created, distributed, and captured. There is a rich research field, about not only servitization or services in manufacturing, but also engineering of supply systems and technological or other systems necessary in production that investigate similar issues than those raised in the aforementioned service research [as an example, see service science by Maglio and Spohrer (2008)].

12.2.4 Services, Technology, and Innovation

The connection between services, technology, and innovation has several dimensions (Miles 1993). The information and communication technology (ICT) development not only created totally new types of services, but it also increased productivity of the old service sectors such as transportation, wholesale and retail, and business services. The outsourcing of services tasks from the manufacturing companies made services visible, but at the same time, the adoption of new technologies allowed the service companies to increase their efficiency and develop new services to their customers. What at first was studied as innovations in and for services, i.e., service companies adopting the new technology, has extended to a wider understanding of the processual nature of service innovations (Coombs and Miles 2000; Gallouj and Savona 2009; Toivonen 2016). Service companies disseminate new technologies as well as the practices that they learn by applying the technology into their customers' contexts. This impact extends across different customer fields such as manufacturing and other services, as well as primary production and public sector—and through the trade of services, from a country to another. For example, the clean tech and green technology solutions for improving the environmental impact or resource efficiency include the infrastructure, its development, and maintenance services. Services are part of the innovation system.

Overall, the interest on services and innovation accumulated in late 1990s. The innovation surveys of the Oslo manual for innovation were also expanded to service sectors in 1997, and marketing and organizational innovation categories were added in the 2005 edition of the manual (OECD and Eurostat 2005). Innovation in services is described as less technological, more incremental in nature, less formally organized, and more difficult to protect with IPRs. Services illustrate more open approaches to innovation: companies offering services interact with their customers, and they gain insights into the customer processes, which they can then utilize to further improve the solution they offer. Alike, the customer improves efficiency if it opens its own processes for developing the necessary supply network, rather than acquiring a specific product or service separately (Boden and Miles 2000; Toivonen 2016).

12.2.5 Technology-Enabled Platforms B2B, B2C, C2B, and C2C

As already highlighted earlier, technology has been an important driver for increasing role of services, particularly, the digital technologies that enable the accumulation of data and more efficient organization of operations. It is worthy to note that these developments do not limit themselves within the manufacturing, processing or the tangible production, but they extend to various forms of market exchanges. Technology-enabled platforms provide new ways to reconnect idle resources: Airbnb became a global accommodation provider without any own hotel infrastructure investments, Uber became a global taxi company without its own car fleet, Facebook became a global media without its own journalists or media production, and so forth. Furthermore, the users can establish their own distribution channels and create added value, which is no more under the control of the traditional producers or service providers. What used to be business-to-business (B2B) and business-to-consumers (B2C) become several forms of transactions, including consumers-to-business (C2B) and consumers-to-consumers (C2C).

Data that the platforms collect on users and their behavior have become a valuable resource on its own. The roles and tasks, as well as the money transactions, become reconfigured; thus, it changes who produces what to whom and what is the basis of the market transactions. What can be provided as a service solution extends to various operations well beyond the digital world; for example, mobility-as-a-service provides a digital platform that connects both public and private transportations as well as different means of mobility into one single service application for the user. What we understand as services and what we expect as service is changing. Technological change is not an add-on to the old operations, but it interlinks with the changing society. Digitalization today is not the same thing as it was in the year 2000. Earlier, it was about e-marketplaces and how marketing of goods and services changes, today it is about how production is organized. At present, the big data, Internet of Things, and robotics are expected to encourage companies to relocate their production and assembly tasks back to the developed countries (Brynjolfsson and McAfee 2014; OECD 2017).

The ongoing industrial revolution is even foreseen to make the division between tangible and intangible production ultimately artificial (OECD 2017; Boden and Miles 2000). Today, it is relatively easy to find examples of the digital platforms as a means of new services. The ICT development led to the emergence of new types of services, and its convergence with other technologies created new operation modes. But in a similar vein, totally new services can be envisaged to emerge from the advancements in bio- and nanotechnologies (Chang et al. 2014; Gallouj et al. 2015).

12.3 FOREST-BASED SECTOR RESEARCH ON SERVICES

12.3.1 Production-Oriented View of Services

The forest-based sector is production oriented, and its market analyses, business development, and innovation models focus on the tangible manufacturing production. The data and metrics have been developed to satisfy the needs of the wood (biomass) production processes. Consequently, services, such as administration, education, research and development, and data and information services, are assessed as necessary inputs in sustaining the production, rather than own business and value creation entities (Hetemäki and Hänninen 2013; Näyhä et al. 2015).

The forest-based sector research has addressed similar topics than were summarized earlier from the services research. However, the investigations have not focused on services specifically, or they have used product-oriented analytic frameworks to also address services (Pelli et al. 2017). Services have been, for example, recognized as one part of the intangible value creation in the analyses of the wood products industries, along with other intangibles such as the product and process qualities, renewable, and traceable raw material, environmental sustainability and corporate social responsibility [e.g., the study by Toivonen et al. (2005)]. New business opportunities have been investigated in nonwood goods and services, thus also including services in the rural development, entrepreneurship, and innovation studies in forestry [summarized by Niskanen et al. (2007) and Weiss et al. (2011)]. Furthermore, similar to the knowledge-intensive business services mentioned earlier, the forest owner services research has addressed questions on outsourcing, innovation and entrepreneurship in forestry, and the changing customer needs (Clark 2005; Anderson 2006; Hull and Nelson 2011; Mattila and Roos 2014).

The research on services that has been carried out in the parallel disciplines may remain less recognized in the forest-based sector research. For example, services business related to forests, such as the nature tourism, has been analyzed in the marketing and business studies and the tourism research. On the other hand, in the engineering literature, the forest-based industry servitization cases are often presented among other manufacturing industries, whether it is the collaboration between the machinery and engineering companies and their customers in the processing industries or between the pulp and paper industries and their customers (Davidsson et al. 2009; Viitamo 2013).

12.3.2   CIRCULAR BIOECONOMY AND CROSS-SECTORAL COLLABORATION

Recent bioeconomy and circular economy strategies have emphasized the need for diversification and cross-sectoral collaboration, thus also highlighting the forest-based sector partnerships extending to energy, textiles, chemicals, food, and other industries (de Besi and McCormick 2015; Hansen 2016; Hetemäki et al. 2017). There is also an increasing focus on the significance of higher value-added products versus bulk products [e.g., the studies by Schipfer et al. (2017) and Toppinen et al. (2017)]. Although the focus in these strategies tends to be in the biomass processes, the development of new biobased materials and their markets requires more close collaboration along the value chains, for example, in wood construction and pulp and paper industry/biorefineries (Näyhä et al. 2014; Toppinen et al. 2017).

Again, the interest of the forest-based sector research and perspectives are not in services specifically, but the substitution strategies and emerging biobased solutions require new thinking about the customer processes and the overall business ecosystem as well as wider societal and environmental challenges, where services also play important roles. Services affect the competitiveness of the extant forest-based businesses, and they are embedded in the renewal processes of the whole forest-based sector, but often, services are not explicated despite being crucial parts in the innovation systems of the evolving bioeconomy (Pelli et al. 2017). It is also worthy to note that different actors and stakeholders understand bioeconomy and forest-based bioeconomy concepts differently, which creates more ambiguities also for discussing new products, processes, and services (Kleinschmit et al. 2014).

An economic paradigm shift calls for connecting the socioeconomic and technology processes with the ecological systems that also directly and indirectly sustain the human systems. The services that forests provide offer diverse and growing economic and employment potential. The forest research has already made first openings toward this aim. Hetemäki et al. (2017) call for science-based decision-making systems and development of tools on how to integrate natural capital in a circular bioeconomy model. Better understanding on the role of natural capital could draw more attention toward services, including ecosystem services, in the bioeconomy strategies, and assist in decoupling economic growth from environmental degradation. On the other hand, Matthies et al. (2016) have elaborated the service-dominant logic conceptualizations from the marketing field and propose the "value in impact" as a conceptual tool for economic and market analysis: the natural ecosystems and the human-based service systems could be brought into the same assessment. Pelli et al. (2017, 2018) seek to address the same question from the perspective of companies and business systems.

12.3.3   Forest-Based Sector and Forest-Based Services

The interest in services in the forest-based sector research is relatively recent. Also, the concepts of what is understood by the forest-based services are still imprecise.

Hetemäki and Hänninen (2013) and Näyhä et al. (2014, 2015) defined three service categories for the forest-based services: (1) forest-related services that are directly related to forests, such as nature tourism and recreation, hunting, mushroom and berry picking, as well as the services that forest produce (ecosystem services, such as soil and water services, and carbon sequestration); (2) forestry-related services, including advisory services, forest management planning, forest inventory, administration, governance, R&D, and education; (3) industry-related services linked to the manufacturing of forest-based products (e.g., R&D, design, production processes, headquarters functions, logistics, and marketing) including not only the forest industry but also the supply and customer industries such as machinery and engineering, energy, and chemicals.

In the forest-based sector strategies, the United Nations Economic Commission for Europe/Food and Agriculture Organization of the United Nations Green Economy action plan (2014) recognized the services that are related to forest products (such as maintenance, planning, servicing, monitoring, programming, patents, R&D, education, and consulting), to forests (tourism, recreation, inventory and monitoring, forestry services, etc.), and to more generally the ecosystem services, such as climate services and protection services. The Forest Europe (2014), in turn, explicated services as a part of the green economy, rural employment, and income: the green jobs in the forest-based sector refer to employment in forests and manufacturing, including R&D, administrative and service activities that contribute to preserving or restoring environmental quality.

Pelli et al. (2017) defined, based on the business research literature, three perspectives for analyzing services in the forest-based sector and its strategic partnerships of the biobased industries, processing industries, manufacturing, energy-efficient buildings, and green vehicles: (1) a production-focused view on services as activities or operations, (2) a product-focused view on services as tradable outputs, and (3) a strategic orientation on service as a business model definition how value is created. From the analyses of the R&D roadmaps, they also derived a fourth category, which was emphasized in the forest-based sector strategies: "services to society" refers to the forest ecosystem services, sustainable management and use of natural resources, and the contribution of the forest-based sector to the rural employment and livelihood.

Compared with the research fields briefly discussed in the Section 12.2, the preceding forest-based services categories mix several concepts that are investigated separately in the services research: services businesses, such as tourism or business services; services produced and offered by the manufacturing companies, such as pre- and aftersale services; services within the primary production such as forest owner services or environmental services for example related to nature conservation; and finally, the ecosystem services that stem from a very different conceptual background than the aforementioned categories. The idea of connecting all these (Figure 12.1) into the evolving new production paradigm or economic model requires new thinking

FIGURE 12.1 What is understood as an increasing role of services differs considerably between disciplines. Evidence bases, tools, and methods are developed in the forest research and, for example, in the engineering or business research fields, but the conceptualizations remain apart (Photo by P. Pelli).

and relating the concepts more closely to that how other disciplines as well as companies, organizations, and various stakeholders already work on services.

This means that without further clarification and further elaboration, it will remain very challenging to deliver the ideas of forest-based services to the decision makers and wider audiences. Services—including the services to society by the forest-based sector—deserve more attention in the evolving economic paradigm. Crucial questions are also what preconditions are needed for developing new or more value-added services, who are the actors providing services and who are the customers or beneficiaries of various servicing processes. The evolving service economy of the twenty-first century is an interesting framework to elaborate these questions from a new perspective, but it requires translating the forest-based services to concepts and models that the actors operating in the present service economy understand.

12.4 FUTURE OPPORTUNITIES AND CHALLENGES

12.4.1 Evolving Forest-Based Sector in a Service Economy Context

As the preceding broad overview illustrates, there is a rich research tradition in several disciplines investigating the increasing role of services and its implications. Macroeconomic analyses as well as services research have been interested in the interdependencies between manufacturing and services (Section 12.2). It has been recognized that the intangible services become more and more difficult to separate from the tangible goods production. In the business and engineering studies, a similar development from goods and services to a synthesis view is perceivable, as well as the trend from analyses of linear value-added chains to analyses of value networks or, further, wider systems. Services such as IT systems, logistics, marketing, R&D, retail/wholesale and various business services are inherent to production processes.

Often, services, such as technical services that assist the customer to use a product or materials, need to be offered for the customer whether they generate separate cash flow or not.

Compared to this, the interlinkages between the primary production and services have been less investigated. Yet similar developments than those recognized in manufacturing can also be found in the primary production: forestry is a knowledge-based activity, where public organizations and private companies utilize technology-enabled services and develop their processes to satisfy the customer needs, for example, the changing needs of forest owners. Wood procurement is not only harvesting and afterharvesting operations, but also various legal, financial, and other services related to the property management are provided for the private forest owners. This development has taken place within the primary production sector, but it connects with the service sector of the economy. In the forest industry, similar to other manufacturing industries, services refer to inputs acquired from external service providers, the increasing capacities built in-house or aimed by joint ventures, as well as services added on the products for the customer.

The notion how the business and engineering sciences have evolved to address the increasing role of services in their investigations also provides new perspective to the evolving forest-based sector: what comes to the fore if we address the changes in production as a more profound system transition, instead of seeking to assess the services separately from the tangible, material processes? Overall, is it possible to assess services and develop the high-value-added services separately from the evolving circular bioeconomy processes? And from the forest-based services conceptualizations (Chapter 12.3.3) point of view, do the services to society, such as sustainable management and use of natural resources, connect with the evolving service economy? The following three forest-based sector contexts illustrate these questions. All examples are from Finland and highlight the recent developments of crossing the traditional sector boundaries.

12.4.1.1   Pulp Industry: From Pulp Mills to Biorefinery Ecosystems

The biggest forest industry investment in Europe in recent years has been the Metsä Fibre bioproduct mill by in Äänekoski, Finland, which started operation in 2017. It has been introduced as a business ecosystem specializing in different bioproducts and services and with several small- and medium-sized companies operating together with Metsä Fibre. Resource efficiency is the key concept of the ecosystem and, in essence, one producer's waste becomes another producer's raw material. The main product of Metsä Fibre is still softwood pulp for paper production, but the company is seeking new uses for this raw material, for example, in the textile industry. The company has participated, together with other large forest industries in Finland, in technology and R&D programs with downstream partners, such as the clothing company Marimekko, in order to support the whole supply chain development from raw materials to the end customers. Even though the pulp companies' position is that of a raw material producer, they seek to enter to more value-added products, to develop a brand name for the new material and to improve their technical services so that the customers could serve their own customers better. The R&D project is still ongoing, but Metsä Fibre has already presented its first pilot garments to raise awareness of the public as well as interest of the potential investors to test and scale up the production for textiles (Palahí and Hetemäki 2017; and further information in http://bioproductmill.com/).

In this example, trying to estimate the value of services separately from the tangible production is very challenging. In fact, many of the services, such as R&D, testing, monitoring, and other expert services, would not exist without the investigations on novel ways to utilize pulp and its production process. Furthermore, the new materials or products of biorefineries are not necessarily directly applicable to the further downstream production but they require also adjustments in the customer industry processes (Bauer et al. 2016). The development of solutions necessitates interaction across several levels of the value networks and deeper understanding of the customer needs throughout the supply chain. Services, such as necessary resource management tools, knowledge flows or data and monitoring systems, develop integrated into the tangible production processes, and they affect both the upstream supply chain and the downstream customers. Processes become reconfigured through resource integration and mutual learning of multiple actors. The higher value-added services are activities embedded in production, and recognizing the opportunities for transferring these skills and knowledge to another use context requires new thinking. Many of the future bioeconomy services do not yet exist because the technologies and applications are currently being developed.

12.4.1.2 Wood Products Industry: From Engineered Wood Products to Building Systems

Examples of the wood products industry leaning toward services are that a company producing wooden frames offers assembly services to its customers or a company producing wooden elements offers the customer with a technical solution including planning, the product, its engineering, and assembly. Large wood product companies that operate in the global markets, such as Stora Enso, have developed building information model tools including country specific requirements, for example, on acoustics and fire safety in their key market areas. These tools assist architects and engineers to design and plan wooden buildings according to the national rules, and thus, support the adoption and use of the engineered wood products, such as cross-laminated timber and laminated veneer lumber. Even if the companies still are suppliers of varied wooden products for the construction industry and prefabrication of wooden elements and modules, they seek to provide solutions that assist both their direct customers and the supply and production networks of these customers (Pelli et al. 2018; further information in http://buildingandliving.storaenso.com /products-and-services/building-systems).

As in the preceding biorefinery example, services do not necessarily generate new cash flow, but they are necessary for the customer adopting wood-based solutions. The wood products offer renewable sustainably sourced materials and novel technical solutions for multistory buildings and large-scale construction, which has traditionally been dominated by the concrete and steel industries (Hurmekoski 2016). The construction industry, in turn, can be defined as a complex products and systems sector, partly manufacturing and partly services. It integrates various supplies, materials, components, and equipment as well as engineering, design, consulting, project management, and financing services. Complex processes, capital intensiveness, and project-based activities lead to an emphasis on cost-efficient processes, risk minimization, and a slow adoption of novel technical solutions (Bröchner 2010). Furthermore, regulations, norms, and standards are local; thus, for example, the planning and engineering services that are necessary for designing the technical solution need to be acquired in the country where the wood products are used. Also, assembly and prefabrication or on-site services are domestically sourced. It is costly to move downstream in the value-added chain, and the construction industry has strong operation modes that dominate the market. Finding higher value added from wood as material would require rethinking the solutions for the whole supply chain as well as for the inhabitants, users, tenants, or businesses operating the buildings. For example, solutions contributing to human health and well-being, better living, or higher employee productivity do not stem from the construction material as such, but the design of the built environment at large. New products and services could be provided, but the established market players are not necessarily the primary source of such thinking to develop the built environment.

12.4.1.3   Forestry: From Forest Data to Big Data

Service platforms, new digital applications, and technology solutions are foreseen to support the implementation of the Finnish bioeconomy strategy by efficient mobilization and use of wood. The forest data, in turn, are developed to ensure more accurate and precision data for the decision-making at different levels. Several lanes of forest big data developments are ongoing and are seeking to combine the data flows from satellite-based data, laser scanning data, and the data collected by harvesters during the forestry operations. Focus is on wood procurement and an efficient biomass supply, but it is foreseen that the digital infrastructure enables also improved biodiversity monitoring and, for example, development of carbon balance-related tasks. The online service "Metsään.fi" for forest owners and service providers (metsaan.fi) was launched by the Finnish Forest Centre in 2011. It provides the forest owners a direct access to data on their own forests. Previously, they could acquire these data only if the forest expert organization had prepared a forest management plan for their property. The goal of the online service is to improve efficient utilization of the data collected by public funding and to support active forestry operations in the private-owned forests. Parallel to this, the large wood procuring companies have developed their own information tools and services. As an example of the evolving market, the online wood procurement platform (kuutio.fi) opened in 2017 in collaboration between the forest owners and wood-procuring industries. The development of these new service platforms is coordinated by the established forest sector organizations (Pelli et al. 2018).

The initiatives of an open forest data strive for platform thinking, thus also opening opportunities for new types of services and new service providers. Pilot ideas are already tested, such as berry data applications or forests for recreation use (http://www.berrymonitor.com), but they are still at an early stage of development. As the examples of digital solutions in other sectors illustrated (Section 12.2.5), the platforms also enable consumer-to-business or citizen-to-administration, citizen-to-citizen, and so forth interactions. We are just in the beginning of these developments:

services as well as the service provisioning and how value becomes created, distributed, and captured are evolving. If somebody has a business idea—or a strong opinion about something—there are new means to turn it into a market action and in a very quick pace test it, multiply it, and sell it further.

12.4.2 NEW PERSPECTIVES: FROM SERVICES TO SERVICE SYSTEMS?

As an overall observation from the preceding forest-based sector, developments can be concluded that the increasing role of services is not only about the services as businesses that we already know how to measure. The phenomenon is also about the qualitative changes within the production processes that do not have concepts and models for describing how the forest-based sector actors benefit their customer industries' processes and how they could improve their service to the customer industries of today as well as the ones in the future.

In marketing language, what companies offer to their customers is called "value proposition." The value proposition answers to questions how the customer benefits from selecting the particular product or service, what is required from the customer to use the product or service, and sometimes, in addition to these, how the proposed solution contributes to the customer processes in a longer term [summarized by Payne et al. (2017)]. Thus, even if there are no explicit services added on a product, the company serves the customer by fulfilling its need and inviting the customer to integrate its resources to create value from the company offering. For example, the forest industry companies offer to their customers the timely delivery of requested quantities and qualities, traceable supply chain to ensure sustainable use of the renewable raw materials, and efficient processes to ensure competitive prices. This value proposition is repeated for products, such as sawn wood or pulp, as well as higher value-added products, such as engineered wood products or pulp for a specific product segment for example in food industry. The new biobased solutions, which are promoted for substitution of materials with the renewable ones, such as biobased fuels for energy, wood for multistory construction, wood-based pulp for textiles or substances for chemicals, and other industries, require a new type of value proposition.

As highlighted in the preceding examples, the new solutions require modifications of the customer industry processes, and the forest industry companies need to think of means to assist their customers to adopt the new products. The means to assist the customers is called service, even if these engineering or other services do not provide cash flow or they would no longer be needed when the customer process is fully accustomed to the new material flow. Value is not created in production, but it requires integration of the customer resources. Service research (Sections 1.2.3 through 1.2.5) is interested in these interactions because they enable continuous process development: innovation does not cease when the customer process is ready to fully utilize the new solution, but the established feedback loop between a producer and its customer generates continuous inputs to further develop both the customer's value network and the producer's supply network.

Today, these interactions are technology enabled [see the algorithm revolution by Zysman et al. (2011)], but the idea is in principle the same in any service exchange. For example, a forest owner is not a target for the offer by the forest service provider, but he/she could be understood as a source of necessary knowledge and skills for the forest management and planning and sustainable use of natural resources. The interaction between the producer and customer, or service provider and user, creates the basis for the future value creation.

The opportunities and challenges by services and service provisioning in a future bioeconomy can no longer be solely assessed with the concepts of past production paradigm. For example, the internationally distributed supply chains where services have played a crucial role during the past decades do not provide a direct analogy to project the development of today's supply chains: enabling technologies are now different, and new applications are already foreseen to create opportunities for new service concepts; resource scarcities have increased the demand for renewable resources as well as for resource efficiency solutions; and what is becoming more and more abundant is data and information, but the rules and practices related to the use of this resource base are developing (OECD 2017). Neither the services businesses can be expected to remain the same: the knowledge-intensive business services as well as the tourism, health, and well-being services will evolve together with the new enabling technologies and changing modes of production and distribution. These developments are taking place parallel to the evolving circular bioeconomy (Figure 12.2)—whether or not the forest-based sector and its organizations are actively developing new services or contributing to the definitions what is understood by service.

In other words, there is an opportunity to rethink what are the services to be developed for future income in a bioeconomy, what is produced as-a-service based on forest resources, or in more general terms, what is provided as service to society by the forest-based sector. Without an overview on the issues, concepts, and models of service research, it is difficult to position the forest-based sector or its services

FIGURE 12.2 Service economy is developing parallel to the evolving bioeconomy. The interdependent developments of technology and society, affected by the increasing resource scarcities, change what we measure as services and what we expect as service. New concepts and models are needed. (Courtesy of Pixabay.com under Creactive Commons CC0.)

into a wider debate about the evolving production modes or what is called as a "service economy." This service economy is under construction. For example, in the research on services, a shift toward systems thinking can be perceived, yet there are several lanes of research on these service systems, such as technological systems or platforms, business-to-business networks or supply systems, markets as institutional systems or the product-service systems, and their impact on sustainability. These conceptualizations provide new perspectives to assess what we understand by increasing role of services, its past developments, as well as its future in the forest-based sector context.

12.5 CONCLUSIONS

This chapter provided a broad overview of the increasing role of services in the forest-based sector. Focus was on trends and drivers detectible in different disciplines investigating services and related developments as well as the forest-based sector research. What was described as a service economy 50 or 20 years ago is not the same ideas of a service economy that are investigated today.

Three forest-based sector developments were described. The examples do not allow projections to be concluded on the macroeconomic developments or, for example, market outlooks of services. Instead, attention was drawn to the observation that the increasing role of services is not only about the services business that we already know how to measure, but also about the qualitative changes within the production processes. The interdependent developments of technology and society, affected by the increasing resource scarcities, change what we measure as services and what we expect as service. These developments are ongoing. The concepts and models developed in several fields of service research could provide means to elaborate renewal of the forest-based sector, including not only the new services businesses, but also more profound transitions toward a circular bioeconomy.

Services—including the services to society promoted in the forest-based sector strategies—deserve more attention both in the forest-based sector research and in the circular bioeconomy strategies. However, the forest-based services require clarification and relating the concepts more closely to how other disciplines as well as companies, organizations, and various stakeholders already work on services. Without further elaboration, it will remain challenging to deliver the ideas of forest-based services to the decision makers and wider audiences.

REFERENCES

Ali-Yrkkö, J. and P. Rouvinen. 2015. Slicing up global value chains: A micro view. *Journal of Industry, Competition and Trade* 15(1):69–85.

Anderson, F. 2006. A comparison of innovation in two Canadian forest service support industries. *Forest Policy and Economics* 8(7):674–682.

Baines, T. S., Lightfoot, H. W., Benedettini, O., and J. M. Kay. 2009. The servitization of manufacturing: A review of literature and reflection on future challenges. *Journal of Manufacturing Technology Management* 20(5):547–567.

Baldwin, R. 2016. *The Great Convergence Information Technology and the New Globalization.* Cambridge, MA: Harvard University Press.

Bauer, F., Coenen, L., Hansen, T., McCormick, K., and V. Y. Palgan. 2016. Technological innovation systems for biorefineries: A review of the literature. *Biofuels, Bioproducts and Biorefining* 11:534–548.

Bitner, M. J. and S. W. Brown. 2008 The service imperative. *Business Horizons* 51(1):39–46.

Boden, M. and I. Miles. 2000. Conclusions: Beyond the services economy. In *Services and the Knowledge-Based Economy*, eds. Miles, I. and M. Boden, 247–264. London and New York: Continuum.

Boehm, M. and O. Thomas. 2013. Looking beyond the rim of one's teacup: A multidisciplinary literature review of "Product-Service Systems in Information Systems, Business Management, and Engineering & Design." *Journal of Cleaner Production* 51:245–260.

Bröchner, J. 2010. Innovation in construction. In *The Handbook of Innovation and Services—A Multidisciplinary Perspective*, eds. Gallouj, F. and F. Djellal, 743–767. Cheltenham: Edward Elgar.

Brynjolfsson, E. and A. McAfee. 2014. *The Second Machine Age: Work, Progress, and Prosperity in a Time of Brilliant Technologies*. New York: W.W. Norton.

Chang, Y.-C., Miles, I., and S.-H. Hung. 2014. Introduction to special issue: Managing technology-service convergence in Service Economy 3.0. *Technovation* 34:499–504.

Clark, P. 2005. The rise of the service provider. *New Zealand Journal of Forestry Science* 9–12.

Coombs, R. and I. Miles. 2000. Innovation, measurement and services: The new problematique. In *Innovation Systems in the Service Economy. Economics of Science, Technology and Innovation*, eds. Metcalfe, J. S. and I. Miles, 18:85–103. Boston MA: *Springer.*

Davidsson, N., Edvardsson, B., Gustafsson, A., and L. Witell. 2009. Degree of service-orientation in the pulp and paper industry. *International Journal of Services Technology and Management* 11:24–41.

de Backer, K., Desnoyers-James, I., and L. Moussiegt. 2015. *Manufacturing or Services—That Is (Not) the Question: The Role of Manufacturing and Services in OECD Economies*. OECD Science, Technology and Industry Policy Papers, No. 19. Paris: OECD Publishing. https://dx.doi.org/10.1787/5js64ks09dmn-en.

de Besi, M. and K. McCormick. 2015. Towards a Bioeconomy in Europe: National, regional and industrial strategies. *Sustainability* 7(8):10461–10478.

Fisk, R. P. and S. J. Grove. 2010. The evolution and future of service—Building and broadening a multidisciplinary field. In *Handbook of Service Science*, eds. Maglio, P. P., Kieliszewski, C. A. and J. C. Spohrer. 643–663. Springer US.

Forest Europe. 2014. Green economy and social aspects of sustainable forest management. Workshop Report, April, 29–30, Santander. http://www.foresteurope.org/sites/default/files/report_WKS_GESocial_final.pdf (Accessed March 27, 2018).

Gallouj, F. and M. Savona. 2009. Innovation in services: A review of the debate and a research agenda. *Journal of Evolutionary Economics* 19:149–172.

Gallouj, F., Weber, K. M., Stare, M., and L. Rubalcaba. 2015. The futures of the service economy in Europe: A foresight analysis. *Technological Forecasting and Social Change* 94:80–96.

Gebauer, H., Fleisch, E., and T. Friedli. 2005. Overcoming the service paradox in manufacturing companies. *European Management Journal* 23(1):14–26.

Hansen, E. 2016. Responding to the bioeconomy: Business model innovation in the forest sector. In *Environmental Impacts of Traditional and Innovative Forest-based Bioproducts*, eds. Kutnar, A. and S. S. Muthu, 227–248. Singapore: Springer.

Hetemäki, L. 2011. Metsäalan palveluvaltaistuminen [Tertiarization of the forest-based sector, in Finnish] In *Murroksen jälkeen—Metsien käytön tulevaisuus Suomessa*, eds. Hetemäki, L., Niinistö, S., Seppälä, R. and J. Uusivuori, 70–76. Hämeenlinna: Metsäkustannus Oy.

Hetemäki, L., Hanewinkel, M., Muys, B., Ollikainen, M., Palahí, M., and A. Trasobares. 2017. *Leading the Way to a European Circular Bioeconomy Strategy: From Science to Policy 5.* Joensuu: European Forest Institute. http://www.efi.int/files/attachments/publications/efi_fstp_5_2017.pdf (Accessed March 27, 2018).

Hetemäki, L. and R. Hänninen. 2013. Suomen metsäalan taloudellinen merkitys nyt ja tulevaisuudessa [The economic impact of the Finnish forest-based sector now and in the future, in Finnish]. *Kansantaloudellinen aikakauskirja* 2/2013:191–208.

Hull, R. B. and K. Nelson. 2011. Wildland-urban interface forest: Entrepreneurs: A look at a new trend. *Journal of Forestry Research* 109(3):136–140.

Hurmekoski, E. 2016. *Long-term outlook for wood construction in Europe: Dissertationes Forestales 211.* Joensuu: University of Eastern Finland, School of Forest Sciences.

Kenney, M. 2013. Where is value in value networks? In *21st Century Manufacturing*, eds. Breznitz, D. and J. Zysman, 13–36. Vienna: United Nations Industrial Development Organization (UNIDO).

Kleinschmit, D., Lindstad Hauger, B., Thorsen Jellesmark, B., Toppinen, A., Roos, A., and S. Baardsen. 2014. Shades of green: A social scientific view on bioeoconomy in the forest sector. *Scandinavian Journal of Forest Research* 29(4):402–410.

Kowalkowski, C., Gebauer, H., and R. Oliva. 2017. Service growth in product firms: Past, present, and future. *Industrial Marketing Management* 60:82–88.

Maglio, P. P. and J. Spohrer. 2008. Fundamentals of service science. *Journal of the Academy of Marketing Science* 36:18–20.

Matthies, B. D., D'Amato, D., Berghäll, S. et al. 2016. An ecosystem service-dominant logic? Integrating the ecosystem service approach and the service-dominant logic. *Journal of Cleaner Production* 124:51–64.

Mattila, O. and A. Roos. 2014. Service logics of providers in the forestry services sector: Evidence from Finland and Sweden. *Forest Policy and Economics* 43:10–17.

Miles, I. 1993. Services in the new industrial economy. *Futures* 25:653–672.

Millennium Ecosystem Assessment. 2005. *Ecosystems and Human Well-being: Synthesis.* Washington, DC: Island Press.

Mont, O. 2002. Clarifying the concept of product-service system. *Journal of Cleaner Production* 10(3):237–245.

Näyhä, A. and H.-L. Pesonen. 2014. Strategic change in the forest industry towards the biorefining business. *Technological Forecasting and Social Change* 81:259–271.

Näyhä, A., Pelli, P., and L. Hetemäki. 2014. Forest-based services outlook. In *Future of the European Forest-Based Sector: Structural Changes Towards Bioeconomy. What Science Can Tell Us 6*, ed. L. Hetemäki, 55–62. Joensuu: European Forest Institute.

Näyhä, A., Pelli, P., and L. Hetemäki. 2015. Services in the forest-based sector—Unexplored futures. *Foresight* 17:378–398.

Neely, A., Benedettini, O., and I. Visnjic. 2011. The servitization of manufacturing: Further evidence. Paper presented at the 18th European Operations Management Association Conference.

Niskanen, A., Slee, B., Ollonqvist, O., Pettenella, D., Bouriaud, L., and E. Rametsteiner. 2007. *Entrepreneurship in the Forest Sector in Europe.* Silva Carelica 52, Joensuu: Faculty of Forestry, University of Joensuu.

OECD (Organisation for Economic Co-operation and Development). 2017. *The Next Production Revolution: Implications for Governments and Business.* Paris: OECD.

OECD and Eurostat. 2005. *Oslo Manual: Guidelines for Collecting and Interpreting Innovation Data*, The Measurement of Scientific and Technological Activities. Paris: OECD Publishing. https://doi.org/10.1787/9789264013100-en.

Oliva, R. and R. Kallenberg. 2003. Managing the transition from products to services. *International Journal of Service Industry Management* 14:160–172.

Ostrom, A. L., Bitner, M. J., Brown, S. W. et al. 2010. Moving forward and making a difference: Research priorities for the science of service. *Journal of Service Research* 13(1):4–36.

Palahí, M. and L. Hetemäki. 2017. Forests and forest-based products. In *Bioeconomy Report 2016.* European Commission JRC Scientific and Policy Report, EUR 28468, eds. T. Ronzon et al. 90–91. Publications Office of the European Union. https://dx.doi .org/10.2760/20166 (online)

Payne, A., Frow, P., and A. Eggert. 2017. The customer value proposition: Evolution, development, and application in marketing. *Journal of the Academy of Marketing Science* 45:467–489.

Pelli, P. 2018. Services and industrial development: Analysis of industrial policy, trends and issues for the forest-based sector. *Journal of Forest Economics* 31:17–26.

Pelli, P., Haapala, A., and J. Pykäläinen. 2017. Services in the forest-based bioeconomy—Analysis of European strategies. *Scandinavian Journal of Forest Research* 7:559–567.

Pelli, P., Kangas, J., and J. Pykäläinen. 2018. Service-based bioeconomy—Multilevel perspective to assess the evolving bioeconomy with a service lens. In *Towards a Sustainable Bioeconomy: Principles, Challenges and Perspectives*, eds. Leal Filho, W., Pociovălişteanu, D., Borges de Brito, P., and I. Borges de Lima, 17–42. World Sustainability Series. Cham: Springer.

Pilat, D., Cimper, A., Olsen, K. B., and C. Webb. 2006. *The Changing Nature of Manufacturing in OECD Economies.* OECD Science Technology and Industry Working Papers, 2006/09. Paris: OECD Publishing. http://dx.doi.org/10.1787/308452426871

Schipfer, F., Kranzl, L., Leclère, D., Sylvain, L., Forsell, N., and H. Valin. 2017. Advanced biomaterials scenarios for the EU28 up to 2050 and their respective biomass demand. *Biomass and Bioenergy* 96:19–27.

Stehrer, R., Baker, P., Foster-McGregor, N. et al. 2014. Study on the relation between industry and services in terms of productivity and value creation. Final report. Study for the European Commission, Directorate-General for Enterprise and Industry. European Consortium for Sustainable Industrial Policy, ECSIP Consortium: The Vienna Institute for International Economic Studies (WIIW), IFO Institute—Leibniz Institute for Economic Research at the University of Munich, and Ecorys Netherlands. Vienna: ECSIP Consortium. https://ec.europa.eu/growth/content/relation-between-industry-and -services-terms-productivity-and-value-creation-0_en (Accessed June 25, 2018).

Sturgeon, T. J., Nielsen, P. B., Linden, G., Gereffi, G., and C. Brown. 2013. Direct measurement of global value chains: Collecting product- and firm-level statistics on value added and business function outsourcing and offshoring. In *Trade in Value Added: Developing New Measures of Cross-Border Trade*, eds. Mattoo, A., Wang, Z., and S.-J. Wei, 289–319. Washington, DC: World Bank.

Taglioni, D. and D. Winkler. 2016. Making global value chains work for development. In *International Bank for Reconstruction and Development.* Washington: World Bank.

Toivonen, M. ed. 2016. Service Innovation—Novel Ways of Creating Value in Actor Systems. Tokyo: Springer.

Toivonen, R., Hansen, E., Järvinen, E., and R.-R. Enroth. 2005. The competitive position of the Nordic wood industry in Germany: Intangible quality dimensions. *Silva Fennica* 39:277–287.

Toppinen, A., Wan, M., and K. Lähtinen. 2013. Strategic orientations in the global forest sector. In *The Global Forest Sector: Changes, Practices, and Prospects*, eds. Hansen, E., Panwar, R., and R. Vlosky, 405–428. Boca Raton, FL: CRC Press.

Toppinen, A., Pätäri, S., Tuppura, A., and A. Jantunen. 2017. The European pulp and paper industry in transition to a bio-economy: A Delphi study. *Futures* 88:1–14.

Tukker, A. 2004. Eight types of product-service system: Eight ways to sustainability? Experiences from SusProNet. *Business Strategy and the Environment* 13(4):246–260.

Tukker, A. 2015. Product services for a resource-efficient and circular economy—A review. *Journal of Cleaner Production* 97:76–91.

United Nations Economic Commission for Europe/Food and Agriculture Organization of the United Nations. 2014. *Rovaniemi Action Plan for Forest Sector in a Green Economy.* UNECE/FAO Forestry and Timber Section. Geneva Timber and Forest Study Paper 35. Geneva: United Nations.

Vandermerwe, S. and J. Rada. 1988. Servitization of business: Adding value by adding services. *European Management Journal* 6:314–324.

Vargo, S. and R. Lusch. 2017. Service-dominant logic 2025. *International Journal of Research in Marketing* 34(1):46–67.

Viitamo, E. 2013. *Servitization as a Productive Strategy of a Firm: Evidence from the Forest-Based Industries.* Research Institute of the Finnish Economy ETLA Reports No 14. Research Institute of the Finnish Economy. https://www.etla.fi/wp-content/uploads/ETLA-Raportit-Reports-14.pdf (Accessed June 25, 2018).

Weiss, G., Pettenella, D., Ollonqvist, P., and B. Slee. 2011. Innovation in Forestry: Territorial and Value Chain Relationships. Wallingford: CAB International.

Wise, R. and P. Baumgartner. 1999. Go downstream: The new profit imperative in manufacturing. *Harvard Business Review* 77(5):133–141.

Zeithaml, V., Parasuraman, A., and L. Berry. 1985. Problems and strategies in services marketing. *Journal of Marketing* 49:33–46.

Zysman, J., Feldman, S., Murray, J., Nielsen, N. C., and K. Kushida. 2011. ICT-enabled services: The implications for manufacturing. In *21st Century Manufacturing*, eds. Breznitz, D. and J. Zysman, 37–52. Vienna: UNIDO.

13 Human Health
A Tertiary Product of Forests

Naomie Herpin-Saunier, Ingrid Jarvis,
and Matilda van den Bosch

CONTENTS

13.1 INTRODUCTION

Forests can be seen as part of "wilderness" and the pristine, great outdoors. These environments are frequently depicted in folklore and art as representing places that are separate from civilization and urban environments. The Food and Agriculture Organisation of the United Nations defines a forest as a "land area of more than 0.5 ha, with a tree canopy cover of more than 10%, which is not primarily under agricultural or other specific non-forest land use" (FAO 2010).

The focus of this chapter is the health benefits of contact with forests. The chapter pays specific attention to urban forests, often referred to as urban green spaces, since much of the world's population live in cities and many of today's public health issues are related to city living and the lack of natural environments. Urban forests are perhaps not the first environment that comes to mind when thinking about a forest; nevertheless, they play an important role for biodiversity, environmental protection, as well as human health. An urban forest has been defined as "the sum of all woody and associated vegetation in and around dense human settlements, ranging from small communities in rural settings to metropolitan areas" (Miller 1997). We follow this definition and include, for example, urban parks, street trees, and other more or less tree-covered urban spaces in cities, in our journey through the topic of forests' impact on human health.

The chapter begins with an overview of the rising global burden of disease and the pathways through which forest environments contribute to improved health—stress relief, physical activity, and social cohesion. The proceeding section outlines various health outcomes that have been correlated with green space accessibility and the use of forest environments as an intervention in therapeutic contexts. Next, the negative effects of forests on human health are reviewed. The chapter then examines how the various ecosystem services provided from forests—provisioning, regulating, and supporting—contribute to human health. The chapter concludes that human health can be maintained and improved through "green" investments, with optimal outcomes dependent on approaches that incorporate multiple systems, including healthcare, urban, and environmental planning. By these kinds of investments and by considering "capitalization" of nature by payment for ecosystem services, it is possible that forests can contribute to a sustainable and improved economic development.

13.2 RISK FACTORS AND THE GLOBAL BURDEN OF DISEASE

In the span of the twentieth century, along with the rapid rates of development and increase in human life expectancy, global disease patterns have radically changed. Up to around a hundred years ago, the greatest dangers to human health were infectious and parasitic diseases. This kept infant and child mortalities high and life expectancies low. However, due to a plethora of medical discoveries, such as antibiotics and vaccines, the average life expectancy in the world has risen from about 31 years of age in 1900 to 71 years today (WHO 2016a).

Nowadays, public health faces new challenges, with the lion's share of morbidity and mortality resulting from chronic, noncommunicable diseases (NCDs), such as cancers, chronic respiratory diseases, obesity, diabetes, mental disorders, and cardiovascular diseases (Vos et al. 2015). By current predictions, heart diseases and depression will be the most prevalent diseases globally by 2020 (WHO 2014).

Many of these chronic diseases depend on unhealthy lifestyles and social and physical living environments. Major risk factors are long-term stress, lack of physical activity, and failing social networks and sense of community (Murray and Lopez 1996).

Another shift in the global burden of disease is an increase in illnesses and deaths related to harmful environmental exposures (Prüss-Ustün et al. 2016). This is mainly due to pollution of air, water, and soil; environmental degradation; and climate change. In addition, the increasing urbanization with a change from rural to urban living environments has an impact on health, with, for example, a higher prevalence of mental disorders. While urban living yields many opportunities, such as access to healthcare, employment opportunities, and cultural diversity, it can also result in challenges related to environmental and social stressors, such as traffic, noise, increased stress, and insecure social networks.

A broader public health perspective allows for a discussion of environmental and behavioral interventions to prevent many of today's most common diseases. Current risk factors and challenges require actions from various sectors, not only healthcare, but also urban planning, environmental management, infrastructure, and social services. This corresponds well to the World Health Organization's (WHO) definition of health as "a state of complete physical, mental and social wellbeing, not only the absence of disease" (WHO 1948). Within this holistic view of health, it becomes clear that access to all kinds of forests can play an important role for health promotion and disease prevention (Figure 13.1).

13.3 FORESTS CONTRIBUTE TO REDUCING RISK FACTORS

In the following, we discuss major risk factors—stress, physical inactivity and social isolation—and provide evidence on the links to forests and urban green spaces.

13.3.1 Stress and Stress Reduction: Risk and Health Factors

Stress is a physiological response to a changing or challenging condition, to which the body must adapt (McEwen 1998). This adaptation is executed by the sympathetic nervous system through a cascade of physiological reactions, which results in a release of hormones and other transmitter substances leading to, for example, increased heart rate and blood pressure, decreased salivation, and increased blood flood to the skeletal muscles. All these reactions are adequate responses to an acute physical stressor. However, if the challenge is not conquered or there is no option for recovery, the stress reactions are maintained, and we may enter a chronic stress state. Chronic stress implies a wear and tear of the body, eventually leading to exhaustion and mental illness (McEwen 2008). Not only mental health is affected. Due to the cascading reactions and the release of various hormones, stress can have an impact on immune system functioning and increase the risk for many other diseases, such as cardiovascular disorders and some cancer forms (McEwen 2008; Peters and McEwen 2015).

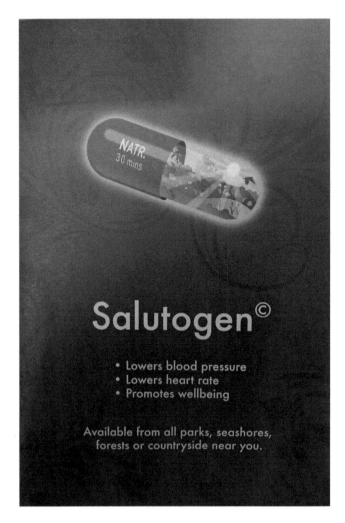

FIGURE 13.1   Nature can act as a resource for health promotion through several pathways. (Courtesy of European Centre of Environment and Human Health, Cornwall.)

As mental disorders and cardiovascular illness are major contributors to the global burden of disease, many lives can be saved by preventing stress and promoting less stressful lifestyles. This is particularly important in a society where social stress is common and increasing. In addition, urban environments appear to make people more vulnerable to stress and less capable of coping (Lederbogen et al. 2011, 2013; Figure 13.2). This has, in part, been attributed to poor connection to forests and other natural environments (Bratman et al. 2015).

Various theories and studies have postulated that forests contain stress reducing features and can consequently act as environments for improved health. The two major theories related to stress reduction were developed in environmental psychology. They are Roger Ulrich's Stress Reduction Theory (SRT) (Ulrich 1983) and

FIGURE 13.2   Individuals often perceive cities as being stressful environments. (Photo *Busy Street Double Exposure* courtesy of Nick Page CC BY 2.0.)

Rachel and Stephen Kaplan's Attention Restoration Theory (ART) (Kaplan 1995; Kaplan and Kaplan 1989). A fundament for both theories is the "biophilia" hypothesis (Wilson 1984; Kellert and Wilson 1995), which argues that humans, because of our evolutionary origins, are innately and instinctively tied to nature. This means that humans carry an inherent love for natural environments, making us feel safe and relaxed.

As a part of his psychoevolutionary theory, Roger Ulrich (1983) analyzed the psychological, as well as physiological effects, of watching natural environments (Ulrich 1984; Ulrich et al. 1991). The SRT claims that humans have subconscious, psychophysical reactions to being in natural spaces with trees and water, causing immediate stress recovery. Several empirical studies have provided support for the theory, by demonstrating, for example, reduced levels of stress hormones in the blood and harmonized heart rate after exposure to nature (Annerstedt et al. 2013; Roe et al. 2013). Recent studies suggest that this stress-reducing impact may contribute to positive health outcomes, such as decreased prevalence of depression and cardiovascular mortality (Reklaitiene et al. 2014; Gascon et al. 2016).

The ART observes that many daily tasks require "direct attention," defined as deliberate concentration on monotonous pursuits, which often results in mental fatigue (Kaplan 1978). The authors postulate that natural environments provide attention recovery and prevent fatigue through four main dimensions in nature:

- Being away: a sense of freedom and escape from daily responsibilities and stress
- Soft fascination: feelings of aesthetic pleasure and astonishment
- Extent: impression of being in another world
- Compatibility: the congruence between a person's aspirations and the sensations offered by the natural scenery

Various types of experimental studies have provided support for these theories (Ulrich 1981; Kaplan and Talbot 1983; Parsons et al. 1998; Kaplan 2001). Research has, for example, demonstrated reduced sympathetic nervous system activity after walking in a forest, but not after a walk in an urban setting (Yamaguchi et al. 2006). Other studies have indicated the same, by measuring various biomarkers of stress, such as cortisol, noradrenalin, adrenalin, progesterone, blood pressure, and heart rate variability (Li et al. 2011; Annerstedt et al. 2013).

Understanding the mechanism through which forests reduce stress and improve cognitive function involves identifying the various sensory stimuli our brain receives while in nature and the physiological and psychological responses they induce. The following sections outline the current state of knowledge in the field.

Researchers have investigated how visual perception of forests affects stress, mental states, and cognitive function. Research on fractal visual input from natural elements analyzes the effects of naturally occurring fractal patterns on human psychophysiology (Taylor et al. 2005). A fractal is a pattern which repeats on a finer and finer scale (Figure 13.3). The patterns forming the leaves of a fern or the branches of a tree are examples of midrange statistical fractals, associated with randomness in scaling properties and a variety of different patterns. Electroencephalography experiments demonstrate that through their combination of predictability and unpredictability, as well as order and disorder, naturally occurring fractal patterns maintain our attention engagement and are easier for our brains to process than artificial shapes (Hägerhäll et al. 2004, 2008). This results in sensations of relaxation and tranquility, associated with positive emotions, stress relief, and attention restoration (Joye 2011).

While much research has focused on visually aesthetic qualities of nature, the sounds of forested environments also have the potential of reducing stress. Auditory stimuli are neurologically closely associated with emotions and mood. This connection carries psychophysiological implications (Kraus and Canlon 2012).

FIGURE 13.3 Cauliflower displaying a typical fractal pattern, with a shape repeating itself at different scales. (Photo *Cauliflower* courtesy of Ian Turk CC BY-NC-ND 2.0.)

FIGURE 13.4   The smell following rainfall on dry soil, geosmin, is recognizable to all human beings. (Photo *Mindful and Resurfacing* courtesy of Justin Kern CC BY-NC-ND 2.0.)

Humans tend to prefer sounds of natural provenance, such as chirping birds, rustling leaves, and croaking frogs, while we dislike sounds associated with technology, such as traffic noise (Benfield et al. 2014).

This relationship has been upheld by research displaying a significant relationship between exposure to sounds of nature and reduction in stress symptoms arising from a stressful task (Alvarsson et al. 2010) as well as reduced stress biomarkers during surgery and anesthesia (Arai et al. 2008; Saadatmand et al. 2013).

Olfaction was a critical sense for our ancestors, who depended upon it for basic survival, allowing them to find food and avoid ingesting harmful substances. Although we rarely utilize olfaction to understand our external world today, we still maintain the ability to detect certain scents, such as earthy "geosmin," which occurs following rainfall on dry soil (Jiang et al. 2007; Figure 13.4). We also have inbuilt neurological systems which are sensitive to olfactory stimuli and help identify harmful toxins and pollutants in our environment. Plants and trees produce toxins with a certain smell to avoid being ingested, as well as pheromones with pleasant odors to attract pollinators. Studies have suggested that some of these smells can generate stress-reducing effects in humans. For example, plants of the cypress family, *Cupressaceae*, can induce physiological and psychological stress recovery responses, such as reduced blood pressure and improved mood (Chen et al. 2015). Other studies have revealed that the perception of nature smells led to improved mood and well-being, while the opposite occurred for synthetic or urban odors (Glass et al. 2014).

13.3.2   PHYSICAL ACTIVITY: AN IMPORTANT HEALTH FACTOR

One of the major lifestyle factors contributing to the increase in the NCD burden is physical inactivity, accounting for 6% of deaths each year, and deemed the fourth leading risk factor for global mortality (WHO 2010). Physical activity improves mental health, general well-being, and neurocognitive development, while preventing, for example, obesity, cardiovascular morbidity, cancer, and osteo-porosis (Cavill et al. 2006; Colditz et al. 2008; Biddle and Asare 2011; Archer and Blair 2012; Brenner et al. 2016). Particularly in cities, physical inactivity results from various factors, such as a deficiency of pedestrian paths, high traffic density, and lack of public open spaces and green spaces (De Nazelle et al. 2011). Green spaces are often preferred for physical activity due to the cleaner air and cooler climate (Figure 13.5).

A growing body of evidence suggests that urban forests may contribute to allevi-ating the health burden of physical inactivity (Sallis et al. 2016). Forested environ-ments have the potential of promoting physical activity by offering a suitable space for exercise. Urban forests provide better air quality and thermal comfort, resulting in an increased willingness to go outside for physical activities (Sugiyama et al. 2013). Further, research suggests that "green exercise" may yield greater benefits than physical activity in other environments (Marselle et al. 2013). For example, a study conducted in Scotland found an association between physical activity in natural environments and reduced risk of poor mental health, while findings of physical activity in other types of environments did not show similar health benefits (Mitchell 2013).

However, not all urban forests are optimal for physical activity. An important prerequisite for physical activity in forested environments is an individual's per-ceived safety (Weimann 2017). For example, urban forests with poorly lit areas and neglected or sparse footpaths are less likely to promote physical activity (Cohen et al. 2016).

Numerous factors determine the relationship between physical activity and urban forests, including proximity, perceived quality, size, biodiversity, and seclusion from noise pollution (Giles-Corti et al. 2005a, 2005b, 2013; Björk et al. 2008; de Jong et al. 2012). Physical activity promoting urban forests can, for example, be defined as

FIGURE 13.5   Forest environments encourage physical activity. (Photo *Vers la lumière* courtesy of Sébastien Launay CC BY 2.0.)

spaces with a high number of recreational amenities, spaciousness, species richness, perceived peacefulness, and wildness (Björk et al. 2008; de Jong et al. 2012). The quality and amenities of an urban forest play an important role in providing experiences that attract people outdoors (Hartig et al. 2014).

The benefits of forests for fostering physical activity likely vary across population subgroups. One review found that ~40% of studies with objective measurements of natural environments demonstrated that park access and degree of vegetation were positively correlated with children's physical activity levels (Ding et al. 2011). For older people, it can be difficult to maintain moderate levels of physical activity and access to urban forests may encourage exercise, even if only at a low level (WHO 2016b). A positive relationship between the greenness of living environment and physical activity levels of older people has been observed, with perceived safety as an important mediating factor (Broekhuizen et al. 2013).

Providing accessible urban forests may encourage individuals to spend more time outdoors and promote physical activity. WHO suggests urban green spaces as a public health indicator and recommends accessibility to green spaces of at least 1–2 ha within a 300 m distance from every residence (Annerstedt van den Bosch et al. 2016; WHO 2016b).

Taking all the evidence together, it is reasonable to suggest that access and exposure to good quality urban and periurban forests may contribute to promoting long-term, healthy behavioral changes for many people in terms of increased and maintained physical activity.

13.3.3   SOCIAL COHESION DETERMINES HEALTH

To have a feeling of being socially connected to other people is of major importance for health. Social isolation is a fatal risk factor, comparable to the hazards of smoking (Pantell et al. 2013). Loneliness increases the risk of premature death and cardiovascular illness and is more common in cities (Holt-Lunstad et al. 2015). Social cohesion is the counterpart to isolation and refers to the population's "sense of community," trust between neighbors and feelings of belonging to a place (Francis et al. 2012).

The presence and quality of urban forest are associated with the surrounding neighborhoods' overall sense of social cohesion (de Vries et al. 2013). Conversely, feelings of loneliness and absence of social support have been linked to a lack of accessible green space (Maas et al. 2009). Urban forests may also promote recovery after traumatic life events by offering safe and nondemanding spaces for social connections (Hordyk et al. 2015).

An important prerequisite for socializing activities is that urban forests are properly maintained and provide amenities that attract people and foster social ties (Figure 13.6). For example, the presence of public events or amenities in urban parks has been shown to draw individuals together and encourage interactions, such as casual conversations and inclusive sports (Peters 2010).

The improved social cohesion has sometimes been attributed as one of the major pathways between urban forests and specific health outcomes (de Vries et al. 2013).

FIGURE 13.6 Amenities of urban parks encourage social interactions among community members, such as these communal chess tables in Timisoara Park, Romania. (Photo *Timisoara Park* courtesy of Cernavoda CC BY-SA 2.0.)

13.4 FORESTS AND THE IMPACT ON DISEASES AND DIRECT HEALTH OUTCOMES

Interactions and synergies between pathways and decreased risk factors associated with forests and urban green spaces likely account for the various health outcomes that have been correlated to green spaces accessibility.

13.4.1 FORESTS DECREASE MORTALITY

Two recent systematic reviews both suggest that higher residential greenness reduces the risk of all-cause and cardiovascular disease mortality (van den Berg et al. 2015; Gascon et al. 2016). This is based on numerous studies with empirical results demonstrating effects. For example, a large study in England showed that the amount of green space has an influence on all-cause mortality, especially in populations of low income and education (Mitchell and Popham 2008). This also contributes to reduced health inequalities due to socioeconomic factors.

13.4.2 FORESTS REDUCE THE PREVALENCE OF HEART DISEASES AND ASTHMA

Research also suggests that urban forests reduce the prevalence of cardiovascular disease (Massa et al. 2016). Particularly, the use of green space seems to be related to a reduced risk of cardiovascular disease (Tamosiunas et al. 2014; Grazuleviciene et al. 2016). Contrary, the loss of trees and green spaces may have a negative influence on health; in a study from the United States, low respiratory tract illness prevalence and mortality due to cardiovascular disease became increased after the loss of around 100 million trees due to the emerald ash borer, an invasive forest pest (Donovan et al.

2013). Similar results have been replicated in later studies, also related to the loss of trees due to the same pest (Donovan et al. 2015).

Several other diseases have been studied in relation to urban green spaces. Recent research has addressed the correlation to asthma and found that children residing in regions with higher surrounding greenness have a lower prevalence of asthma (Lovasi et al. 2008; Dadvand et al. 2014; Sbihi et al. 2015, 2016).

13.4.3   CAN FORESTS REDUCE ALLERGY? THE IMMUNE SYSTEM BOOST FROM NATURE

Somewhat surprising, residential greenness seems to be associated with both increased and decreased prevalences of allergic rhinitis (Fuertes et al. 2016). Studies in support of the protective effects of urban forests on allergies suggest that the exposure to biodiverse organisms in nature supports immunoregulatory mechanisms that protect against autoimmune diseases, such as allergies (Rook 2013). This has been supported by studies demonstrating that children growing up on farms or other environments with a higher exposure to biodiverse microorganisms are at lower risk of developing allergic diseases (Hanski et al. 2012) as well as infectious diseases (Keesing et al. 2010; Keesing and Ostfeld 2015).

13.4.4   FORESTS IMPROVE MENTAL HEALTH AND COGNITIVE DEVELOPMENT

Systematic reviews indicate an association between forests and urban green spaces and improved mental health (Gascon et al. 2015; Mitchell et al. 2015; van den Berg et al. 2015). For instance, higher urban street tree density has been associated with reduced rates of antidepressant prescriptions (Taylor et al. 2015) and reduced reports of physical and mental health complaints (de Vries et al. 2013). We also know from brain-imaging studies that symptoms of depression are reduced after a walk in a green setting, as indicated both through self-reports and by changes in brain structures and function. These changes did not occur after a walk in a built environment (Bratman et al. 2015).

The effectiveness of natural settings at improving mental states is likely to vary depending on variables such as age, gender, socioeconomic status, cultural belonging, and familiarity with nature (Kloek et al. 2013; Baran et al. 2014). Especially children, suffering from cognitive or behavioral disorders such as attention deficit hyperactivity disorder, seem to significantly benefit by being in nature (van den Berg and van den Berg 2011; Amoly et al. 2014; Markevych et al. 2014a). There is also evidence that children's cognitive development and functioning is positively influenced by access and exposure to urban green spaces (Dadvand et al. 2015). Green space usage improves behavioral development, including reduced behavioral difficulties, emotional symptoms, and peer relationship problems (Amoly et al. 2014).

13.4.5   FORESTS IMPROVE THE CONDITIONS FOR MOTHERS AND INFANTS

Studies across the world have shown that the access and proximity to green space is associated with positive pregnancy outcomes, including maternal peripartum depression, and increased birth weight (Dadvand et al. 2012; Donovan et al. 2011, 2012; Dzhambov et al. 2014; Hystad et al. 2014; Markevych et al. 2014b; Banay et al. 2017). Birth weight is in turn an important indicator of an infant's immediate health, as well as the subsequent health and development throughout their life course (Black et al. 2007). Proximity appears to be an important determinant of the positive effect of greenness on pregnancy outcomes. For example, Grazuleviciene et al. (2015) found a negative association between distance to park from residential location and risk of preterm birth.

This emerging body of evidence provides opportunities for healthcare professionals and public health workers to develop preventative measures as well as therapeutic treatments for improved public health through exposure to forested environments.

13.5   FORESTS AS AN INTERVENTION OPTION

Nature-based therapy is the umbrella term for interventions that utilize natural elements either as essential components of the treatment or as a setting for any treatment to take place (Annerstedt and Währborg 2011). Nature-based or nature-assisted therapies aim to treat patients through the engagement with plants, natural materials, and/or natural environments. This can be through horticultural activities or nature conservation or by just dwelling in a forest environment. The type of conditions treated through nature-based therapies varies from depression, anxiety, and posttraumatic stress disorder to rehabilitation after stroke (Annerstedt and Währborg 2011).

Shinrin-yoku, taking in the forest atmosphere or "forest bathing," is a specific treatment founded in Japan, which aims to improve an individual's mental and physical health by dwelling in a forest (Park et al. 2010; Figure 13.7). Studies of *shinrin-yoku* conclude that phytoncides, aromatic volatile substances released from trees, may increase the activity and number of natural killer (NK) cells and anticancer proteins that are critical for our immune system function and protection against malignant tumor development (Li et al. 2006, 2008, 2011; Li 2010). A comparative study indicated that walking trips in unforested areas did not increase the activity or number of NK cells, indicating that the results found for trips to forested areas were not due to the physical activity itself, but rather the forest environment (Li et al. 2008). Results suggest that the effect of phytoncides on human immune system responses may last at least 7 days to 1 month. *Shinrin-yoku* also has the potential of lowering cortisol concentrations, pulse rate, heart rate variability, blood pressure, and other biomarkers of sympathetic nervous activity, all of importance for several chronic diseases (Park et al. 2010; Tsunetsugu et al. 2013; Lee and Lee 2014).

Shinrin-yoku is gaining increasing attention as a potential aid in reducing stress, enhancing mental restoration and improving many other health conditions.[1]

Forest therapy, a type of nature-based therapy, is gaining recognition for its therapeutic health effects among people with, for example, depressive exhaustion

FIGURE 13.7   Visitors taking in the forest atmosphere at Arashiyama bamboo forest in Kyoto, Japan. (Photo *Arashiyama* courtesy of Catherine Shyu CC BY-ND 2.0.)

syndrome and personality and behavioral disorders (Eikenæs et al. 2006). According to Korea's Forestry Culture and Recreation Act, forest therapy is the practice of promoting health and immune function through various activities incorporating the fragrance and scenic view of the forest (Jung et al. 2015). It is important to note that forest therapy is more accurately classified as preventive medicine than as a medical intervention (Ohe et al. 2017). Evolved from *Shinrin-yoku*, forest therapy has been developed by researchers and healthcare professionals to include not only walking or "bathing" in the forest, but also a wealth of structured activities and cognitive–behavioral therapy, such as meditation; acupuncture; massages, aerobics; and even practicing hobbies such as chess, drawing, sewing, listening to music, and photography (Kim et al. 2009). Different types of forest therapies exist, and in some cultures, they are presented as "wilderness" or adventure programs or therapies. To promote a more holistic and longer-lasting experience, accommodation and food are often included in forest therapy treatments, and according the Beijing Forests and Parks Department of International

Cooperation's recommendations, endeavor to use locally sourced materials and ingredients, stimulating local economies.

Forest therapy sites often encompass several different facilities, leading to "forest experience centres," which employ people for construction and maintenance, visitor accommodation and hospitality, child environmental education, as well as forest resource use and protection (Figure 13.8).

Consequently, while evidence points to forest therapy and *shinrin-yoku* potentially alleviating the disease burden of various physical and psychological ailments previously mentioned, the establishment of forest therapy sites also appears to entail environmental and economic benefits (Cho et al. 2014; Zhou et al. 2015). For instance, a South Korean study examined three forest therapy bases and concluded that they generated a revenue of 689 billion wons (approximately US$607 million) and 8,176 new jobs (Cho et al. 2014) annually. See also Box 13.1.

FIGURE 13.8 Mural in Badaling's newly-constructed "Forest Experience Centre," showing a family leaving the cityscape to experience the forest. The centre contains a museum for children, dining area, housing facilities, and forest therapy facilities. (Photo by Naomie Herpin-Saunier.)

**BOX 13.1   PROMOTING HEALTH, ENVIRONMENTALISM,
AND TOURISM THROUGH FOREST THERAPY IN BEIJING**

With its 20 million residents, Beijing is a megacity plagued by environmental health concerns such as air pollution resulting from its rapid development and poor urban planning (Yang et al. 2005), as well as health concerns arising from a stressful city lifestyle (Zhou et al. 2015). While it is most prominent in China's neighbors South Korea and Japan (Zhou et al. 2015), forest therapy has been proposed as a solution to some of the environmental and lifestyle-related health problems in several Chinese cities and is now in its initial stages of development in China.

- Amidst outrage over the city's poor environmental quality, the government and public have taken many steps to improve the city's greenness over the last 60 years, including afforestation, reforestation, and forest resource management projects, and Beijing now possesses a forest cover of 36.7%, with 843,000 ha of publicly managed forest (Zhou et al. 2015), creating opportunities for construction of forest therapy and forest experience zones.
- Furthermore, with the Beijing population's average income per capita reaching 40,321 RMB (approximately US$5,898) in 2013 and the tertiary sector accounting for 75.7% of the city's GDP (Zhou et al. 2015), Beijing is considered one of the most developed and prosperous cities in China, increasing the likelihood for its residents to afford engagement in periods of recreational and therapeutic outdoor activities.

Finally, a number of the city's universities, research institutes and nongovernmental organization have engaged in forest therapy research, which is now accumulating and demonstrating some promising results (Zhou et al. 2015). There are also initiatives for the promotion of forest therapy as a tourism industry such as China's first Forest Therapy and Medical Tourism Forum, held in Beijing in December 2016 (State Forestry Administration of the People's Republic of China 2016).

There are now five forest therapy centers undergoing construction around Beijing: Songshan, Badaling, Shichangyu, Baiwangshan (Figures 13.9 and 13.10), and Xishan (Liu and Jin 2017) which have led to the employment of 40,000 rural workers (Zhou et al. 2015).

The forest therapy sites contain several facilities such as log cabins for accommodation, forest boardwalks, meditation sites, camping sites, and forest bathing sites (Figures 13.9 and 13.10). Forest therapy in Beijing remains in its initial stages, but the potential positive economic and health ramifications are evident. Through cooperation with international partners, Beijing is seeking to become a hub for a new type of tourism industry, which promotes environmental education and appreciation, as well as mental and physical well-being.

FIGURE 13.9   Wooden outdoor dining area at the Badaling Forest Experience Centre. (Photo by Naomie Herpin-Saunier.)

FIGURE 13.10   Wooden boardwalk leading to a viewpoint in the Badaling Forest Experience Centre. (Photo by Naomie Herpin-Saunier.)

13.6   FORESTS AS DANGEROUS PLACES

Sometimes the negative effects of forests are mentioned as ecosystem "disservices" (Von Döhren and Haase 2015). This refers to, for instance, fear of nature, pollen allergies, vector-borne diseases, and falling branches with potential damage on people and properties. While these nuisances are reality, they are still a minor problem in comparison to the negative health effects that are the result of today's lifestyles and changing environment. In addition, many of these obstacles can easily be prevented by accurate approaches, such as protective clothing, rather than considering the removal of forests or urban green spaces. It is also worthwhile to consider that many of the issues have been worsened by human behavior in the first place; allergies are exponentially increasing partly due to our disconnection from nature and impaired immune systems as well as a consequence of climate change and spread of allergenic species and prolonged pollen seasons. The augmenting spread of vector-borne diseases is also a result of global warming and changing environmental conditions (UNEP/UNECE 2016).

Altogether, existing evidence strongly support the notion that the health benefits related to forests and urban green spaces by far outweigh the potentially negative effects (Wolf and Robbins 2015; van den Bosch and Nieuwenhuijsen 2017).

13.7   ECOSYSTEM SERVICES FROM FORESTS

Considering human health as a tertiary product of forests leads us to discuss the various services provided from forests to people. In this context, we use the concept of ecosystem services (MA 2005) referring to the provisioning, regulating, supporting, and cultural services from nature.

The concept of ecosystem services has also been used as a way to "capitalize" forests and other ecosystems. By putting a price on nature, its value for human beings and the development of society may become more obvious or potentially easier to take account for in policy making. The capitalizing of nature has mostly been done in areas of provisioning services, where quantitative measures of, for instance, price for timber can be relatively easily estimated. Later attempts to assess economic benefits by the health services from nature have used methods such as willingness to pay for or travel cost methods. These methods basically determine how much an individual is willing to pay for preserving a forest for recreational use or how far the person is ready to travel to visit a recreational site. The time and travel costs are then used as a proxy for the "price" of the environment.

Another study looked at health perception and possible savings from reduced healthcare (Kardan et al. 2015). The researchers found that increasing the street tree density with 10 trees in a city block would correspond to an annual income increase of US$10,000 per household. As the average cost for establishing and maintaining street trees is estimated to be around US$300–5,000 per year, this would provide a cost-efficient method for improved health. The results were further supported by additional analyses of cardiometabolic health, which demonstrated that 11 more trees in a city block decreases these conditions in ways comparable to an increase in annual personal income of US$20,000.

13.7.1   PROVISIONING ECOSYSTEM SERVICES

Let us begin by providing a brief overview of the provisioning ecosystem services from forests with a direct impact on human health and survival, simply through the products of trees and other organisms from a forest environment.

Forests represent renewable sources of valuable nutritive, pharmaceutical, and various other materials for sustaining health, particularly in low- and middle-income countries (FAO 2010). Throughout human evolution and development, these products have been essential to livelihoods, well-being, and traditional medicine. Today, many forest products are increasingly exploited for their value in the pharmaceutical and nutraceutical industries, and pharmaceuticals, such as fungicides and antibiotics of natural origin, are sold for many billion dollars each year. This provides a fundamental basis for protecting areas of naturally occurring forests to preserve various health-promoting products and values. Especially in low- and middle-income countries, accessibility to healthy forest ecosystems with nutrients and medicinal resources helps combat malnutrition and increase resilience to infectious diseases such as human immunodeficiency virus (HIV)/acquired immune deficiency syndrome (AIDS) (Box 13.2) and malaria.

13.7.1.1   Food in the Forest

Forests provide food in various forms and properties. As one of the most diverse ecosystems on the planet, forests host a variety of types of animals, plants, and fungi, which hold nutritional value for humans (Colfer 2006; Johns and Eyzaguirre 2006). Forest foods also have a role to play in social activities and well-being in certain regions, such as in Scandinavia, where foraging for berries and mushrooms is a social activity essential to the rural way of living (Nilsson et al. 2010).

Forest foods provide essential nutrients such as carbohydrates, fat, vitamins, and minerals.

The following list provides a few examples of forest foods, related nutrients, and respective health effects (Nilsson et al. 2010):

- Nuts, such as walnuts, almonds, and macadamia, found in temperate forests, are rich in omega-3 and monounsaturated fatty acids, which reduce risks of cardiovascular and other NCDs.
- *Andansonia digitata* (baobab) leaves, widely found in forests of sub-Saharan Africa, are not only rich in protein and carbohydrates, but are also a source of xanthophylls, which are conducive to optimal vision.
- Wild mushrooms, such as *Cantharellus ciparius*, are high in protein, low in fat, and have antimicrobial and antifungal properties.

13.7.1.2   Medicine in the Forest

Forests encompass a rich reserve of bioactive compounds, such as polyphenols and carotenoids, with medicinal properties (Karjalainen et al. 2010). Trees, plants, and other life forms that naturally occur in forests have been exploited for medicinal properties throughout history and are still utilized in the production of many pharmaceuticals (Karjalainen et al. 2010). It is estimated that 80% of the world's population depends on plant-derived material for their healthcare needs (Nilsson et al. 2010).

One pertinent example of a medical product, originally derived from forests, is quinine, which is extracted from the bark of cinchona trees (Cechinel-Filho 2002). Quinine is used for treating malaria and is considered among the most effective and safe medicines needed in a health system.

Other examples of pharmaceutical components derived from forests are the following:

- Taxol, from the bark of *Taxus* spp., an anticancer agent found in drugs for breast and ovarian cancers (Karjalainen et al. 2010; Figure 13.11)
- Hydroxymatairesinol, found in the lignin of Norway spruce (*Picea abies*), which has strong antioxidant, anti-inflammatory, and antitumor properties (Bylund et al. 2005)
- Pilocarpine, found in *Pilocarpus jaborandi*, which is used for treating glaucoma (Fifer 2008).

BOX 13.2   FOREST RESOURCES AS COPING MECHANISMS FOR RURAL MALAWIAN HIV/AIDS-AFFECTED HOUSEHOLDS (FROM STUDIES BY THE AFRICA FORESTS INITIATIVE ON CONSERVATION AND DEVELOPMENT (AFRICAD) BASED AT THE UNIVERSITY OF BRITISH COLUMBIA

HIV/AIDS is the sixth most common cause of death worldwide, resulting in 2 million deaths in 2004 (WHO 2004). The disease affects underdeveloped countries most severely and, within them, the most economically valuable members of society; according to the WHO's 2004 Global Burden of Disease update, HIV/AIDS is the main cause of adult mortality in Africa, causing 35% of adult deaths in 2004. As a result, the epidemic is responsible for the negation of several decades of human and economic development (Timko 2011). While enduring the physical and emotional hardships the disease entails, households affected by HIV/AIDS—those led by a widower whose spouse has passed away from HIV/AIDS or where a prime-age adult is infected by the disease—suffer radical changes in their livelihoods. This includes a decrease in their human capital, modification of their existing labor structures, and degeneration of their family–community networks. Consequently, households find themselves unable to earn income and maintain assets (McPherson 2005; Slater and Wiggins 2005). In an effort to reduce the socio-economic burden of the disease, households rely on freely available products from trees and nontimber forest resources. HIV/AIDS infection rates have been found to correlate with deforestation in countries such as Zambia, Tanzania, and Malawi (Barany et al. 2005; Frank and Unruh 2008), supporting observed trends of unsustainable resource use by affected households and raising concerns about their access to essential coping strategies. When these families lose access to crucial forest resources due to unsustainable harvesting, it undermines their poverty-alleviation coping strategies, increasing their vulnerability to shocks and exposure to risks (Hunter et al. 2008).

Malawi is an ideal instance for observing the relationships between rural HIV/AIDS-affected households and forest resources due to its relatively high HIV/AIDS infection rate of 12% (National AIDS Commission 2008) and its prevalence of rural land, with 61% of the country devoted to agriculture and 38% dominated by forests. Furthermore, the government's "Forestry and HIV and AIDS Strategy" (Malawi Government 2007) demonstrates the country's awareness and openness to the importance of forests for the HIV/AIDS community.

In sub-Saharan Africa, medicine is a primarily forest-based practice (Barany et al. 2005). It is estimated that between 70% and 95% of primary healthcare is provided by traditional healers who source themselves from forest products (Colfer et al. 2006). Furthermore, forest resources are responsible for the subsistence and supplementary income of over two-thirds of Africa's 600 million people (Timko 2013; CIFOR 2005). A balanced diet is crucial to the improvement and maintenance of the human immune system and resisting opportunistic disease, especially for HIV/AIDS sufferers; they have been found to have a 15% higher calorie requirement than nonsufferers, as well as a 50% higher demand for protein (Kaschula 2008). Some key nutrients satisfied by forest foods include protein, fat, iron, zinc, and vitamins A and C (Barany et al. 2004). Furthermore, forest foods and other resources have been associated with health improvement and detoxification of drug treatments for AIDS patients (ABCG 2002).

A study conducted by Timko (2013, 2014) sought to identify the most important forest-resource-related coping strategies for families affected by HIV/AIDS in rural Malawi.

FIREWOOD

Predominantly collected by women, firewood is used for the cooking of calorie-dense foods such as cassava, and provides a supplementary income to purchase medication as well as a heat source for affected family members.

The vast majority of affected households reported difficulties accessing firewood resources, which they withstood by burning other materials such as rubbish, travelling further to attain collection sites or purchasing firewood, which diminishes their income. Other reactions include eating less or eating uncooked food and, in the most desperate cases, taking children out of school to assist with collection. To ease the burden of the diminished availability of firewood, the researchers recommended several forest-related actions:

- Directly supplying firewood to households or providing communities with seedlings of fast-growing firewood trees such as eucalyptus to be managed in home gardens or communal lands
- Reforesting surrounding degraded areas with firewood species
- Decreasing the pressure on scarce forest resources by offering households fuel-efficient stoves, such as locally produced clay-based *Chitetzo mbaula* stoves or alternative fuel types.

MEDICINAL PLANTS

Especially for morbidity-affected households, medicinal plants are used for treating the opportunistic diseases that frequently befall sufferers, such as diarrhea and shingles (Barany et al. 2005). Most of the affected participants reported decreasing availability of medicinal plants, possibly due to overharvesting. In response to this, family members walk further to collect plants, dry them into powders or seek treatment from traditional healers for a price. Based on the participants' demands and ideas, the recommendations made by the study to counter this problem include

- Ensuring traditional healers are constantly supplied with sufficient medicinal plants
- Equipping communities with a medicinal plant community herbarium
- Providing communities with training on sustainable harvesting methods.

THATCH GRASS

Grass is collected seasonally and primarily for roofing. Many affected respondents reported a lack of available thatch grass and therefore used other materials such as plastic, leading to rat problems, or banana leaves and reeds, which were less effective as water-proof roofing materials. Some also walked longer distances or purchased grass, and HIV-affected households especially reported a shift in the labor burden of roofing houses toward women, which reduces the time women can spend on other important tasks including cooking, child-rearing, and firewood collection. Innovations suggested by participants included the following:

- Acquiring fiber-cemented roofs
- Planting vertivier and other local grass species

OTHER RESOURCES

While firewood, medicinal plants, and thatch grass were the most mentioned forest products, other important forest-based amenities were noted in the study, such as fruits, bushmeat, and honey, which were recounted to be scarce or difficult to acquire. In order to address the affected households' need for these products, the following strategies were recommended:

- Fruits: Invest on agroforestry projects (Jayne et al. 2005) and provision of domesticated fruit tree seedlings such as *Hacontia indica* for planting in gardens or public lands.
- Bushmeat: Reforest degraded areas and revitalize wild local populations and encourage and train in animal husbandry and fish farming to fulfill protein requirements of HIV/AIDS sufferers.
- Honey: Provide beehives and other necessary materials to households for subsistence or income generation.

Timko's research agrees with previous assertions that forest resource use is tightly related to the stage of the disease. Necessity appears to be relatively low in the pre-HIV phase, generally increasing during the morbidity phase, and diminishing postmortality. Exceptions to this include thatch grass and firewood, whose demand increases after mortality. The recommendations made by the study have the potential to instigate positive feedback loops, which allow households to escape the poverty cycle and achieve higher living and health standards. For instance, acquiring more fuel-efficient stoves leads to a decreased labor burden on women collecting firewood, subsequently allowing them to devote more time to income generation and withstand the economic impacts of the disease. They can then invest in medicine, which allows the affected member to have a higher quality of life. The intimate relationship between forest products and communities affected by AIDS/HIV demonstrated in this study manifests the importance of effective management of forest resources in addressing the public health crises of the developing world.

13.7.2 Regulating Services from Forests

Regulating ecosystem services refer to environmental functions that regulate various conditions, such as temperature, humidity and soil fertility. By regulating these factors, they contribute to moderating potential environmental threats and challenges and can thus prevent diseases related to harmful environmental exposures, such as pollution or excessive heat.

13.7.2.1 Air Pollution Reduction

Air pollution is one of the greatest environmental threats to human health (Brauer et al. 2016; Figure 13.12) with harmful impacts both locally and globally (Zhang et al.

FIGURE 13.11 Collection of yew size residues from *Taxus* in container yards in Belgium. (Courtesy of Christine Farcy.)

2017). The WHO (2014) estimated the total air pollution-related deaths at approximately seven million in 2012. Air pollution is, for example, associated with increased mortality, lung cancer, respiratory infections, and cardiovascular morbidity and mortality (Burnett et al. 2014). The major anthropogenic sources of air pollution are fossil-fuel powered industrial facilities and motor vehicle exhaust. Like most health risk factors, air pollution impacts low- and middle-income countries, as well as disadvantaged populations disproportionally (Brauer et al. 2016).

Air pollutants with the heaviest health impact include solid sources, such as small particulate matter (PM) that is formed as a result of chemical reactions between different pollutants. These particles can be inhaled and affect the heart and lungs and cause serious health effects. Depending on particle size, they are often denoted as either PM_{10}

FIGURE 13.12   Air pollution in Delhi, India. (Photo courtesy of Jean-Etienne Minh-Duy Poirrier CC BY-SA 2.0.)

(diameter 10 μm and smaller) or $PM_{2.5}$ (diameter 2.5 μm and smaller). Other common pollutants with serious health impact are gaseous sources such as carbon monoxide (CO), ground-level ozone (O_3), sulfur dioxide (SO_2), and nitrogen dioxide (NO_2).

Current strategies to reduce the health impact of air pollution mostly include reducing emissions, increasing atmospheric dispersion, and relocating high pollution emitters away from populated areas. While these are probably the most efficient strategies, forest vegetation and soil may potentially contribute to removing some atmospheric contaminants. Gaseous polluting agents are absorbed by leaf stomata and throughout the plant surface and then diffuse into intercellular spaces where they may be stored in water films being involved in metabolic and cellular processes (Nowak et al. 2014). Trees also intercept particles on their surfaces, and while they may be resuspended by windy or dry conditions, they are often absorbed into the tree or transferred to the ground due to rain or leaf and twig fall (Nowak et al. 2014). Several factors affect the efficacy of air pollution abatement by forests and urban green space and must be considered for effective management. These factors often vary both temporally and spatially. In general, the removal increases with tree cover, height, and foliage density (Paoletti et al. 2011).

Through an investigation conducted in Santiago, Francisco Escobedo et al. (2008) observed that urban forests were as cost-effective in reducing PM_{10} pollution as other abatement policies and technologies. Other estimations by David Nowak et al. (2014) reported that the removal of 17.4 t of air pollution in the conterminous United States in 2010 by forests and trees corresponded to US$6.8 billion in avoided health expenses. This is a direct example of how forests can contribute to tertiarization of the economy, by reducing expenses for healthcare related to diseases caused by air pollution. Although the exact quantification remains challenging due to the complexity and the multitude of variables involved, research documenting health effects from air pollution reduction by forests and green space is increasingly emerging (Zupancic et al. 2015a, 2015b).

Studies have shown that the implementation and management of urban forestry and green infrastructure could be a viable method to be used in combination with existing strategies to mitigate the health effects of air pollution (TNC 2016). For instance, the replacement of certain roads with pedestrian green infrastructure provides the population the compounded health benefit from the reduction of transportation pollution emissions, exercise, recreation, and mental well-being resulting from walking in a green area, as well as the pollution-abating properties of trees (Zupanic et al. 2015b).

13.7.2.2   Heat Reduction

Heat reduction is an important health-related, regulating ecosystem service provided by forests and urban trees. Disease and mortality associated with extreme temperatures represent a significant public health burden globally, which is bound to increase with continued global warming (Matthews et al. 2017). Exposure to high temperatures ($\geq 35^\circ$C) can have serious acute physiological effects. This leads to heat stress, which encompasses dehydration, heat exhaustion, heat collapse, and finally heatstroke. Excess heat can also cause the exacerbation of preexisting disorders, such as cardiovascular, renal, and respiratory diseases (Basagaña et al. 2011). This serious health issue has been brought to the forefront of environmental health research by the health effects and death tolls of extreme heat events, such as the European heat wave of 2003, which is estimated to have caused an extra 40,000 deaths (Garcia-Herrera et al. 2010). Three global trends are likely to increase the severity of this issue: climate change, population ageing, and urbanization (IPCC 2014). Urgent mitigation and adaptation strategies to this phenomenon are thus ever more important.

Urbanization is a particular issue in this context. Cities are additionally vulnerable to the health impacts of extreme heat, because of high population, infrastructure density, and characteristics of the built environment that contribute to the "urban heat island" (UHI) effect (Oke 1973). This is a phenomenon wherein urban areas experience a microclimatological "island" of temperatures up to 7–12°C higher than the surrounding countryside. This microclimate provides little relief from daytime heat, as materials, such as concrete and asphalt, trap heat overnight. One of the fundamental causes of the UHI is the supplanting of natural vegetation with heat-absorbing, impervious surfaces that inhibit evapotranspiration.

Trees, urban forests, and shrubs affect temperature in various ways. They generate shade on the ground below and cause cooling through evapotranspiration from leaf surfaces. The greening of cities has proven to be among the most cost-efficient ways of cooling urban areas (TNC 2016). The heat-reducing impact of trees is confirmed by a review and meta-analysis of literature conducted by Diana Bowler et al. (2010). The review concluded that urban parks are an average of 1°C cooler than their surrounding areas, with temperature impacts reaching up to 1 km from their boundaries. Another review by Shishegar (2014) provided evidence for the impact of parks, street trees, and green roofs on the UHI, suggesting increased urban greening as a viable mitigation strategy now and in the future.

The extent of this "air-conditioning" ecosystem service depends on several characteristics of the green area, such as size, vegetation type, and total vegetation cover. A study from Taipei by Chi Ru Chang et al. (2007) showed that parks greater than

3 ha were consistently cooler than surrounding areas, whereas smaller parks were less reliable in cooling outcomes. Trees seem to be in general more efficient in reducing heat, in comparison to herbaceous vegetation (Edmondson et al. 2016).

Urban forests and trees can also indirectly contribute to reduced energy consumption, thereby potentially preventing the underlying cause of excessive urban heat by reducing global warming. The presence of greenery moderates the consumption of air-conditioning, which reduces not only household costs but also greenhouse emissions. This has long-term implications for reduced effects of climate change, promoting both health and environmental sustainability (van den Bosch and Nieuwenhuijsen 2017).

Given the magnitude of the health problems related to excessive heat, trees, forests, and urban green spaces may prove to be an efficient prevention method saving many lives. This will be particularly noticeable in areas with hot climates and many people involved in outdoor physical labor (Dash and Kjellstrom 2011; Crowe et al. 2015). As often is the case, it is the most vulnerable—children, elderly, disadvantaged, chronically ill—who are most exposed to the risk. This means that by investing in green infrastructure and proper forest management, health promotion can be provided for those at highest needs.

13.7.2.3 Role of Forests in Prevention of and Responses to Natural Hazards and Disasters

Natural hazards and disasters are a significant threat for global public health. It is estimated that about 217 million people were affected by natural hazards every year since 1990, with health effects ranging from immediate death and injury, to outbreaks of infectious diseases, to long-term mental and emotional trauma (Leaning and Guha-Sapir 2013). While low- and middle-income countries are disproportionately affected, the Great East Japan earthquake and tsunami of March 2011 sent a clear message that also high-income countries and populations are vulnerable to severe natural disasters. Unsustainable development practices, deforestation, ecosystem degradation, poverty, as well as climate variability and extremes have led to an increase in both natural and human-made disaster risk at a rate that poses a significant threat to lives and development efforts (UNISDR 2012). The scope and severity of hazard events and disasters has been projected as increasing over time (Leaning and Guha-Sapir 2013).

In order to reduce the negative human health consequences of natural disasters, disaster risk management approaches should be incorporated in land use and urban and spatial planning, particularly in vulnerable areas. This should include aspects of community resilience and how to rebuild societies in the aftermath of a disaster. In the book *Greening in the Red Zone: Disaster, Resilience and Community Greening*, Keith Tidball and Marianne Krasny (2014) explain how urban forests and community gardens are conducive to community resilience in the face of environmental catastrophes and argue that they should be integral to disaster preparedness plans.

Studies suggest that forests play a role in reducing vulnerability of certain hazards, as well as reinforcing social–ecological resilience after disasters have occurred (Tidball 2012). The United Nations Office for Disaster Risk Reduction's urban resilience campaign cites several examples of forests being effectively used as "protective ecosystem services," simultaneously integrating risk reduction, community resilience, and sustainable development (Johnson and Blackburn 2014).

There are three levels at which forests may contribute to assuaging the negative effects on public health caused by natural hazards and disasters:

- Forests reduce the vulnerability of communities to, for example, flooding, tsunamis, and landslides.
- Forests promote resilience that allows communities to cope with the ongoing social and ecological effects of hazards.
- Forests act as a sustainable solution for development and urban planning in disaster-prone regions.

Overarching these levels is the notion that forests should be maintained and managed in a manner so that the disaster risk is diminished in first place.

In the following, we briefly outline the mitigation effects of forests on tsunamis and floods.

13.7.2.3.1 Tsunamis

The most convincing evidence for protecting people from natural hazards and potential disasters lies in the case of coastal forests as a defense against tsunamis. Harada and Imamura (2005) identified and tested four different functions that coastal forests serve in reducing the occurrence of tsunami disasters. These functions include the following:

- Trap drifts and ships carried by waves on their way to populated areas
- Reduce tsunami velocity by creating drag as water flows past their surfaces
- Form natural sand dunes which act as a structural barrier
- Prevent people from being carried into the ocean with receding waves.

Research in Chile (Rodríguez et al. 2016) found that stands of *Pinus radiata* and *Cupressus macrocarpa* shelterbelts along the coast of the Biobío region were effective in protecting the infrastructure and populations of coastal villages. This was demonstrated by their capacity to reduce the velocity and height of a tsunami in February 2010. The study also revealed important management implications for protective coastal forests. Due to the proper management at place, trees incurred minimal stem breakage and uprooting, reducing their own potential for damage after the impact, thanks to the diversity of the horizontal structure and diameters of the forests' trees. Forests also have the potential to reduce the duration and severity of flooding following tsunamis.

13.7.2.3.2 Flooding

Floods are some of the most common natural hazards. They can arise from abnormally heavy precipitation, dam failures, rapid snow melts, river blockages, or even burst of water mains. While floods are necessary to maintain river ecosystems, extreme floods can cause death by drowning and injury, increase the risk of infectious diseases by disrupting basic services such as electricity and sanitation, and destroy infrastructure and livelihoods (Noji 2000). Urbanization contributes to urban flooding as roads and buildings prevent infiltration of water, so the run-off forms artificial streams. In addition, deforestation and removal of root systems increase runoff.

Trees and forests reduce the risk for flooding and following fatalities by improving drainage, trapping water droplets on their surfaces, and absorbing water from the ground and releasing it to the air through evapotranspiration. Urban green spaces improve storm water runoff and subsequently the severity, frequency, and duration of flooding and its impacts (Laurance 2007).

Deforestation has in many areas resulted in substantial increases in flooding in terms of both frequency and duration (Laurance 2007). The issue is largest in low- and middle-income countries. This has led many countries to adopt afforestation and forest protection strategies upstream of flood-prone regions as a means of protecting vulnerable areas and their inhabitants from flooding (WHO 2017).

13.8   CONCLUSIONS

Forests protect human health through several pathways, both in urban settings and on the countryside. Many of the pathways relate to lifestyles—physical activity, stress, and social networks—where particularly urban forests have an important role to play. What is important to remember is that although definite evidence may not yet be at hand for certain outcomes, the chronic diseases that are the results of unhealthy lifestyles are increasingly common across the globe. This means that also seemingly small interventions, such as establishing a park in a vulnerable area, can have a major health impact on a larger population health level, even if the individual effect is small.

While the health benefits from forests do not (yet) carry a direct market value, attempts are ongoing to also quantify the value in economic terms. A few of the existing scientific examples are described in this chapter; otherwise, most assessments appear in the gray literature (Wolf and Robbins 2015). The costs and saving estimates are preliminary and vary by region and country. In United States, The Trust for Public Land (2013) calculated that physical activity in public parks across 10 US cities corresponded to avoided costs of healthcare ranging between US$4 million and 69.4 million per year. We can conclude that while scientific attempts to quantify the economic value of health benefits from forests are nascent, there are ample opportunities for developing this field by creating new interdisciplinary approaches that integrate environmental and health economics. For instance, cost-effectiveness and cost-minimization methods can be useful for evaluating tradeoffs between public health outcomes and the costs of creating or improving urban forests. In this context, it is also important to consider cobenefits, meaning that benefits to human health by urban forests may also provide advantages for the environment itself by promoting biodiversity and sustainable management.

Particularly encouraging is that "green" investments provide most benefits in disadvantaged and vulnerable groups, groups that are also disproportionally affected by chronic disorders. Intriguingly, deprived areas in cities usually have a poor accessibility to green spaces and especially green spaces of a high quality. If the goal is to achieve equal health and opportunities for all, a lot can be gained by establishing and maintaining high-quality urban forests in areas where they are needed the most (Figure 13.13).

Another important aspect is that many of these disorders are difficult to treat and cure for people who already suffer from them. Would it not be better to prevent people from the need to search healthcare in the first place? If investments in urban forests and other natural environments can protect health and reduce suffering, it is likely that the costs for healthcare and hospitalization could also be reduced.

However, in order to reach these kinds of cost-efficient, "nature-based" solutions, another approach to both healthcare and urban and environmental planning must be implied. In many areas, the traditional kind of "silo-thinking" still prevails. This is insufficient and often inadequate in today's increasingly complex world, where various systems interact, and the health of the environment and the health of humans are strongly interconnected. This displays in various forms, such as how urbanization contributes to the increasing burden of chronic NCDs and how social and physical environments shape lifestyles and thereby disease patterns. It is perhaps even more evident in the case of anthropogenic impact on climate change and environmental degradation, which is eventually striking back and influencing human health, often in devastating ways.

And ultimately, human survival depends on functional interactions with healthy ecosystem for the sake of food, shelter, basic products, as well as cultural and recreational experiences.

We, as a species, must to a much higher extent recognize that we are but a part of a larger ecosystem, and we cannot thrive unless our surrounding environment is thriving. From this perspective, the fundamental importance of forests for human health should be evident.

FIGURE 13.13 Young community members of Oakland, California, prepare the soil for tree planting. The tree plantings represent an initiative to establish and maintain urban green spaces in areas of the region that have little greenery. (Photo IMG_0531 courtesy of US Department of Agriculture, Washington, DC, CC BY 2.0.)

REFERENCES

ABCG (Africa Biodiversity Collaborative Group). 2002. HIV/AIDS and Natural Resource Management Linkages. Workshop proceedings Africa Biodiversity Collaborative Group, Nairobi.

Alvarsson, J. J., Wiens, S., and M. E. Nilsson. 2010. Stress recovery during exposure to nature sound and environmental noise. *International Journal of Environmental Research and Public Health* 7(3):1036–1046.

Amoly, E., Dadvand, P., Forns, J. et al. 2014. Green and blue spaces and behavioral development in Barcelona schoolchildren: The BREATHE Project. *Environmental Health Perspectives* 122(12):1351–1358.

Annerstedt, M. and P. Währborg. 2011. Nature-assisted therapy: Systematic review of controlled and observational studies. *Scandinavian Journal of Public Health* 39(4): 371–388.

Annerstedt, M., Jönsson, P., Wallergård, M. et al. 2013. Inducing physiological stress recovery with sounds of nature in a virtual reality forest—Results from a pilot study. *Physiology and Behavior* 118:240–250.

Annerstedt van den Bosch, M., Mudu, P., Uscila, V. et al. 2016. Development of an urban green space indicator and the public health rationale. *Scandinavian Journal of Public Health* 44:159–167.

Arai, Y. C., Ushida, T., Matsubara, T. et al. 2008. Intra-operative natural sound decreases salivary amylase activity of patients undergoing inguinal hernia repair under epidural anesthesia: 123. *Regional Anesthesia and Pain Medicine* 33(5):e234.

Archer, E. and S. N. Blair. 2012. Physical activity, exercise and non-communicable diseases. *Research in Exercise Epidemiology* 14(1):1–18.

Banay, R. F., Bezold, C. P., James, P., Hart, J. E., and F. Laden. 2017. Residential greenness: Current perspectives on its impact on maternal health and pregnancy outcomes. *International Journal of Women's Health* 9:133–144.

Baran, P. K., Smith, W. R., Moore, R. C. et al. 2014. Park use among youth and adults: Examination of individual, social, and urban form factors. *Environment and Behavior* 46(6):768–800.

Barany, M., Hammett, A. L., Stadler, K. E., and E. Kengni. 2004. Non-timber forest products in the food security and nutrition of smallholders afflicted by HIV/AIDS in Sub-Saharan Africa. *Forest, Trees and Livelihoods* 14:3–18.

Barany, M., Holding-Anyonge, C., Kayambazinthu, D., and A. Sitoe. 2005. Firewood, food, and medicine: Interactions between forests, vulnerability and rural responses to HIV/AIDS. Paper presented at IFPRI conference.

Basagaña, X., Sartini, C., Barrera-Gómez, J. et al. 2011. Heat Waves and cause-specific mortality at all ages. *Epidemiology* 22(6):765–772.

Benfield, J. A., Taff, B. D., Newman, P., and J. Smyth. 2014. Natural sound facilitates mood recovery. *Ecopsychology* 6(3):183–188.

Biddle, S. J. H., and M. Asare. 2011. Physical activity and mental health in children and adolescents: A review of reviews. *British Journal of Sports Medicine* 45(11):886–895.

Björk, J., Albin, M., Grahn, P. et al. 2008. Recreational values of the natural environment in relation to neighbourhood satisfaction, physical activity, obesity and wellbeing. *Journal of Epidemiology and Community Health* 62(4):e2.

Black, S. E., Devereux, P. J., and K. G. Salvanes. 2007. From the cradle to the labor market? The effect of birth weight on adult outcomes. *Quarterly Journal of Economics* 122(1):409–439.

Bowler, D. E., Buyung-Ali, L., Knight, T. M., and A. S. Pullin. 2010. Urban greening to cool towns and cities: A systematic review of the empirical evidence. *Landscape and Urban Planning* 97(3):147–155.

Bratman, G. N., Hamilton, J. P., Hahn, K. S., Daily, G. C., and J. J. Gross. 2015. Nature experience reduces rumination and subgenual prefrontal cortex activation. *Proceedings of the National Academy of Sciences* 112(28):8567–8572.

Brauer, M., Freedman, G., Frostad, J. et al. 2016. Ambient air pollution exposure estimation for the global burden of disease 2013. *Environmental Science and Technology* 50(1):79–88.

Brenner, D. R., Yannitsos, D. H., Farris, M. S., Johansson, M., and C. M. Friedenreich. 2016. Leisure-time physical activity and lung cancer risk: A systematic review and meta-analysis. *Lung Cancer* 95:17–27.

Broekhuizen, K., de Vries, S. I., and F. H. Pierik. 2013. Healthy aging in a green living environment: A systematic review of the literature: Summary, vol. TNO R10154. Leiden: the Netherlands Organisation for Applied Scientific Research (TNO).

Burnett, R. T., Pope III, C. A., Ezzati, M. et al. 2014. An integrated risk function for estimating the global burden of disease attributable to ambient fine particulate matter exposure. *Environmental Health Perspectives* 122(4):397–403.

Bylund, A., Saarinen, N., Zhang, J. X. et al. 2005. Anticancer effects of a plant lignan 7-hydroxymatairesinol on a prostate cancer model in vivo. *Experimental Biology and Medicine* (Maywood) 230(3):217–223.

Cavill, N., Kahlmeier, S., and F. Racioppi. 2006. *Physical Activity and Health in Europe: Evidence for Action*. Copenhagen: World Health Organization (WHO) Europe.

Cechinel-Filho, V. 2002. *Plant Bioactives and Drug Discovery: Principles, Practise and Perspective*. Hoboken, NJ: John Wiley & Sons.

Chang, C.-R., Li, M.-H., and S.-D. Chang. 2007. A preliminary study on the local cool-island intensity of Taipei city parks. *Landscape and Urban Planning* 80(4):386–395.

Chen, C.-J., Kumar, K. J., Chen, Y.-T. et al. 2015. Effect of Hinoki and Meniki Essential oils on human autonomic nervous system activity and mood states. *Natural Product Communications* 10(7):1305–1308.

Cho, T. H., Lee, Y., and S. M. Kim. 2014. The economic spillover effects of forest therapy projects in Korea. *Journal of Korean Forest Society* 103(4):630–638.

CIFOR (Centre for International Forestry Research). 2005. Contributing to African development through forests: Strategy for engagement in Sub-Saharan Africa. Bogor: CIFOR.

Cohen, D. A., Han, B., Derose, K. P. et al. 2016. The paradox of parks in low-income areas: Park use and perceived threats. *Environment and Behavior* 48(1):230–245.

Colditz, G. A., Dart, H., and C. T. Ryan. 2008. Physical activity and health. In *International Encyclopedia of Public Health*, ed. K. Heggenhougen, 102–110. Oxford, UK: Academic Press.

Colfer, C. J. P., Sheil, D., Kaimowitz, D., and M. Kishi. 2006. Forests and human health in the tropics: Some important connections. *Unasylva* 57(224):3–10.

Crowe, J., Nilsson, M., Kjellstrom, T., and C. Wesseling. 2015. Heat-related symptoms in sugarcane harvesters. *American Journal of Industrial Medicine* 58(5):541–548.

Dadvand, P., Sunyer, J., Basagana, X. et al. 2012. Surrounding greenness and pregnancy outcomes in four Spanish birth cohorts. *Environmental Health Perspectives* 120(10):1481–1487.

Dadvand, P., Villanueva, C. M., Font-Ribera, L. et al. 2014. Risks and benefits of green spaces for children: A cross-sectional study of associations with sedentary behavior, obesity, asthma, and allergy. *Environmental Health Perspectives* 122(12):1329.

Dadvand, P., Nieuwenhuijsen, M. J., Esnaola, M. et al. 2015. Green spaces and cognitive development in primary schoolchildren. *Proceedings of the National Academy of Sciences* 112(26):7937–7942.

Dash, S. K. and T. Kjellstrom. 2011. Workplace heat stress in the context of rising temperature in India. *Current Science* 101(4):496–503.

de Jong, K., Albin, M., Skärbäck, E., Grahn, P., and J. Björk. 2012. Perceived green qualities were associated with neighborhood satisfaction, physical activity, and general health: Results from a cross-sectional study in suburban and rural Scania, southern Sweden. *Health and Place* 18(6):1374–1380.

De Nazelle, A., Nieuwenhuijsen, M. J., Antó, J. M. et al. 2011. Improving health through policies that promote active travel: A review of evidence to support integrated health impact assessment. *Environment International* 37(4):766–777.

de Vries, S., van Dillen, S. M. E., Groenewegen, P. P., and P. Spreeuwenberg. 2013. Streetscape greenery and health: Stress, social cohesion and physical activity as mediators. *Social Science and Medicine* 94(0):26–33.

Ding, D., Sallis, J. F., Kerr, J., Lee, S., and D. E. Rosenberg. 2011. Neighborhood environment and physical activity among youth: A review. *American Journal of Preventive Medicine* 41(4):442–455.

Donovan, G. H., Michael, Y. L., Butry, D. T., Sullivan, A. D., and J. M. Chase. 2011. Urban trees and the risk of poor birth outcomes. *Health and Place* 17(1):390–393.

Donovan, G. H., Butry, D. T., Michael, Y. L. et al. 2013. The relationship between trees and human health: Evidence from the spread of the emerald ash borer. *American Journal of Preventive Medicine* 44(2):139–145.

Donovan, G. H., Michael, Y. L., Gatziolis, D., Prestemon, J. P., and E. A. Whitsel. 2015. Is tree loss associated with cardiovascular-disease risk in the women's health initiative? A natural experiment. *Health and Place* 36:1–7.

Dzhambov, A. M., Dimitrova, D. D., and E. D. Dimitrakova. 2014. Association between residential greenness and birth weight: Systematic review and meta-analysis. *Urban Forestry and Urban Greening* 13(4):621–629.

Edmondson, J. L., Stott, I., Davies, Z. G., Gaston, K. J., and J. R. Leake. 2016. Soil surface temperatures reveal moderation of the urban heat island effect by trees and shrubs. *Scientific Reports* 6:33708.

Eikenæs, I., Gude, T., and A. Hoffart. 2006. Integrated wilderness therapy for avoidant personality disorder. *Nordic Journal of Psychiatry* 60(4):275–281.

Escobedo, F. J., Wagner, J. E., Nowak, D. J. et al. 2008. Analyzing the cost effectiveness of Santiago: Chile's policy of using urban forests to improve air quality. *Journal of Environmental Management* 86(1):148–157.

FAO (Food and Agriculture Organization of the United Nations). 2010. *Global Forest Resources Assessment* 2010. FAO Forestry Paper. Rome: FAO.

Francis, J., Giles-Corti, B., Wood, L., and M. Knuiman. 2012. Creating sense of community: The role of public space. *Journal of Environmental Psychology* 32(4):401–409.

Frank, E. and J. Unruh. 2008. Demarcating forest, containing disease: Land and HIV/AIDS in southern Zambia. *Population and Environment* 29:108–132.

Fuertes, E., Markevych, I., Bowatte, G. et al. 2016. Residential greenness is differentially associated with childhood allergic rhinitis and aeroallergen sensitization in seven birth cohorts. *Allergy* 71(10):1461–1471.

Garcia-Herrera, R., Díaz, J., Trigo, R. M., Luterbacher, J., and E. M. Fischer. 2010. A review of the European summer heat wave of 2003. *Critical Reviews in Environmental Science and Technology* 40(4):267–306.

Gascon, M., Triguero-Mas, M., Martínez, D. et al. 2015. Mental health benefits of long-term exposure to residential green and blue spaces: A systematic review. *International Journal of Environmental Research and Public Health* 12(4):4354.

Gascon, M., Triguero-Mas, M., Martínez, D. et al. 2016. Residential green spaces and mortality: A systematic review. *Environment International* 86:60–67.

Giles-Corti, B., Broomhall, M. H., Knuiman, M. et al. 2005a. Increasing walking: How important is distance to attractiveness and size of public open space? *American Journal of Preventive Medicine* 28:169–176.

Giles-Corti, B., Timperio, A., Bull, F., and T. Pikora. 2005b. Understanding physical activity environmental correlates: Increased specificity for ecological models. *Exercise and Sport Sciences Reviews* 33(4):175–181.

Giles-Corti, B., Bull, F., Knuiman, M. et al. 2013. The influence of urban design on neighbourhood walking following residential relocation: Longitudinal results from the RESIDE study. *Social Science and Medicine* 77:20–30.

Glass, S. T., Lingg, E., and E. Heuberger. 2014. Do ambient urban odors evoke basic emotions? *Frontiers in Psychology* 5:340.

Grazuleviciene, R., Danileviciute, A., Dedele, A. et al. 2015. Surrounding greenness, proximity to city parks and pregnancy outcomes in Kaunas cohort study. *International Journal of Hygiene and Environmental Health* 218:358–365.

Grazuleviciene, R., Vencloviene, J., Kubilius, R. et al. 2016. Tracking restoration of park and urban street settings in coronary artery disease patients. *International Journal of Environmental Research and Public Health* 13(6):550.

Hägerhäll, C. M., Purcell, T., and R. Taylor. 2004. Fractal dimension of landscape silhouette outlines as a predictor of landscape preference. *Journal of Environmental Psychology* 24(2):247–255.

Hägerhäll, C. M., Laike, T., Taylor, R. P. et al. 2008. Investigations of human EEG response to viewing fractal patterns. *Perception* 37(10):1488–1494.

Hanski, I., von Hertzen, L., Fyhrquist, N. et al. 2012. Environmental biodiversity, human microbiota, and allergy are interrelated. *Proceedings of the National Academy of Sciences* 109(21):8334–8339.

Harada, K. and F. Imamura. 2005. Effects of coastal forest on tsunami hazard mitigation—A preliminary investigation. In *Advances in Natural and Technological Hazards Research*, vol 23, ed. K. Satake, 279–292. Dordrecht: Springer.

Hartig, T., Mitchell, R., de Vries, S., and H. Frumkin. 2014. Nature and health. *Annual Review of Public Health* 35(1):207–228.

Holt-Lunstad, J., Smith, T. B., Baker, M., Harris, T., and D. Stephenson. 2015. Loneliness and social isolation as risk factors for mortality: A meta-analytic review. *Perspectives on Psychological Science* 10(2):227–237.

Hordyk, S. R., Hanley, J., and É. Richard. 2015. Nature is there; it's free: Urban greenspace and the social determinants of health of immigrant families. *Health and Place* 34(0):74–82.

Hunter, L. M., Twine, W., and A. Johnson. 2008. *Adult Mortality and Natural Resource Use in South Africa: Evidence from the Agincourt Health and Demographic Surveillance Site*. Institute of Behavioural Science Working Paper EB2005–0004. Boulder, CO: University of Colorado.

Hystad, P., Davies, H. W., Frank, L. et al. 2014. Residential greenness and birth outcomes: Evaluating the influence of spatially correlated built-environment factors. *Environmental Health Perspectives* 122(10):1095–1102.

IPCC (Intergovernmental Panel on Climate Change). 2014. *Climate Change 2014: Impacts, Adaptation, and Vulnerability*. Working Group II Contribution to the IPCC 5th Assessment Report, Final draft. IPCC.

Jayne, T. S., Villarreal, M., Pingali, P., and G. Hemrich. 2005. HIV/AIDS and the agriculture sector: Implications for policy in eastern and southern Africa. *Journal of Agricultural and Development Economics* 2(2):158–181.

Jiang, J., He, X., and D. E. Cane. 2007. Biosynthesis of the earthy odorant geosmin by a bifunctional Streptomyces coelicolor enzyme. *Nature Chemical Biology* 3(11):711–715.

Johns, T. and P. B. Eyzaguirre. 2006. Linking biodiversity, diet and health in policy and practice. *Proceedings of the Nutrition Society* 65(2):182–189.

Johnson, C. and S. Blackburn. 2014. Advocacy for urban resilience: UNISDR's making cities resilient campaign. *Environment and Urbanization* 26(1):29–52.

Joye, Y. 2011. A review of the presence and use of fractal geometry in architectural design. *Environment and Planning B: Planning and Design* 38(5):814–828.

Jung, W. H., Woo, J. M., and J. S. Ryu. 2015. Effect of a forest therapy program and the forest environment on female workers' stress. *Urban Forestry and Urban Greening* 14(2):274–281.

Kaplan, S. 1978. Attention and fascination: The search for cognitive clarity. In *Humanscape: Environments for People*, eds. R. Kaplan and S. Kaplan, 84–90. Ann Arbor, MI: Duvbury Press.

Kaplan, S. 1995. The restorative benefits of nature: Toward an integrative framework. *Journal of Environmental Psychology* 15(3):169–182.

Kaplan, R. 2001. The nature of the view from home psychological benefits. *Environment and Behavior* 33(4):507–542.

Kaplan, S. and J. F. Talbot. 1983. Psychological benefits of a wilderness experience. *Human Behavior and Environment: Advances in Theory and Research* 6:163–203.

Kaplan, S. and R. Kaplan. 1989. *The Experience of Nature: A Psychological Perspective.* New York: Cambridge University Press.

Kardan, O., Gozdyra, P., Misic, B. et al. 2015. Neighborhood greenspace and health in a large urban center. *Scientific Reports* 5:11610. http://www.nature.com/articles/srep11610 #supplementary-information.

Karjalainen, E., Sarjala, T., and H. Raitio. 2010. Promoting human health through forests: Overview and major challenges. *Environmental Health and Preventive Medicine* 15(1):1–8.

Kaschula, S. A. 2008. Wild foods and household food security responses to AIDS: Evidence from South Africa. *Population and Environment* 29:162–185.

Keesing, F. and R. S. Ostfeld. 2015. Is biodiversity good for your health? *Science* 349 (6245):235–236.

Keesing, F., Belden, L. K., Daszak, P. et al. 2010. Impacts of biodiversity on the emergence and transmission of infectious diseases. *Nature* 468(7324):647–652.

Kellert, S. R. and E. O. Wilson. 1995. *The Biophilia Hypothesis.* Washington DC: Island Press.

Kim, W., Lim, S. K., Chung, E. J., and J. M. Woo. 2009. The effect of cognitive behavior therapy-based psychotherapy applied in a forest environment on physiological changes and remission of major depressive disorder. *Psychiatry Investigation* 6:245–254.

Kloek, M. E., Buijs, A. E., Boersema, J. J., and M. G. C. Schouten. 2013. Crossing borders: Review of concepts and approaches in research on greenspace, immigration and society in northwest European countries. *Landscape Research* 38(1):117–140.

Kraus, K. S. and B. Canlon. 2012. Neuronal connectivity and interactions between the auditory and limbic systems: Effects of noise and tinnitus. *Hearing Research* 288(1–2):34–46.

Lattin, D. L. and E. Fifer. 2008. Drugs affecting cholinergic neurotransmission. In *Foye's Principles of Medicinal Chemistry*, eds. D. A. Williams, W. O. Foye, and T. L. Lemke, 373–374. Philadelphia, PA; Baltimore, MD: Lippincott Williams & Wilkins.

Laurance, W. F. 2007. Environmental science: Forests and floods. *Nature* 449(7161):409–410.

Leaning, J. and D. Guha-Sapir. 2013. Natural disasters, armed conflict, and public health. *New England Journal of Medicine* 369(19):1836–1842.

Lederbogen, F., Kirsch, P., Haddad, L. et al. 2011. City living and urban upbringing affect neural social stress processing in humans. *Nature* 474(7352):498–501.

Lederbogen, F., Haddad, L., and A. Meyer-Lindenberg. 2013. Urban social stress—Risk factor for mental disorders: The case of schizophrenia. *Environmental Pollution* 183(0):2–6.

Lee, J. Y. and D. C. Lee. 2014. Cardiac and pulmonary benefits of forest walking versus city walking in elderly women: A randomised, controlled, open-label trial. *European Journal of Integrative Medicine* 6(1):5–11.

Li, Q. 2010. Effect of forest bathing trips on human immune function. *Environmental Health and Preventive Medicine* 15(1):9–17.

Li, Q., Nakadai, A., Matsushima, H. et al. 2006. Phytoncides (wood essential oils) induce human natural killer cell activity. *Immunopharmacology and Immunotoxicology* 28(2):319–333.

Li, Q., Morimoto, K., Kobayashi, M. et al. 2008. Visiting a forest, but not a city, increases human natural killer activity and expression of anti-cancer proteins. *International Journal of Immunopathology and Pharmacology* 21(1):117–127.

Li, Q., Otsuka, T., Kobayashi, M. et al. 2011. Acute effects of walking in forest environments on cardiovascular and metabolic parameters. *European Journal of Applied Physiology* 111:2845–2853.

Liu, J. and L. Jin. 2017 April. *Practice of Forest Therapy in Beijing.* Trees, People and the Built Environment 3 Conference, Institute of Chartered Foresters, Birmingham.

Lovasi, G. S., Quinn, J. W., Neckerman, K. M., Perzanowski, M. S., and A. Rundle. 2008. Children living in areas with more street trees have lower asthma prevalence. *Journal of Epidemiology and Community Health* 62(7):647–649.

MA (Millenium Assessment). 2005. *Millennium Ecosystem Assessment: Ecosystems and Human Well-Being.* Washington, DC: Island Press.

Maas, J., van Dillen, S. M. E., Verheij, R. A., and P. P. Groenewegen. 2009. Social contacts as a possible mechanism behind the relation between green space and health. *Health and Place* 15(2):586–595.

Malawi Government. 2007. *Forestry and HIV and AIDS Strategy 2007–2011.* Lilongwe: Ministry of Energy and Mines.

Markevych, I., Tiesler, C. M. T., Fuertes, E. et al. 2014a. Access to urban green spaces and behavioural problems in children: Results from the GINIplus and LISAplus studies. *Environment International* 71:29–35.

Markevych, I., Fuertes, E., Tiesler, C. M. T. et al. 2014b. Surrounding greenness and birth weight: Results from the GINIplus and LISAplus birth cohorts in Munich. *Health and Place* 26(0):39–46.

Marselle, M., Irvine, K., and S. Warber. 2013. Walking for well-being: Are group walks in certain types of natural environments better for well-being than group walks in urban environments? *International Journal of Environmental Research and Public Health* 10(11):5603–5628.

Massa, K. H. C., Pabayo, R., Lebrão, M. L., and A. D. P. Chiavegatto Filho. 2016. Environmental factors and cardiovascular diseases: The association of income inequality and green spaces in elderly residents of São Paulo, Brazil. *BMJ Open* 6(9):e011850.

Matthews, T. K. R., Wilby, R. L., and C. Murphy. 2017. Communicating the deadly consequences of global warming for human heat stress. *Proceedings of the National Academy of Sciences* 114(15):3861–3866.

McEwen, B. S. 1998. Stress, adaptation, and disease: Allostasis and allostatic load. *Annals of the New York Academy of Sciences* 840(1):33–44.

McEwen, B. S. 2008. Central effects of stress hormones in health and disease: Understanding the protective and damaging effects of stress and stress mediators. *European Journal of Pharmacology* 583(2–3):174–185.

McPherson, M. F. 2005. Asset preservation in African agriculture in the face of HIV/AIDS: The role of education. *American Journal of Agricultural Economics* 87(5):1289–1297.

Miller, R. W. 1997. *Urban Forestry: Planning and Managing Urban Greenspaces.* 2nd ed. Englewood Cliffs, NJ: Prentice Hall.

Mitchell, R. 2013. Is physical activity in natural environments better for mental health than physical activity in other environments? *Social Science and Medicine* 91:130–134.

Mitchell, R. and F. Popham. 2008. Effect of exposure to natural environment on health inequalities: An observational population study. *Lancet* 372(9650):1655–1660.

Mitchell, R. J., Richardson, E. A., Shortt, N. K., and J. R. Pearce. 2015. Neighborhood environments and socioeconomic inequalities in mental well-being. *American Journal of Preventive Medicine* 49(1):80–84.

Murray, C. J. L. and A. D. Lopez. 1996. The Global Burden of Disease: A Comprehensive Assessment of Mortality and Disability from Diseases, Injuries, and Risk Factors in 1990 and Projected to 2020. Cambridge, MA: Harvard University Press, World Health Organization and the World Bank.

National AIDS Commission. 2008. *HIV and Syphilis Sero—Survey and National HIV Prevalence and AIDS Estimates Report for 2007.* Lilongwe: Ministry of Health, Department of Preventive Health Services.

Nilsson, K., Sangster, M., Gallis, C. et al. 2010. *Forests, Trees and Human Health.* Dordrecht: Springer.

Noji, E. K. 2000. The public health consequences of disasters. *Prehospital and Disaster Medicine* 15(4):147–57.

Nowak, D. J., Hirabayashi, S., Bodine, A., and E. Greenfield. 2014. Tree and forest effects on air quality and human health in the United States. *Environmental Pollution* 193:119–129.

Ohe, Y., Ikei, H., Song, C., and Y. Miyazaki. 2017. Evaluating the relaxation effects of emerging forest-therapy tourism: A multidisciplinary approach. *Tourism Management* 62:322–334.

Oke, T. R. 1973. City size and the urban heat island. *Atmospheric Environment (1967)* 7(8):769–779.

Pantell, M., Rehkopf, D., Jutte, D. et al. 2013. Social isolation: A predictor of mortality comparable to traditional clinical risk factors. *American Journal of Public Health* 103:2056–62.

Paoletti, E., Bardelli, T., Giovannini, G., and L. Pecchioli. 2011. Air quality impact of an urban park over time. *Procedia Environmental Sciences* 4:10–16.

Park, B., Tsunetsugu, Y., Kasetani, T., Kagawa, T., and Y. Miyazaki. 2010. The physiological effects of Shinrin-yoku (taking in the forest atmosphere or forest bathing): Evidence from field experiments in 24 forests across Japan. *Environmental Health and Preventive Medicine* 15(1):18–26.

Parsons, R., Tassinary, L. G., Ulrich, R. S., Hebl, M. R., and M. Grossman-Alexander. 1998. The View From the road: Implications for stress recovery and immunization. *Journal of Environmental Psychology* 18(2):113–140.

Peters, K. 2010. Being together in urban parks: Connecting public space, leisure, and diversity. *Leisure Sciences* 32(5):418–433.

Peters, A. and B. S. McEwen. 2015. Stress habituation, body shape and cardiovascular mortality. *Neuroscience and Biobehavioral Reviews* 56:139–150.

Prüss-Ustün, A., Wolf, J., Corvalán, C., Bos R., and M. Neira. 2016. *Preventing Disease through Healthy Environments—A Global Assessment of the Burden of Disease from Environmental Risks.* Geneva: World Health Organization (WHO).

Reklaitiene, R., Grazuleviciene, R., Dedele, A. et al. 2014. The relationship of green space, depressive symptoms and perceived general health in urban population. *Scandinavian Journal of Public Health* 42(7):669–676.

Rodríguez, R., Encina, P., Espinosa, M., and N. Tanaka. 2016. Field study on planted forest structures and their role in protecting communities against tsunamis: Experiences along the coast of the Biobío Region, Chile. *Landscape and Ecological Engineering* 12(1):1–12.

Roe, J., Thompson, C., Aspinall, P. et al. 2013. Green space and stress: Evidence from cortisol measures in deprived urban communities. *International Journal of Environmental Research and Public Health* 10(9):4086–4103.

Rook, G. A. 2013. Regulation of the immune system by biodiversity from the natural environment: An ecosystem service essential to health. *Proceedings of the National Academy of Sciences* 110(46):18360–18367.

Saadatmand, V., Rejeh, N., Heravi-Karimooi, M. et al. 2013. Effect of nature-based sounds' intervention on agitation, anxiety, and stress in patients under mechanical ventilator support: A randomised controlled trial. *International Journal of Nursing Studies* 50(7):895–904.

Sallis, J. F., Cerin, E., Conway, T. L. et al. 2016. Physical activity in relation to urban environments in 14 cities worldwide: A cross-sectional study. *Lancet* 387(10034):2207–2217.

Sbihi, H., Tamburic, L., Koehoorn, M., and M. Brauer. 2015. Greenness and incident childhood asthma: A 10-year follow-up in a population-based birth cohort. *American Journal of Respiratory and Critical Care Medicine* 192(9):1131–1133.

Sbihi, H., Koehoorn, M., Tamburic, L., and M. Brauer. 2016. Asthma trajectories in a population-based birth cohort: Impacts of air pollution and greenness. *American Journal of Respiratory and Critical Care Medicine* 195(5):607–613.

Shishegar, N. 2014. The impact of green areas on mitigating urban heat island effect: A review. *International Journal of Environmental Sustainability* 9(1):119–130.

Slater, R. and S. Wiggins. 2005. Responding to HIV/AIDS in agriculture and related activities. *Natural Resource Perspectives 98*, Overseas Development Institute, London.

State Forestry Administration of the People's Republic of China. 2016, December 24. *China's First Forest Therapy and Medical Tourism Forum Held in Beijing [Press Release]*. http://english.forestry.gov.cn/index.php?option=com_contentandview=article andid=1388:china-s-first-forest-therapy-and-medical-tourism-forum-held-in-beijing andcatid=25andItemid=163 (Accessed November 2, 2017).

Sugiyama, T., Giles-Corti, B., Summers, J. et al. 2013. Initiating and maintaining recreational walking: A longitudinal study on the influence of neighborhood green space. *Preventive Medicine* 57(3):178–182.

Tamosiunas, A., Grazuleviciene, R., Luksiene, D. et al. 2014. Accessibility and use of urban green spaces, and cardiovascular health: Findings from a Kaunas cohort study. *Environmental Health* 13(1):20–31.

Taylor, R. P., Spehar, B., Wise, J. A. et al. 2005. Perceptual and physiological responses to the visual complexity of fractal patterns. *Nonlinear Dynamics, Psychology, and Life Sciences* 9(1):89–114.

Taylor, M. S., Wheeler, B. W., White, M. P., Economou, T., and N. J. Osborne. 2015. Research note: Urban street tree density and antidepressant prescription rates—A cross-sectional study in London, UK. *Landscape and Urban Planning* 136:174–179.

Tidball, K. G. 2012. Urgent biophilia: Human-nature interactions and biological attractions in disaster resilience. *Ecology and Society* 17(2):5.

Tidball, K. G. and M. E. Krasny. 2014. *Greening in the Red Zone: Disaster, Resilience and Community Greening*. Amsterdam: Springer.

Timko, J. A. 2011. *HIV/AIDS, Forests and Futures in Sub-Saharan Africa*. STEPS Working Paper 43. Brighton: STEPS Centre.

Timko, J. A. 2013. Exploring forest-related coping strategies for alleviating the HIV/AIDS burden on rural Malawian households. *International Forestry Review* 15(2):230–240.

Timko, J. A. 2014. Exploring the links between HIV/AIDS and forests Malawi: Morbidity, mortality, and changing dependence on forest resources. In *Ecological Health: Society, Ecology and Health* (Advances in Medical Sociology, Volume 15), ed. Gislason, M. K., 147–171. Bingley: Emerald Group Publishing.

TNC (The Nature Conservancy). 2016. *Planting Healthy Air: A Global Analysis of the Role of Urban Trees in Addressing Particulate Matter Pollution and Extreme Heat*. eds. R. McDonald, T. Kroeger, T. Boucher et al. Arlington, VA: The Nature Conservancy. https://thought-leadership-production.s3.amazonaws.com/2016/11/07/14/13/22/685dccba-cc70–43a8-a6a7-e3133c07f095/20160825_PHA_Report_Final.pdf (Accessed April 1, 2017).

Trust for Public Land. 2013. *The Economic Benefits of Parks*. San Francisco, CA. Trust for Public Land, Center for City Park Excellence Bibliography. https://www.tpl.org/center-city-park -excellence-bibliography#sm.001ojlywb14anf0fps41o16l5t1jz (Accessed March 16, 2017).

Tsunetsugu, Y., Lee, J., Park, B.-J. et al. 2013. Physiological and psychological effects of viewing urban forest landscapes assessed by multiple measurements. *Landscape and Urban Planning* 113:90–93.

Ulrich, R. S. 1981. Natural versus urban scenes: Some Psychophysiological Effects. *Environment and Behavior* 13(5):523.

Ulrich, R. S. 1983. Aesthetic and affective response to natural environment. *Human Behavior and Environment: Advances in Theory and Research* 6:85–125.

Ulrich, R. S. 1984. View through a window may influence recovery from surgery. *Science* 224(4647):420.

Ulrich, R. S., Simons, R. F., Losito, B. D. et al. 1991. Stress recovery during exposure to natural and urban environments. *Journal of Environmental Psychology* 11(3)201–230.

UNEP (United Nations Environment Programme)/UNECE (United Nations Economic Commission for Europe). 2016. *GEO-6 Assessment for the Pan-European Region*. Nairobi: UNEP.

United Nations Office for Disaster Risk Reduction (UNISDR). 2012. *Disaster Risk and Resilience*. Thematic Think Piece, UN System Task Force on the Post-2015 UN Development Agenda. Geneva: United Nations Office for Disaster Risk Reduction. http://www.un.org/en/development/desa/policy/untaskteam_undf/thinkpieces/3_disaster _risk_resilience.pdf (Accessed May 12, 2017).

van den Berg, A. E. and C. G. van den Berg. 2011. A comparison of children with ADHD in a natural and built setting. *Child Care Health and Development* 37(3):430–439.

van den Berg, M., Wendel-Vos, W., van Poppel, M. et al. 2015. Health benefits of green spaces in the living environment: A systematic review of epidemiological studies. *Urban Forestry and Urban Greening* 14(4):806–816.

van den Bosch, M. and M. Nieuwenhuijsen. 2017. No time to lose—Green the cities now. *Environment International* 99:343–350.

Von Döhren, P. and D. Haase. 2015. Ecosystem disservices research: A review of the state of the art with a focus on cities. *Ecological Indicators* 52:490–497.

Vos, T., Barber, R. M., Bell, B. et al. 2015. Global, regional, and national incidence, prevalence, and years lived with disability for 301 acute and chronic diseases and injuries in 188 countries, 1990–2013: A systematic analysis for the Global Burden of Disease Study 2013. *Lancet* 386(9995):743–800.

Weimann, H., Rylander, L., Annerstedt van den Bosch, M., Albin, M., Skärbäck, E., Grahn, P. et al. 2017. Perception of safety is a prerequisite for the association between neighbourhood green qualities and physical activity: Results from a crosssectional study in Sweden. *Health and Place* 45:124–130.

WHO (World Health Organization). 1948. *Preamble to the Constitution of the World Health Organization as Adopted by the International Health Conference*, New York, June 19–22, 1946. New York: World Health Organization.

WHO. 2010. *Global Recommendations on Physical Activity for Health*. Geneva: World Health Organization.

WHO. 2014. *Global Status Report on Noncommunicable Diseases 2014*. Geneva: World Health Organization.

WHO. 2016a. *World Health Statistics 2016: Monitoring Health for the SDGs*. Geneva: World Health Organization.

WHO. 2016b. *Urban Green Spaces and Health—A Review of Evidence*. Copenhagen: World Health Organization, European Regional Office.

WHO. 2017. *Preventing Drowning: An Implementation Guide*. Geneva: World Health Organization.

Wilson, E. O. 1984. *Biophilia: The Human Bond with Other Species.* Cambridge, MA: Harvard University Press.

Wolf, K. L. and A. S. Robbins. 2015. Metro nature, environmental health, and economic value. *Environmental Health Perspectives* 123(5):390.

Yamaguchi, M., Deguchi, M., and Y. Miyazaki. 2006. The effects of exercise in forest and urban environments on sympathetic nervous activity of normal young adults. *Journal of International Medical Research* 34(2):152–9.

Yang, J., McBride, J., Zhou, J., and Z. Sun. 2005. The urban forest in Beijing and its role in air pollution reduction. *Urban Forestry and Urban Greening* 3(2):65–78.

Zhang, Q., Jiang, X., Tong, D. et al. 2017. Transboundary health impacts of transported global air pollution and international trade. *Nature* 543(7647):705–709.

Zhou, C., Zhang, F., Feng, D., Ma, H., Zou, D., and H. Nan. 2015. Research on promoting public health by forest therapy in Beijing: Beijing Forestry and Parks Department of International Cooperation. *Journal of Beijing Forestry University* 14:2.

Zupancic, T., Westmacott, C., and M. Bulthuis. 2015a. *The Impact of Green Space on Heat and Air Pollution in Urban Communities: A Meta-Narrative Systemic Review.* Vancouver: David Suzuki Foundation.

Zupancic, T., Kingsley, M., Jason, T., and R. Macfarlane. 2015b. *Green City: Why Nature Matters to Health—An Evidence Review.* Toronto: Toronto Public Health.

ENDNOTE

1. https://www.youtube.com/watch?v=yMczMjZBA7I by World Economic Forum.

14 Framing Investments in Forest Services

Patrice Harou

CONTENTS

14.1   INTRODUCTION

One of the main works undertaken by forest economists is sector analysis to help identify the most important investments in forest. This chapter, relying on forest economics pillars and principles, so mainly production-oriented, anticipates how the analysis of the sector and of forest investments could evolve while the tertiarization of the economy increases the importance of services, in particular those directly related to forests (see Chapter 12). The chapter also pertains to the philosophy of the International Union of Forest Research Organizations Working Group New Frontiers in Environmental Economics (Kant et al. 2013). The term forestry sector is used here, not the classical forest sector of the forest economist, since it also includes in the analysis all the environmental and other services a forest cover can provide. The chapter is organized linearly for clarity even though the forestry sector and the entire economy systems are recognized as entirely nonlinear (Figure 14.1).

First, one reviews the analysis of the forestry sector duly considering the evolving global and economy-wide policies in a new environmental and anthropogenic context. Today, any sector works must envisage a new global future in which the rural area is impacted by global trade, macroeconomic policies, environmental concerns, the service economy, and urbanization. These trends could modify the forest area and its management and thus the supply of timber in the analysis of the forestry sector. Increased demand for forest-related services impacting indirectly the forest area use for timber production and so the future supply of timber and directly the service values of the forest and so the value of forest land.

Second, the analysis of investments in forest goods and services with an emphasis on the latter is presented. The services and amenities provided by the forest cover are

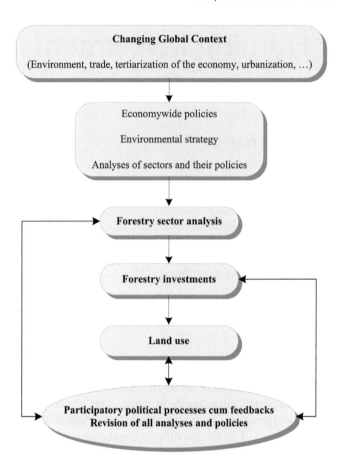

FIGURE 14.1   Chapter linear thoughts of an entirely nonlinear forestry system.

demanded by an expanding urban population and a growing economy of services. The increased value of these forest services should change the profitability of forest investments and could impact land use over time.

The chapter then presents a case for an integrative forest policy instrument, in particular land use planning, and develops principles for its implementation.

As the services provided by the forests gain increasing relevance for society, the social dimension of forests increases. The analyses of the forestry sector, of its investments and land use, should involve the stakeholders and be reactive to changes in values and norms. Citizens should increasingly be involved in scientific research, the valuation of services, the identification of forest management alternatives, and the monitoring of land use.

We conclude with the need to revisit continuously and participatively these analyses and plans to adapt them to an ever changing and more uncertain context, not an easy feat for foresters used to clear and immutable planning.

14.2   FRAMING THE FORESTRY SECTOR

Framing the forestry sector is needed to help prioritize forest investments. The forestry sector is influenced by the new global context. Nationally, the sector is impacted by macroeconomic policies, both monetary and financial, by other sectors of the economy, by environmental policies, and by the strength of the institutions in place. These policies are, for the most part, set outside the forestry sector. The projections of the production and consumption of forest goods and services are made more complex because of the interactions of many factors exogenous to the forestry sector and that change more rapidly. The estimation of the future supply and demand of forest goods and services and of their prices are challenging given the long-term horizon of forestry in a fast-changing world.

The overall socioeconomic context influencing forest policies and investments is represented schematically in Figure 14.2 (World Bank 1998). In a new global context, the ellipse of the organigram contains the economy-wide policies influencing the different steps in the analysis of forest investments. The new global context (box 1), the macroeconomic policies (box 2), and different sectors policies (box 3) impinge on prices throughout the economy. It also influences the price of time, the discount rate, and inflation (monetary policies), and taxes (fiscal policies). The sector analysis forecasts the supply and demand of timber to help set the sectoral policies, taking macroeconomic policies as given. The same analyses of supply and demand are made for all the sectors of the economy, e.g., the agriculture sector. The environment (box 4), which is not a sector, should be mainstreamed in all sectors of the economy including the forestry sector. The strength of the institutions in charge of the annual budgets (box 5) reflects a cultural dimension in the run of the economy. It is another factor to take into consideration because it influences the implementation of forest policies.

Economy-wide and sectoral policies should provide the sustainable development context (SD; box 6) under which forestry projects are appraised (project cycle; box 7). Indeed, prices, exchange rates, subsidies, taxes, environmental regulations, and the institutions implementing policies will impact the analysis of investments across all sectors, including the forestry sector. These policies have been established for the entire economy. It is possible that some of these economy-wide policies could not only favor the forestry sector, such as the creation of a carbon market, but could also hinder the forestry sector such as property taxes or subsidies to another sector, e.g., agriculture, which could push the forestland out of production with negative environmental impacts. In case of a negative impact, the forestry sector should propose countermeasures to guarantee that forest options are efficiently considered in the overall economy. Forestry sectoral policies should rely on a realistic model in which services needs to be duly considered.

A word of caution is necessary as we talked about sector modeling particularly in the long-time frame of forestry investments. A physicist and later Wall Street modeler, Emanuel Derman, (2011) in his book *Models Behaving Badly*, makes the differences between theories, models, and intuition. Theories tell you what something is while models merely tell you what something is like. The first is useful in physics in attempting to discover the principles that drive the world. Models are just metaphors

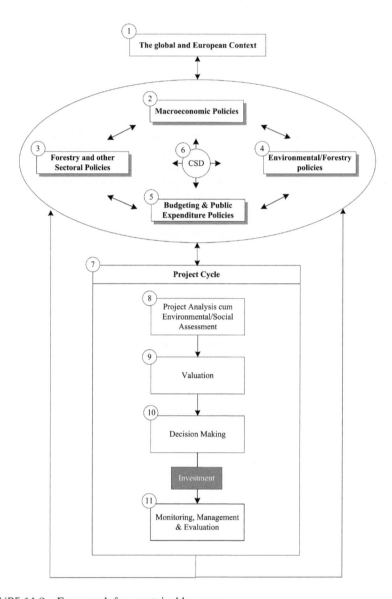

FIGURE 14.2 Framework for a sustainable economy.

comparing the object of their attention, here the supply and demand for timber, to something else that just resembles it but without reflecting all the intricacies of real life such as Wall Street dealings or the politics of trade. Intuition unifies the subject with the object. It allows envisaging theories that need to be proven and to recognize the limits of models. The outputs of models should be checked with intuition and experience.

The global partial equilibrium models of supply and demand of forest products such as the global forests products model (Buongiorno et al. 2003; Buongiorno 2014) are useful to understand the direction of forest products prices and quantity traded with changes in policies, technologies, and trade. However, most services provided by forests, our focus here, are not tradable. Trade is also submitted to the vagaries of new future policies. While high value-added products will probably continue to be regularly traded, the environmental impact and costs of transporting heavy goods with low value-added should be minimized over a long distance in the future. Trade can also be constrained by politics, e.g., the decades-long trade war on forest products between Canada and the United States. The impact of the transatlantic treaty on forest products trade simulated in 2014 (Buongiorno et al. 2014) should now consider different US trade views and Brexit.

In our new fast-changing world and the decreasing carrying capacity of Mother Earth to produce what we need to survive decently, the environment will need to be carefully managed and in more details locally. This is particularly important in developing countries where population increase during the last decades has been unprecedented. For all these reasons, we refer here to a simple gap model of consumption and production (Harou 1992) to identify and rank forest investments including in services. This simple approach allows incorporating more easily the impact of urbanization and the new emphasis on services with direct link to the management of distinct landscapes and timbersheds. This, in turn, helps integrate more efficiently forestry and the forest products industry into sustainable regional land use.

The sector analysis attempts to quantify the gap between aggregate consumption and production for forest goods and services at a given price level. With aggregates, there are no relationships with price as in a partial equilibrium model between supply and demand for forest products. For services, the gap model should follow the same process as for forest products but with the extra difficulty to standardize services provided by forests. Once these aggregates are compared and the gap between production and consumption estimated, policies and investments are proposed to close that gap. The gap model will be periodically revised. The policy to close the gap to maintain prices at present level, i.e., avoiding price increase, is to encourage wood consumption considered a more environmentally friendly and healthy construction material. The investments to close the gap can be in changing policies, such as increasing forest plantation or regeneration on private lands by providing subsidies or in a public extension program. It could also provide green infrastructures increasing both timber and amenities in conurbations. Alternatively, a partial equilibrium model at a national or supranational (European Union) level could be necessary if the trade in and out of the region under study is necessary to fill the gap when the landscape under study is already managed at its more efficient and sustainable level.

Macroeconomic policies have precedence over sectoral policies at the national level. If these policies negatively affect forestry, it is for the sector policies to rectify negative impacts and seek investment support through the budgeting process. The macroeconomic policies impact investments in the entire economy and in the forestry sector. Countries without a proper macroeconomic framework will not attract new forestry investments (Kanieski da Silva et al. 2017). A simple comparison of production and consumption of forest products can help direct policies and investments at the land use level. At that level, the demand of services provided by the forest is easier to identify, to quantify, and to value. The response to this increased demand of services is to efficiently invest in their production. This is the object of the following section.

14.3   INVESTMENTS IN SERVICES PROVIDED BY FORESTS

The analysis of investments in forest goods and services goes well beyond a simple accounting exercise (box 7 of Figure 14.2). While the financial analysis of investments is well known by foresters who master the Faustmann formula, the investments in services made from the point of view of the society require a far more complete analysis of their environmental and social aspects (see box 8 of Figure 14.2). The values of the services provided by the forest cover, in most circumstances, must be estimated since markets rarely exist for them. If they do exist, the specificity of a forest habitat precludes generalization. The urbanizing and tertiarization trends are rapidly changing the nature and value of these forest services.

What are those services? While still using the wording "forest services" in this chapter, we refer to the concept of ecosystem services and their economic values as described in "The Economics of Ecosystems and Biodiversity (TEEB)" reports in 2008 and 2010 [United Nations Environment Programme (UNEP)]. TEEB describes ecosystems services as "the direct and indirect contributions of ecosystems to human-well-being." These services are further divided into provisioning, regulating, supporting (habitat), and cultural services. Forest ecosystems provide food, raw material, and medicines. They regulate clean air and water. They support habitat for a diverse range of species. Finally, they hold spiritual and recreational values for mental and physical health. A meaningful description of forest services following that UNEP classification is provided by Sing et al. (2015).

The ability of forests to explicitly provide these different benefits depends on where they are located and how they are managed. A proper land use maintains or creates forests where it is ecologically appropriate and economically efficient for society. The objective for a given forest plan must direct the management regime for that forest while ensuring the possible adaptation of not only that management to potential changes in the future demand for timber but also of forest services. By maintaining biodiversity, hence, a diversity of services over time, the forest should be more resilient to climate change. Forest valuation and forest investments must consider these possible futures running the alternative test periodically (Harou et al. 2014).

How is the profitability of forest investments analyzed? For private investments, foresters are used to preparing a cash flow table for a forest stand with a specific site index once the growth function has been simulated for a given forest management. The cost function for managing that stand is estimated. The profitability of the investments is calculated with the Faustmann formula. The financial profitability of forestry investments is estimated for making decision in the private sector.

The appraisal of the same investments from a societal point of view, called here the economic analysis of the investment by opposition to the financial analysis just described earlier, requires a more elaborated analysis especially if there is a coproduction of forest services from the management of that forest stand

(Harou 1987). The production function estimated in the financial analysis suffers from the "wake effect," i.e., does not include services, and so needs to be complemented by an environmental and a social assessment. Both assessments provide supplementary information on the inputs and outputs related to the public investment. These input–outputs could enter in the investment cash flows if they have a market or can be shadow priced as explained in the following. This may require to value, to shadow price, inputs and outputs that may not have a market (valuation, box 9 in Figure 14.2). If environmental externalities and social considerations cannot enter the cash flow table, the information from the environmental and social assessments should be jointly considered with the economic analysis in a multiple objective decision-making process (box 10 in Figure 14.2).

What is a shadow price? Shadow pricing corrects market prices, or estimates nonmarket prices, when market, policy, or institutional failures exist, i.e., externalities. The basic principle of shadow pricing is that the inputs and outputs entering the investment analysis must reflect their real cost of opportunity to society. If a market exists and is competitive, the market price is used in the economic analysis of the investment. However, if the market is imperfect, i.e., in situations of oligopoly or monopoly, shadow prices should be estimated assuming (almost) perfect competition. Since most forest services are characterized by externalities, no market prices exist for them and so shadow pricing is necessary. The valuation of nonmarket services should become much more common in the analysis of public forest investments. The shadow prices could also be used for private investments to decide on public policies aiming at increasing timber production, the production of forest services, or the protection of forest ecosystems on private lands, as explained in the following.

A series of potential valuation techniques exist to shadow price for imperfect and nonexistent market. These techniques are presented in Figure 14.3. If markets exist and were perfect for all inputs and outputs underlying the economic analysis, there would be no need to shadow price since market prices would reflect their true costs of opportunity. However, the perfect competition model is just a model, not the reality and prices, especially labor price, and will need to be adjusted to reflect their opportunity costs to society (Harou 1987). If macro, sectoral, environmental policies have corrected some of the market failures, we could eventually take these prices observed in the corrected market. However, even if macro and sectoral policies have mitigated some of the market, policy, and/or institutional failures, some important inputs and outputs prices could still need further review.

The market for forest services often does not have market at all. In this case, the analyst can follow Figure 14.3 and use direct proxy techniques, which are usually cost based, or indirect proxy techniques, for example, by revealing embedded environmental values from property assets, such as land or houses prices. When no proxy is available for shadow pricing, we are forced to survey the willingness to pay (WTP) directly such as in the contingent valuation method. A more detailed coverage of these techniques, including their limits, is out of the scope of this chapter but can be found in a book by Markandya et al. (2002).

Another important distinction between the economic and financial analyses of forestry investment is the price of the most important inputs of forest production,

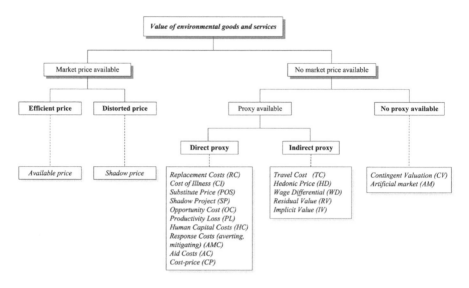

FIGURE 14.3 Shadow pricing forest goods and services.

time. The discount rate in the Faustmann formula used for private forest investments is the market alternative rate of return of the investors. For society, the discount rate has an intergenerational dimension of the flow of costs and benefits of the investment over time. The economic analysis uses a social discount rate which implies a value judgment and is given to the analyst by the politics. It is often given to the analyst by the forest administration. The literature of the proper discount rate for forestry is ample and out of our scope (Harou 1987).

Another distinction between the private and public investment analysis needed to appraise forests services related investments are the transfer payments. In the financial analysis, the taxes and eventual subsidies are entered in the cash flow tables as costs or benefits. For the investment analysis made from the point of view of society, the economic analysis, taxes, and subsidies are just transfer among people but do not change the overall wealth in the economy, rather the distribution of that wealth. The transfers are thus omitted from the economic analysis (Harou 1987).

Services provided by the forests when they have no market but are desired by society will become a political issue. Should the politics decided to correct market failures by increasing services to the citizens, many different policy instruments are available in forestry (Cubbage et al. 2007) and specifically for forest plantation (Zhang et al. 2015). The most economically efficient instrument can be identified by a dual financial–economic analysis (Harou et al. 2013). If the present value in the financial analysis of a private forest is negative for a given forest management enhancing the desired production of a mix of services, the private owner would not adopt that type of management. However, if the economic analysis is profitable, society could consider providing the instruments required to motivate private forest owners to manage for obtaining these desired services (see Chapter 15). One of the most encompassing instruments for managing forest landscapes to produce a mix of forest services and timber and at the same time try to smooth the sprawling of urban centers is certainly land use planning to which we now turn.

The approach to invest in forestry for both the private and public sectors using the financial and economic analysis cum environmental and social assessments is illustrated using a stylized case in Box 14.1. The example is used to define the forestry instrument to be used to entice a private landowner to undertake an investment useful to society but not profitable for the private forest owner (Harou et al. 2013).

BOX 14.1   SIMPLIFIED ILLUSTRATION OF THE FINANCIAL AND ECONOMIC ANALYSIS OF A FORESTRY INVESTMENT CUM ENVIRONMENTAL AND SOCIAL IMPACTS

The investment is in a plantation of maritime Pine (*Pinus pinaster*) in a very specific landscape in Aquitaine, an area of close to 1 million ha in the southwest part of France, with sandy and acid soils which varies from rendzinas limestone to podzolic soils. The mean annual temperature ranges from 11°C to 14°C. Precipitation is well distributed all year around from 600 to 1200 mm per year, with the possibility of drought periods between spring and summer.

The example is for investing in 1 ha of standard characteristics and site index. The forest yield, the production function, has been modeled using the Capsis platform (http://capsis.cirad.fr/capsis/models) and average 11 m³/ha and per year. The soil is poor, but genetic improvements have allowed a doubling of the yield in the last 50 years. The financial and economic cash flows of that investment provided elsewhere is summarized here to illustrate the approach just described.

The private forest owner considered here has an opportunity cost of capital of 3%. If she does not invest in the forestry plantation and sell the land, she can get a 3% on that capital. The financial cash flow shows that the investment is not profitable at a 3% discount rate, which corresponds to the landowner opportunity cost of capital. The internal rate of return is 2.6%, i.e., below the acceptable rate of return. In that case, the landowner could neglect forest management on that land as what happens often on the small forest estates encountered in Europe. Note that in Aquitaine, the medium property size of 8 ha is well above the average forest area size of 3 ha in France. The owner could also change the use of that land if allowed or sell it.

If the society wants the owner to plant so to supply the numerous forest products mills of the region, an instrument is needed to entice that private investment. An economic analysis of the same investment is now done to gauge the profitability of the investment but from a societal standpoint this time. The social discount rate is given at 2%. From the society's point of view, the internal rate of return (IRR) resulting from the economic analysis is now 4.2%. The investment is clearly profitable for the society.

The difference between the two analyses also resulted from shadow pricing labor, because of the social impact assessment showing elevated level of structural unemployment in the region. It was estimated that the true opportunity cost of the unskilled worker used in forest plantation could be half the wage paid in the private sector which is the minimum wage.

The difference of profitability also resulted from the price of the carbon fixed in the tree plantation as identified in the environmental assessment. A value is attached to the fact that the extra carbon fixed by the plantation is worth approximately €30/ha/year.

The financial analysis has thus showed that the investment by the private owner is clearly not profitable in this case. However, from a society standpoint, it is a worthwhile investment. Using this dual analysis, the analyst can build an argument to provide an instrument to entice forest investments by the private sector. Which instruments to choose is another matter. As said earlier, a panoply of possible instruments exists. We must choose the one that is most efficient, easy to manage and equitable. A CBA for each possible instrument must be run and their net present worth (NPW) must be compared. The instrument with the highest NPW *ceteribus paribus* is retained. A payment for fixing carbon has been found the most appropriate and efficient instrument in this case. A payment of € 30/ha and per year to fix the carbon on 1 ha was used. Using this shadow price for carbon as the incentive to the private forest owner, the financial analysis shows the investment profitable with an IRR (as defined earlier) well above 3%.

However, is this instrument, consisting of a payment of €30/ha/year for carbon fixing, efficient? To be efficient, the private forest investment in the financial analysis should just break even with the carbon payment and no more. Any payments over and above that point represent an inefficient transfer since it was not necessary to entice the private owner to invest. It was found that a carbon payment of half that amount, or €15/ha/year, was just enough to break even. The instrument recommended is thus a payment for carbon of just €15/ha/year.

14.4 LAND USE PLANNING FOR PROVIDING FOREST SERVICES

To cope with urbanization and the tertiarization of the economy, a connected phenomenon, and to organize the production of forest goods and services required by these contemporary trends, a better planning of the use of the land is paramount. The main task of land use planning[1] is the coordination of the different land-use interests to use land sustainably and efficiently for the general good while simultaneously establishing an enabling environment for investment.

Urbanization should be part of a proper land use. Urban sprawl is the result of a lack of planning duly incorporating not only ecological but also social factors.

Appropriate urban planning reflects the social cohesion of the people living in the city region. The well-planned city is seen today as part of a regional planning duly incorporating all the aspects of a sustainable economy (Calthorpe 2001). This new regional structure is an interconnectedness of networks of transports, communities, open space, economic systems, and cultures. As for the Internet, these networks need to smoothly link for the entire system to properly work.

The transport network is particularly important in the regional plans. It should serve the creation of green tree corridors and green infrastructure, ensuring the continuity of ecocorridor needed to preserve biodiversity and to adapt to climate change. Some principles given for transit-oriented development by Calthorpe (1993) organize economic growth on a regional level to be compact and transit supportive; place commercial, mix housing, jobs, parks, and civic uses within walking distance of transit stop; create pedestrian/bike friendly street/path networks connecting local destination; and preserve sensitive habitat, riparian zones, and quality open space.

In effective regional planning, urban centers smoothly transit to rural landscape in which agriculture and forestry predominate and is well integrated with the urban region center. The transition is more clear-cut where people density and social cohesiveness are high. The participatory process in designing these plans is needed to ensure a sustainable land use that maximizes social welfare and to assuage the Nimbys (Not in My Backyard) and even the Bananas (Build Absolutely Nothing Anywhere Near Anything) of this world (Duany et al. 2001).

Rural areas should be planned at the landscape, ecological level. Integrated landscape management scales and leverages land for agriculture, forestry, and water management in such a way that the whole ecological and economic gains are greater than the sum of its sectors taken separately. The key actions to operationalize land use planning in this way are (Eco-Agriculture Partners 2013): (1) plan for multiples objectives at different scales; (2) manage ecological, social and economic interactions to optimize synergies; (3) plan for adaptability and so resilience; and (4) involve all the stakeholders combining bottom-up and top-down participation to ensure land use buy-in, detailed monitoring, and political and administrative support.

Land use is directly influenced by economic forces as depicted in Figure 14.4. The use of the land reflects its biophysical characteristics and the socioeconomic and cultural context in which the different sectors compete for the use of the land at a given time. The suitability of land for forestry, as for agriculture or any other uses, is based not only on its biophysical characteristics but also on economic and political factors influencing investments toward a particular use. The macroeconomic and sector policies mentioned earlier provide the framework explaining the current and future prices of forest goods and services as well as prices in other sectors. The price changes resulting from new policies and demand will be reflected in the changing profitability of a certain use of the land, hence the value of that land. The land use pattern shifts. These links between integrated land use and changes in economic policies is well documented for forestry (Harou and Essmann 1990).

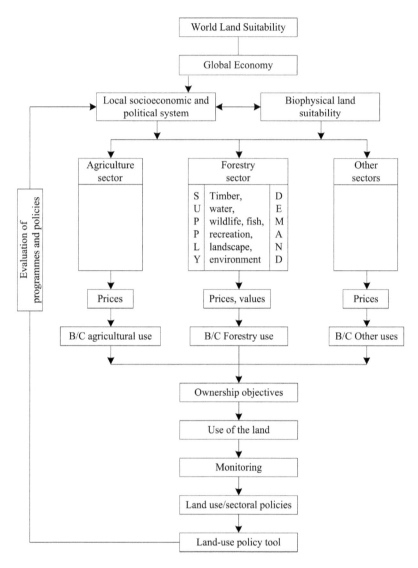

FIGURE 14.4 Economic linkages to land use. (From Harou, P. A. and H. Essmann, Integrated land uses and forest policies: A framework for IUFRO Research, in *Proceedings of the IUFRO World Congress*, 188–197, Montréal, Wissenschaftsverlagvauk, 1990.)

Land use management consists of directing the use of the land to its most economically efficient and sustainable use. A sustainable use of the land is duly incorporating its ecological and social dimensions as explained earlier in the analysis of sustainable investments. If it is not the case, for example, when marginal lands are used for agriculture, one must question the policies that brought that unsustainable use of the land, such as agriculture subsidies. Likewise, if a biodiverse private forest is transformed in a parking lot, it is the service of the forest cover that has not been reckoned at its just value. In that case, a proper land use instrument should be proposed to bring about the proper use of the land, as a not net loss of wetlands for instance.

A range of instruments exists to direct the use of the land to its most sustainable and efficient use (Harou 1990). Some influence market behavior, e.g., land titling, others affect the land management process through improved regulations, subsidies, or provision of information. These instruments fall broadly into six main categories: regulatory, economic, land acquisition, property rights, provision of infrastructure, and information.

- Regulatory instruments include zoning, subdivision regulations, transfer of development rights, and other types of land use controls designed to protect sensitive land resources, public interests, and environmental and cultural values.
- Economic instruments include economic incentives such as preferential taxation schemes, transfer and development taxes, and subsidies and contracts, all of which can be used to encourage developers and landowners to develop (or keep in natural state) land in accordance with environmental objectives.
- Land acquisition alternatives include various types of land acquisition approaches, such as voluntary sales, expropriation, easements, and land exchanges, including land readjustment which will enable urban land managers to correct sprawling (UNHABITAT 2016).
- Property rights approach involves the provision of secure land tenure to promote investment in land and infrastructure improvement and is particularly important in developing countries (UNHABITAT 2012; FAO 2016).
- Government provision of infrastructure involves the provision of appropriate infrastructure, such as roads, to guide development as well as to serve the special needs of land resources or hazard-prone areas.
- Information and education provides methods for expanding knowledge of the issues, land conditions, and the environmental implications of various types of development on environmentally sensitive or hazard-prone lands. They can be used to support land use decisions and to encourage landowners and public authorities to carry out voluntary conservation. They include land information systems, various types of assessments, and public information.

No single instrument will be effective in achieving land management with multiple objectives as is the case in forestry. An effective participatory process is needed to balance these objectives. A mix of instruments is useful to properly integrate forestry in a continuous landscape of urban, regional, and rural areas extending throughout national boundaries.[2] The instruments can also help specific forest project from a forest of production to a forest of services or a mix of them. The most efficient instrument, as for forestry instruments, must be identified using cost–benefit analysis (CBA) (Harou et al. 2013).

Forestry is thus becoming an important player within an emerging field of work of environmental economic geography that provides services inputs and service activities in green economic development (Jones et al. 2016). The green economy has been defined by UNEP (2011): "In its simple expression, a green economy is low-carbon, resource-efficient, and socially inclusive. In a green economy, growth in income and employment are driven by public and private investments that reduce carbon emissions and pollution, enhance energy and resource efficiency, and prevent the loss of biodiversity and ecosystem services".

Today, consulting activities in forestry cover a broad range of expertise important to switch to a green economy of services: mapping soil and vegetation, timber certification, marketing of wood products, wood architecture, wood products manufacturing, carbon offsetting, urban greens and smart cities, agroforestry, green finance, forest valuation, forest products certification, environmental auditing, and other green knowledge-intensive business support services (Bryson and Daniels 2015). Providing all these services to other sectors in addition to forestry should become a useful source of employment for foresters. The suggestion made here to use land use as an important instrument to adapt forestry to changes fits well in this new field of environmental economic geography.

14.5 CONTINUOUS DECISION-MAKING PROCESS

Given the fast changes in our new economies, there is a need to not only continuously review the sector work, which is usually done periodically, but also to monitor the analysis of the forest investments using the alternative test (AT), which is much less frequent (Harou 1987). Adapting a forest investment once it has started is rarely done because of not only inertia but also the high sunk costs involved. The necessity to manage forest ecosystems for services in addition to timber production requires for certain forests, particularly public forests, a multi-objective and adaptable silviculture. It is sometimes called continuous cover forestry, e.g., Pro-Sylva, type of forest management, and it favors an ecosystem more resilient to climate and socioeconomic changes. This should not preclude to designate also in the land use some intensive monofunctional plantations to produce timber and wood fiber monitored with the AT too.

Sector analysis cum investments appraisal has for finality to help prioritize the actions to be undertaken in the forestry sector at the local, regional, and national scales, duly considering the global context and adapting to its evolution. While the economic analysis is not a finality in and out itself, it should help clarify the tradeoffs involved, structure the participatory process, and help make decisions on the investments and policies required to reach individual and society objectives of sustainable land use management and economic development. The major limitation of the economic analysis is that equity is not considered. Eventually, some weights can be placed on the benefits accruing to different income beneficiaries (Harou 1987), but this is rarely done.

The question of allocating the territory to different uses, including forests, becomes then a question of political economy. A certain land use also reflects a status quo of power, and any change in land use involves a play of influences. To manage changing behavior for ensuring the best land use and effective participation in the decision-making process is not an easy task in developed as in developing economies. The higher the social cohesion, the better the participatory process and the results of that process are. For not only developing, but also developed countries, the changes need to come from social activism. To be successful, the movers would learn a lot by reading Duncan Green's (2016) new book *How Change Happens*, which sets out a power and system approach, which tries to adapt to the erratic behavior of complex system following an evolutionary approach rather than a linear one.

The movers should bring along with them the policy makers and get them out of their short-termism. Their presentism bias in policy making decision is well known by foresters and environmentalists who work for the long term and future generations. In his book *Governing for the Future*, Boston (2017) proposed an approach including: the establishment of future-focused advisory institutions to ensure that the policy debates take next generations' interests into consideration and to embed long-term considerations in day-to-day policy making; greater thoughts in the foresight of the sector work and a stronger integration of that work in policy making processes; due considerations of the social and environmental impacts in the economic analyses of investments and the development of green accounting at the macro level; more extensive participatory processes to help nurture shared values and greater political trust; and enhancement of the private sector long-term decision-making through better governance, management, and accounting practices including green accounting at the firm level.

14.6   CONCLUSION

For clarity, the chapter structure has followed a line of thoughts (Figure 14.1) of a totally nonlinear forestry system in which all its components interact among themselves and with land use. The approach proposed here to accommodate today's changes affecting forestry and its increased demand for services, even more than for tangibles ones and specific to ecological niches, relies on constant feedbacks on the sector, investment, and land use analyses within a moving economic system. This cybernetic approach to forest system management did consider economy-wide policies, forestry sector foresight, sequential investment analysis using the AT, and an integrative land use instrument cum participation and constant monitoring. With the help of the citizenry and its movers and by coopting the policy makers, this approach should be evolutionary to better manage uncertainty inherent to the long-term forestry sector. A society with more social cohesion will have institutions which react more readily to any types of change.

REFERENCES

Boston, J. 2017. *Governing for the Future: Designing Democratic Institutions for a Better Future*. Emerald Volume 25 of the Public Policy and Governance series. Bingley: Emerald Group Publishing Ltd.

Bryson, J. R. and P. W. Daniels. 2015. *Handbook of Service Business*. Cheltenham: Edward Elgar.

Buongiorno, J., Zhu, S., Zhang, D., Turner, J., and D. Tomberlin. 2003. *The Forest Products Model: Structure, Estimation and Applications*. Amsterdam: Academic Press/Elsevier.

Buongiorno, J. 2014. Global Modelling to predict timber production and prices: The GFPM approach. *Forestry* 88:291–303.

Buongiorno, J., Rougieux, P., Barkaoui, A., Zhu, S., and P. A. Harou. 2014. Potential impact of a transatlantic trade and investment partnership on the global forest sector. *Journal of Forest Economics* 20:252–266.

Calthorpe, P. 1993. *The Next American Metropolis: Ecology, Community and the American Dream.* New York: Princeton Architectural Press.

Calthorpe, P. 2001. *The Regional City: Planning for the End of Sprawl.* Washington: Island Press.

Cubbage, F., Harou, P. A., and E. Sills. 2007. Policy instruments to enhance multi-functional forest management. *Forest Policy and Economics* 9:833–851.

Derman, E. 2011. *Models Behaving Badly.* New York: Free Press.

Duany, A., Plater-Zyberg, E., and J. Speck. 2001. *Suburban Nation: The Rise of Sprawl and the Decline of the American Dream.* New York: North Point Press.

Eco-Agriculture Partners. 2013. *Defining Integrated Landscape Management for Policy-Makers.* Eco-agriculture Policy Focus 10. Washington, DC: Eco-Agriculture Partners. http://www.ecoagriculture.org/documents/files/doc_547.pdf (Accessed February 28, 2014).

FAO (Food and Agriculture Organization of the United Nations). 2016. *Responsible Governance of Tenure and the Law.* Rome: FAO.

Green, D. 2016. *How Change Happens.* Oxford, UK: Oxford University Press.

Harou, P. A. 1987. *Essays in Forestry Economics: Appraisal and Evaluation of Forestry Investments, Programs and Policies.* Kiel: Wissenschaftsverlag Vauk.

Harou, P. A. 1990. Forestry taxes and subsidies for integrated land use. In *Land Use for Agriculture, Forestry and Rural Development*, eds. Whitby, M. C. and P. J. Dawson, 85–94. Newcastle: Tyne University Press.

Harou, P. A. 1992. Minimal review of the forestry sector. In *Proceedings of IUFRO working groups Integrated land use and forest policy (S6.12-03) and Forest Sector Analysis (S6.11-00), 10th Forestry World Congress*, 85–94.

Harou, P. A. and H. Essmann. 1990. Integrated land uses and forest policies: A framework for IUFRO Research. In *Proceedings of the IUFRO World Congress*, 188–197, Montréal: Wissenschaftsverlagvauk.

Harou, P. A., Rose, D., and A. Lobianco. 2013. Cost-benefit of forestry instruments. In *Socio-Economic Analysis of Sustainable Forest Management*, 70–80, eds. Sisak, L., Dudek, R. and M. Hirb. Vienna: International Union of Forest Research Organizations.

Harou, P. A., Zheng, Ch., and D. Zhang. 2014. The alternative test approach in forestry. *Forest Policy and Economics* 34:41–46.

Jones, A., Strom, P., Hermelin, B., and G. Rusten. 2016. Services and the green economy. London: Macmillan.

Kanieski da Silva, B., Cubbage, F. W., Estraviz, L. C. R., and C. N. Singleton. 2017. Timberland investment management organizations: Business strategies in forest plantations in Brazil. *Journal of Forestry* 115:95–102.

Kant, S., Wang, S., Deegen, P. et al. 2013. New frontiers of forest economics. *Forest Policy and Economics* 35:1–8.

Markandya, A., Harou, P. A., Bellu, L., and V. Cistulli. 2002. *Environmental Economics for Sustainable Growth—A Handbook for Practicioners.* Camberley Surrey: Edward Elgar Publisher.

Sing, L., Kevin, R., and K. Watts. 2015. *Ecosystem Services and Forest Management.* Forestry Commission Research Note. https://www.forestry.gov.uk/pdf/FCRN020.pdf/$FILE/FCRN020.pdf (Accessed January 18, 2018). Camberley: Edward Elagar.

UNEP (United Nations Environment Programme). 2008. *The Economics of Ecosystems and Biodiversity.* Nairobi: UNEP. http://www.teebweb.org/ (Accessed January 18, 2018).

UNEP. 2010. *Mainstreaming the Economics of Nature: A Synthesis of the Approach, Conclusions and Recommendations of TEEB*. Nairobi: UNEP. http://doc.teebweb.org /wp-content/uploads/Study%20and%20Reports/Reports/Synthesis%20report/TEEB %20Synthesis%20Report%202010.pdf (Accessed January 18, 2018).

UNEP. 2011. *Towards a Green Economy: Pathway to Sustainable Development and Poverty Eradication*. Nairobi: UNEP. https://sustainabledevelopment.un.org/index.php?page =view&type=400&nr=126&menu=35 (Accessed July 15, 2017).

UNHABITAT. 2012. *Handling Land—Innovative Tools for Land Governance and Secure Tenure*. Nairobi: UNHABITAT.

UNHABITAT. 2016. *Remaking the Urban Mosaic—Participatory and Inclusive Land Readjustment*. Nairobi: UNHABITAT. https://gltn.net/home/2016/10/10/remaking-the -urban-mosaic-participatory-and-inclusive-land-readjustment/ (Accessed January 18, 2018).

World Bank. 1998. Political Economy of the Environment. Training material in environmental economics and policy. Mimeo: World Bank Institute.

Zhang, D., Stenger, A., and P. A. Harou. 2015. Policy instruments for developing planted forests: Theory and practices in China, USA, Brazil and France. *Journal of Forest Economics* 21:223–237.

ENDNOTES

1. The concept is different from landscape approach shortly presented in Chapter 3. Here it is a planning process relying on the study of the network of all sectors of the economy in a given area.

2. As, for instance, the initiative to bring ecological tourism in the Ardennes region of Europe including Belgium, Luxembourg, and France (Ardenne Grande Région, Eco-Tourisme et Attractivité AGRETA); http://collin.wallonie.be/la-grande-r-gion -moteur-de-d-veloppement

15 Motivation of Forest Service Producers

Philippe Polomé

CONTENTS

15.1 INTRODUCTION

15.1.1 TERTIARIZATION

Tertiarization implies the idea that the forestry sector has moved, partly, from the first sector (essentially, extractive production of wood or other forest products) to the third: production of services, in the sense of not only nonmaterial goods, e.g., tourism, outdoor recreation, and scenic beauty, but also more importantly, "functional" ecosystem services associated with water, soil, biodiversity, protection, and cultural values (Farcy et al. 2016). Forestry has been, possibly since time immemorial, mainly considered a first-sector activity, although one could argue that the forest has always held religious or magical symbols. Accordingly, forest management has been primarily focused on wood production (or in some cases by-products such as rubber). Even when other goals were acknowledged, they were implicit or subordinate, that is, it was assumed that optimizing wood production would also optimize these other goals. Consequently, education, ministries, associations, and even the legal system have been designed according to this goal (Nair 2004). With the realization that forests supply a large number of services to human societies, this focus has to shift, from a material goods-oriented policy to a much less material services-oriented policy.

Conceptually, forest-related regulations everywhere have either a main focus on forest and, in that case, target essentially extractive use, possibly sustainably, or a main focus on related issues such as rural development, climate, biodiversity, energy, and targeting services. These policies are not necessarily coordinated and may hamper each other (see Chapter 20). For example, management experiences exist, particularly in Latin America (Villalobos et al. 2012), that aim for higher sustainability of extractive resources and, in the process, increase some environmental and social outcomes (or avoid their deterioration). But these outcomes do not form part of the institutional agreements that essentially aim at market or marketable products. Nonmarket benefits remain side benefits: gratefully acknowledged, but unsought and uncontracted.

A key issue in the tertiarization process is that a number a forest services are not marketable or are difficult to market. Some services are "public goods."[1] Carbon sequestration and landscape come to mind. Other services are "common-pool resources."[2] That would be the case for recreation (e.g., mushroom collection) or water supply. Governance issues are compounded with the latter, especially when the resource does not fall under a single jurisdiction.

Some services can be "marketable." Recreation, for example, can be appropriated by tourism, at least partly. Other services face technical difficulties. Water supply and filtration, for example, could, in principle, be reflected in a water bill, but it is technically difficult to assign a particular value to a particular management of an area of the forest. Forest protection versus landslides and mudslides is implicitly accounted for in house prices and possibly in insurance contracts premiums, but that is not appropriable by the owner or manager of the forest. Carbon fixing might be paid to forest owners, or even to countries, but it is not done as the technical difficulties are appalling (Börner et al. 2017). The Food and Agriculture Organization of the United Nations (2014) recognizes the importance of such services and even classify some of them, such as water supply, into the basics needs, but does not include them into forest benefits because they are deemed too difficult to estimate. Indeed, the issue here is about a flow of nonappropriable services.

15.1.2 FOREST OWNERS AS SERVICES PRODUCERS

The shift from goods to services is therefore major from the point of view of the producers of those services: public and private forest owners, as well as local communities who manage forest. The present chapter is primarily concerned with behaviors and motivations of (nonindustrial) individual private forest owners (PFOs), either as a PFO or as an individual in a community, in this transition from first to third sector. Generally, fragmentation of forest ownership implies an important heterogeneity of owners, with regard to both their properties and their socioeconomic characteristics. This heterogeneity means that different owners will react differently to incentives (or other aspects of a policy) and have different objectives for their forest. In certain circumstances, it is imaginable that tertiarization renews the interest of small or medium forest owners for managing their properties, not only for services, but also for wood production. Services and wood production are not necessarily substitutes, and they may be complements, particularly in a context of sustainable forest management.

Several contemporary factors may lead to a renewal of forest management arrangements. Climate change is certainly one of them as forests almost everywhere are affected, and managers have to consider how new or evolving local conditions of temperature and amount of rain modify the ecosystem. Biological invasions may also be a driver for change in some areas. But possibly more important is the gradual realization that forest ecosystems provide unmarketable services to the community. These services may be essential (e.g., water supply) or amenities, but their provision varies depending on the forest owners management choices.

Consequently, these choices are a legitimate concern for communities who want to steer management decisions toward their preferred outcomes. The private vs. public ownerships or management regimes are not the only arrangements that are possible. As demonstrated by the classic work of Ostrom (1990), there exist many forms of alternative managements, with various types of collective properties that may be the product of a long history or of new arrangements. There are abundant well-documented examples [e.g., the study by Villalobos et al. (2012)]. Public policies include "command-and-control" or norms instruments, such as natural areas and generally limits to forestry practices, and voluntary approaches, in particular payment for ecosystem (or environmental) services (PES) that may contract on outcomes (e.g., deforestation) or on actions (e.g., good practices plans), but can rarely contract on actual environmental services due to observability (Börner et al. 2017). Numerous instances of PES exist. The European Union Agrienvironment measures, dating back to the 1980s, are broad examples, primarily compensation of costs (e.g., of set asides or environmentally friendly practices), not fee-for-service. Costa Rica famously implemented a national payment scheme in 1997 to maintain and enhance environmental service provision in the forestry sector. Following Cardenas (2004), this shift to services might concern even those local owners who sustainably manage the provision of ecosystem benefits, because there might be a need for introducing mechanisms that address externalities, that is, outside of the property.

15.1.3 MOTIVATIONS AND BEHAVIORS

It is classic economic textbook that, failing to provide incentives for the provision of public goods or common-pool resources, will result in underprovision, as the private incentive does not coincide with the public value. Following Ryan and Deci's (2000a) self-determination theory, incentives (also called motives in the psychology literature) can be intrinsic or extrinsic. Extrinsic ones lead to what is termed *controlled* actions, that is, actions motivated by a pressure that is perceived as external to the individual and can be more or less accepted ("integrated") by that individual. Such pressures are akin to penalties or rewards and encompass economic or financial incentives (fines or payments and legal obligation) to include threats, peer pressures, and feelings of moral obligation (more generally: social motives).[3] Other extrinsic motives may also exists (see Table 15.1).

Intrinsic motivations, on the other hand, lead to actions that are called "autonomous" or "self-determined," in the sense that individuals carry them out for themselves, without any perceived external pressure. Examples of such intrinsic motives include identification with the action, emotional attachement to their forest (so that

TABLE 15.1

Motivations of Service Producers

	Extrinsic		Intrinsic
Economic	Social	Others	
Financial incentives	Reputation	Utilitarian	Ethics and beliefs
Tax break	Peer pressure	Recreation	Attachment
Higher prices	Peer recognition	Ecosystem service	Control aversion
Time savings	Networking	("Warm glow")	Stewardship
Cost cutting	Emulation		Philanthropy
Fines and penalties			Reciprocity
Complexity			Information

owners will strive to preserve it in a certain state), or concern about biodiversity (so that owners manage their forest accordingly). The satisfaction (or utility) of an action is deemed higher when it is intrinsically motivated than by an extrinsic motivation. Intrinsic motives can be enhanced by fostering a feeling of competence and autonomy (through, e.g., participation or personal choices or meaningful feedback or skills training) or of relatedness (through networks).

The diversity of motives, in silvicultural and harvesting behaviors, is well documented for nonindustrial PFOs, e.g., Karppinen (1998) or Kendra and Hull (2005). These different motivations may offer additional levers for public policy beyond monetary fee-for-service or cost coverage. In terms of public policy, there are two major issues associated with individual motivations.

First, they may not coincide with what society as a whole desires. Since these services are by definition not in a market, its "clients" have little means to make their preferences known to the PFOs, who, in turn, receive little reward to cater to them. This is heterogeneity, as different PFOs may hold different ideas about the services they want to provide. The level of the incentive is a value cue: the payment made by the public agency may appear small or large compared to the other motive(s) originally felt. Such value cue may also apply to the general public. If individual values are uncertain (to the individual themselves), maybe because they are not used to think of nonmarket goods in terms of value, then any stated value might influence their own value (Crutchfield 1953).

Second, the nature of the incentive(s) that a community could offer to PFOs may change their motivation(s) for supplying the service. For example, if PFOs carry certain management actions to preserve biodiversity, because they think it is the greater good or by attachment, when a financial incentive is introduced, the nature of the management actions may evolve to become a service in exchange for money or, when an economic incentive is introduced, PFOs may find that the payment is small comparative to their intrinsic motive; they may come to realize that society, expressing itself by the public agency, hold a smaller value than they are holding. In these cases, the introduction of an economic incentive may or may not crowd out other (previous) motivations, a phenomenon that has been abundantly described since Titmuss (1970).

The existence of a crowding out effect implies that some monetary incentives may not reach their effects if they conflict with nonmonetary motivations.

Crowding out has generated much research in psychology, summarized by Deci and Ryan (1985a) and in economics, e.g., Frey (1992) and summarized by Frey and Jegen (2001). Bénabou and Tirole (2006) explore possible crowding out of economic incentives on social motivations. When economic incentives are introduced, even if intrinsic motives are not crowded out, they may alter the social image of the PFOs, e.g., their social network may now think that they are acting for the money instead of for the greater good. The perception of the owner in the eyes of others may change-and that may alter the owner's motivation to act. Festré and Garrouste (2015) intend to unify the prevailing theories in economics and psychology. Börner et al. (2017) acknowledge such crowding out in a broad literature review on PESs.

Intrinsic motives may also be affected by specific elements of a policy. For example, Cetas and Yasué (2017) indicate that when the terms of the incentives are dictated by the public agency, they may alter the owner's feeling of "sovereignty," or mastery, over their forest, thus reducing the PFO intrinsic motive. Börner et al. (2017) state that excessive control or penalty can reduce compliance or participation in PES schemes. Generally, Ryan and Deci (2000b), summarizing a body of research in psychology, suggest that policies that foster feelings of competence and of autonomy (through, e.g., participation or personal choices or by meaningful feedback and skills training) and/or of relatedness (through networks) enhance intrinsic motives.

The self-determination theory advances that any motivation for any particular action can be gradually internalized ("appropriated"), that is, the motivation can become more intrinsic as the individual identifies more with it. Following Ryan and Deci (2000a), extrinsic motivations can be internalized either because of their social component or because of self-endorsement (possibly in association with the person's other goals). Based on such internalization ideas, relying on extrinsic motivations alone is not efficient from an economic point of view insofar as policies that build environmental awareness may gradually turn external motives into internal ones, obviating the need of economic incentives.

Internalization also opens the idea of behavior persistence (Gneezy and Rustichini 2000) over a longer time period as well as policy acceptability and, in the long run, policy sustainability. In particular, it is of importance to find out under what condition(s), if any, an extrinsic motivation may be appropriated (become intrinsic), making the behavior change permanent, or on the contrary, the conditions under which the behavior reverts once the incentive disappears. Disney et al. (2013) reviewed a few real-world environmental policies as showcases for persistence and nonpersistence. Pagiola et al. (2016) found that silvopastoral practices were maintained years after a 4-year PES scheme terminated. Experimental economics has investigated this issue (Chaudhuri 2011), but it is difficult to assess how relevant the results are to PFOs because experiments are very short term and often involve students, while forestry decisions typically span and commit more than one generation.

For the rest of this chapter, we briefly summarize the research most related to the forestry context. While much effort has been dedicated to individual motivations in general context public good situations or social dilemmas, less effort has been dedicated to environmental or conservation issues, and within those much less to forestry. Section 15.2 is methodological, and it presents the two lines of evidence on the motivation of service providers. Section 15.3 synthesizes the motives, and Section 15.4 concludes with policy recommendations.

15.2 LINES OF EVIDENCE ON MOTIVATION

In this section, different lines of evidence are presented on how motivations affect PFOs' provision of nonmarket services. Relatively little of the evidence has been strictly focused on forests, as research is usually by geographical area or policy program or human behavior. The literature covers two broad lines of evidence. One is primarily concerned by individuals, centered on economics and psychology, and intends to elicit motivations using experiments and surveys. The other line of evidence approaches the issue by programs, e.g., natural areas, and is centered on conservation sciences.

15.2.1 INDIVIDUAL-CENTERED EVIDENCE

Rode et al. (2015) focus on 18 peer-reviewed empirical articles for evidence in which the use of economic incentives undermine ("crowd out") or reinforce ("crowd in") people's intrinsic motivations to engage in biodiversity and ecosystem conservation. Across the 18 articles, they identified seven psychological mechanisms leading to crowding out, including Bénabou and Tirole's (2006) crowding out of the social motivations, and four mechanisms to crowding in, albeit for two of these; it is the social motives that lead to crowd in the intrinsic motive, e.g., when social recognition leads to higher intrinsic values, or through the idea of making examples. Ten of these articles concern forest cases, out of which five are "framed field,"[4] while the other five are "natural" experiment; they are either Latin American or sub-Saharan.

In framed field experiments, subjects are invited to participate in interactive tasks that represent forest management, usually a type of contextualized resource extraction social dilemma. These are essentially short-run games played by local owners or users, sometimes with students. This is in sharp contrast with forestry decisions, which are deeply related to the growth of trees and may only show their full effects on the next generation. Although the stakes and time scale are much smaller than with actual forest decisions, the literature seems to consider that such field experiments still provide valid insight into real-scale forest decisions. There is a very large empirical literature in experimental or behavioral economics and in psychology dedicated to understanding the crowding out and crowding in effects in general contexts. Festré and Garrouste (2015) present a detailed review and intend to unify the results. This literature appears to be mainly focused on understanding the crowding out of intrinsic motivations by economic incentives; it is beyond the scope of the present chapter, in particular because the frame might be quite different from a forest context.

In the so-called natural experiments, the owners are part of a policy, e.g., being included in a natural park, in which there is, at least, an economic incentive. In other words, natural experiments correspond to actual policy interventions involving economic incentives. In essence, individuals here are surveyed on their motivations.[5]

Over the frame field and natural experiments, Rode et al. (2015) conclude that "there is a growing body of empirical studies that supports the hypothesis that economic incentives can impact on intrinsic motivations for engaging in biodiversity and ecosystem conservation." However, the evidence is not clear-cut as there are several cases in which no statistically significant effect can be found. It is possible that when the economic incentive is too small, it does more harm than good (Kerr et al. 2012), but beyond that, the conditions under which crowding out may take place remain elusive.

García-Amado et al. (2013) interviewed 731 households who participated in (primarily lowland deciduous or dry tropical forest) conservation programs in Mexico, more specifically in "La Sepultura" Biosphere Reserve. They were asked about the benefits derived from living in a Biosphere Reserve and about motive conservation. Positive answers were grouped in monetary, "utilitarian" and intrinsic reasons. Utilitarian referred here to the clean water ecosystem service, an extrinsic motive. Negative answers were "restrictions," lack of funding, and "diverse." "Restrictions" refers to control, possibly unaccepted, a factor that Ryan and Deci (2000a) cite as undermining intrinsic motivations. Such motives can be interpreted in the framework of the self-determination theory.

Polomé (2016) describes a sample of 627 PFOs from a French regional natural area. They were asked whether they had adopted any of the biodiversity-related programs available to PFOs in this area. Relatively few respondents (22%) participate in any program while less than 2% participated in more than one. The strong negative correlation between participation decisions indicates that programs are competing with each other in terms of the PFOs' time and resources. For each of those programs, respondents were queried on their motives. Table 15.2 presents all the motives that were suggested; these motives proceed from meetings with stakeholders and were later grouped into economic, social, and intrinsic.

Although the intrinsic motive did not appear to play as significant a role as the social and economic motives, there was significant crowding out between the economic and intrinsic motives. In the present context, the intrinsic motive primarily refers to feelings of attachment to the forest and concerns about the mastery of one's own practices (concerns about excessive control). Contrarily to the hypothesis in Bénabou and Tirole (2006), there is no crowding out between economic and social motives. Therefore, social motives might appear in a privileged position for being used as leverages of public programs that would focus on the social motivations through the use of nudges, see, e.g., the book by Oliver (2013). Social motives have been shown to be strong (Freeman 1996), but in this particular sample, they have been displayed by few respondents, contrary to the economic and intrinsic motives. This relative scarcity of social motives in a forestry context also occurs in the study by Rode et al. (2015) and the survey by García-Amado et al. (2013).

Côté et al. (2015) surveyed the nonindustrial PFOs of Québec on their motivations for managing their forest property. Reviewing previous literature, they broadly categorize motivations into external, corresponding mostly to financial incentives, and internal, corresponding to the PFO personal or family values, identity, cultural background, or "knowledge," a concept close to the idea of curiosity, an intrinsic motivation in the self-determination theory. Internal motivations appear to be the primer driver for management. Québec PFOs conform with these motivations;

TABLE 15.2

Detailed Motives for Adoption (or Not) of a Conservation Policy

Adoption	Nonadoption
Economic	
Economic interest	Lack of economic interest
Time savings	Time consuming
Fiscal advantages	Too complicated
Compensation for game damages	Commitments are too long or uncertain
Better insurance contracts	Owner's practices already include the program
Owner believes program is legally mandatory	Owner believes program is not legal in his/her case
Social	
Recommended by a friend/family/colleague	Lack of commitment of the other owners
Do as the others	The program does not have social recognition
Make an example	Lack of dialogue with forest owners on this program
Enter a network (professional or not)	
Intrinsic	
Attachment to forest	Owner wishes no intervention in his/her practices
Desire to bequest a better forest	Incompatible with owner ecological or management
Improving the woods	actions
Personal belief	Control by administration is too uncertain or
Protect certain species	complicated
Information need (technical, management...)	Lack of information on the program
Curiosity	

Source: Polomé, P., *Journal of Environmental Management*, 183, 212–219, 2016.

they are active in forest management activities and in forest management programs intended to encourage silvicultural work, wood certification, and conservation. Such management practices are found to be predominantly motivated by intrinsic motives (forest improvement, pleasure of working in the woods, etc.) or utilitarian (spending free time) and to a lesser degree by economic incentives (direct income from wood sale or firewood collection); social motives appear absent.

15.2.2 Conservation Program Evidence

Although not exclusively in forestry, another line of research on motivations addresses conservation projects of protected areas directly. Cowling (2014) advocates conservationists to seek insights into behavioral economics, on the ground that, at least outside of protected areas, conservation essentially intends to influence human choices toward biodiversity considerations.

Börner et al. (2017) review a large empirical and comparative literature on the effectiveness of PES schemes, possibly nested in a policy mix. Most PES schemes appear to occur in forest contexts, with some notable exceptions in the European Union and the United States where the context is more agricultural. Intrinsic motivations are not precisely defined but appear to be occasionally crowded out or crowded in by the PES scheme depending on a number of reasons. Such effects may sometimes spill over to nonparticipants in the PES. Crowding in may occur when the scheme is perceived as supportive or fair to the applicants (schemes striving for effectiveness might target specific PFOs only) and/or when there are associated development objectives in the policy mix. When a PES crowds out intrinsic motivations, it is said that the program effectiveness is lowered; unfair schemes or excessive control are cited as sources of crowding out. PES schemes do not seem to cause much social effects, although Börner et al. indicate that the number of studies directed at measuring such effects is quite small.

Cetas and Yasué (2017) consider conservation projects as a mix of policy instruments, which, in turn, can be categorized as relying more on extrinsic or intrinsic motivations. For example, regulation, alternative livelihood, and direct payments, three of the most common instruments, mostly rely on extrinsic motivations; on the other hand, education and monitoring instruments are based on intrinsic motivations. Conservation projects are also characterized by their outcomes, which can be categorized into ecological, economic, or social. The links between policy instruments and outcomes can be statistically assessed sufficiently using many conservation projects. On this basis, Cetas and Yasué (2017) estimate that the proportion of intrinsic instruments significantly and positively influence the likelihood of social, ecological, and economic success, in which success is determined on the basis of each project's self assigned goal(s).

The study by Mayer and Tikka (2006) is a qualitative review of some biodiversity management programs for PFO through economic incentives only. Using a sample of voluntary incentive programs for private forests in Europe and North America, they characterize the economic, social, and ecological dimensions of these programs and their associated forests. Although not explicitly referring to the self-determination theory, "ecological" characteristics of the program appear to hold intrinsic values for the owners, although they refer only to biodiversity and not to the various other intrinsic motives. The "social" characteristics are defined by acceptability and not the full range of network effects. Similar to Cetas and Yasué (2017), Mayer and Tikka evaluate the success of these programs with respect to their stated goals and find that important drivers of program success include an allowance for some economic productivity, a long period since program inception, and little interference from other incentive programs.

Souto et al. (2014) review a number of conservation cases and categorize their objectives according to the origin of the motivation: internal or external to the local community. Although not cast in the self-determination theory, internal motivation is associated with linking human needs with biodiversity conservation, while external motivation is associated with requests from outside the community, that need to be addressed with economic incentive and are, broadly, less likely to succeed. An interpretation of the review by Souto et al. is that conservation policies will be more likely to succeed if local communities internalize their objectives. Such internalization seems deeper when communities have a larger share in setting conservation objectives and means; this appears similar to appropriation in the self-determination theory in which PFOs might appropriate an extrinsic motive when they are more autonomous and their competence recognized.

TABLE 15.3

Policy Instruments and Associated Motives

Instruments	Intrinsic	Extrinsic
Alternative livelihoods	Designed by the community	Established by external players
Community monitoring	By local people using traditional knowledge	By park staff
Direct payments	Ability to opt out	Coercion into program
	Payments made locally	Design by external players
Education	Locally led	Ignores traditional ecological knowledge
	Empowering	
Regulations	Created and enforced locally	Imposed

Source: Cetas, E. R. and Yasué, M., *Conservation Biology*, 31, 203–212, 2017.

A feature of the conservation literature is to consider that policy instruments are not wholly associated with intrinsic or extrinsic motivations, but that each instrument generates motivations that may differ depending on the social context. For example, the "alternative livelihood" instrument may be deemed extrinsic in some conservation project when it is predetermined and developed by external players, but may be deemed intrinsic when it has been designed through stakeholder meetings. Table 15.3 summarizes the elements that help categorize an instrument as intrinsic or extrinsic. It is apparent that instruments are categorized as intrinsic if they foster competence, autonomy, and self-relatedness of the stakeholders, corresponding with the expectations of the self-determination theory.

15.2.3 DIFFICULTY OF MEASURING THE INFLUENCE OF MOTIVATION

Cetas and Yasué (2017) indicated that they "could not conclusively determine whether enhanced intrinsic motivation caused greater success." The difficulty, from a statistical point of view, is that policy instruments are categorized as intrinsic primarily on the basis that they have been designed by stakeholders (Cetas and Yasué 2017; Table 15.2). If the hypothesis being tested was instead that it is such stakeholder design that leads to the success of a policy, the conclusion would likely be the same as with the intrinsic characteristic. Thus, causality cannot be easily established. This is an issue of endogeneity. Testing causality could still be possible, using instrumental variable or other techniques as reviewed by Börner et al. (2017).

Similarly, in the framed field experiments reviewed by Rode et al. (2015), since participation is voluntary, participants may or may not have a particular profile, and this profile may be correlated with their motivations. That would be the source of a so-called selection bias in the classical sense of Heckman (1979), but it appears to be generally untested.

Natural experiments are also not free from such a selection issue since the policy under consideration has not been randomly implemented, e.g., natural parks are established in carefully selected areas, making it difficult to evaluate the effect of such policy (Ferraro 2009). Framed field experiments might also be subject to a similar selection issue, depending on how the study site is selected. This is an issue similar to evaluating the success of a drug by looking only at the treated. Failing a control group, to guarantee consistency of an estimate of a policy effect, it may be necessary, but not sufficient, to understand the reason each program has been established (or each area has been made eligible to the program). Generally, endogeneity issues may cause inconsistencies in interpreting the observed, possibly spurious, correlations, whether the analysis uses quantitative statistical tools or qualitative ones [see, e.g., the book by Wooldridge (2012)]. Börner et al. (2017) review similar issues in program evaluation.

15.3   TYPES OF MOTIVATIONS OF SERVICE PRODUCERS

Table 15.1 presents a number of motives that were collected through the literature for forestry cases. Some motives can incite for action (participation) while others cannot, but several motives act both ways, e.g., financial incentives may be deemed insufficient at low levels and thus inhibit action, but become attractive at higher levels. The economic motives not only include the classical financial incentives, e.g., PES, but may also take the form of tax breaks, higher selling prices (e.g., with labeled forest products), time or cost savings, or some form of cognitive burden ("complexity").

Social motives, although frequently detailed in the literature (Freeman 1996; Bénabou and Tirole 2006), do not seem to find such a clear-cut echo in forest cases (Rode et al. 2015; Polomé 2016; Cetas and Yasué 2017). Social motives such as peer recognition, peer pressure, or reputation are sometimes classified as intrinsic (Rode et al. 2015); however, since Ryan and Deci's (2000b) self-determination theory defines an extrinsic motive as a reward which is not the activity itself, they are classified as extrinsic here.

Other extrinsic motives are sometimes suggested, such as "utilitarian" (García-Amado et al. 2013), in which PFOs engage in conservation activity to collect recreational benefits from the managed forest or benefit from ecosystem services, e.g., protection from flood or mudslides. The famous "warm glow" of impure altruism suggested by Andreoni (1990), in which individuals participate in the common good to feel a "warm glow," is an extrinsic motive that does not fall in the other categories, but that does not appear cited in relation to forest. Ryan and Deci (2000b) introduce gradients of appropriation of extrinsic motives. Table 15.1 merely enumerates extrinsic motives that may be more or less appropriated by PFOs in any practical situation.

Intrinsic motives are diverse; d'Adda (2011) suggests two broad groups: prosocial and proenvironmental. The former relates essentially to humans, with motivations including all sorts of philanthropy (altruism and bequest), control aversion (including a sense of lower autonomy), and reciprocity. Proenvironmental motives include the attachment to the land (Polomé 2016; Figure 15.1) and stewardship (concern for wildlife well-being). Ethics (norms, including guilt), personal beliefs (beauty of the environment and existence value), and reciprocity might constitute a separate category or be considered prosocial. The quest for information (curiosity) is one of the early instances of recognized intrinsic motivation (Ryan and Deci 2000a).

FIGURE 15.1   Colorful camping tents in a forest. Extrinsic motivation, e.g., recreation, may foster intrinsic motivation, e.g., attachment. (Courtesy of Harrison Misfeldt, Creative Commons Zero (CC0) license, available at http://snapwiresnaps.tumblr.com/post /167727798156/harrison-misfeldt-free-under-cc0-10.)

15.4   CONCLUSIONS ON THE DESIGN OF FOREST POLICIES IN THE CONTEXT OF TERTIARIZATION

Tertiarization implies that PFOs are increasingly called for to make management choices that favor conservation and biodiversity or, more generally, to supply nonmarket services that have characteristics of public or collective goods. Such public component causes tertiarization to be the object not only of many public policies, but also of local arrangements. An increasingly popular policy instrument in such context is the PESs, seen as a way to motivate PFOs. Although economic incentives are undoubtedly an important, statistically significant motivation for some PFOs, and may be expected to increase their supply of the targeted ecosystem services [the study by Börner et al. (2017) is a review of their effectiveness], other PFOs, particularly smaller ones, might find them relatively unattractive. In some cases, economic incentives may crowd out (undermine) or even crowd in intrinsic motivations; the precise interaction mechanism remains unclear (Rode et al. 2015). More generally, there is abundant evidence that PFOs have diverse motives to supply ecosystem services, a diversity that may lead to refine the use of direct payment as a policy instrument.

While social motives appear to concern few PFOs, the role of intrinsic motivations in the sustainable provision of ecosystem services has been acknowledged by numerous authors. De Young (2000) argues that intrinsic motivations can be leveraged (e.g., by forest managers) to promote environmentally responsible behavior and consequently, long-term effectiveness of environmental policies. DeCaro and Stokes (2013) argue that public participation to environmental institutions can be improved by increasing acceptance of such institutions through intrinsically motivated behaviors. Dedeurwaerdere et al. (2016), in a nonforestry context, advances that owners are willing to organize themselves and with the public authority and the civil society, in response to local or global threats to biodiversity. Such forms of organizations may rely on extrinsic motivations (not only economic incentive but also regulations and generally command-and-control instruments) and on intrinsic motivations in a "mixed governance" scheme. While command-and-control instruments might achieve faster results, instruments based on intrinsic motivations may further long-term commitment. Certainly, the current understanding of such mixed governance is an active field of research. From an economic efficiency perspective, a mixed governance policy, based on intrinsic and economic motives, might reach a given objective at a lower cost than if it was based only on economic motives. Since extrinsic motives can be gradually appropriated and turned into intrinsic ones, in the long run, this might be the most efficient option.

How can a conservation policy rely more on intrinsic motivation? There are two issues: avoid undermining and fostering. A policy may undermine intrinsic motives if it contains extrinsic incentives, particularly economic ones that crowd out intrinsic ones. The precise mechanisms of crowding out do not seem clearly understood yet (Rode et al. 2015). Other elements of a policy may also undermine intrinsic motives, such as excessive control. Research in conservation studies (Cetas and Yasué 2017) advocates that participatory policies (e.g., involving the participants in collective action) will foster intrinsic motives. Börner et al. (2017) note that PES schemes that are designed not only for the actual payment of the service(s), but also for development goals (e.g., poverty alleviation), may crowd in intrinsic motivations for conservation, in participants and nonparticipants. Conversely, targeting the payment for maximum environmental effectiveness might result in perceived unfairness (e.g., by targeting only large owners or protected areas), in which case the PES might crowd out intrinsic motivations and may not be politically sustainable.

More generally, following the cognitive evaluation theory (Deci and Ryan 1985b), contexts (e.g., rewards, communications, feedback) that generate feelings of competence for an action can increase intrinsic motivation for that action. Such a feeling of competence may be brought about by elements of the policy, e.g., feedbacks that promote exploring and influencing one's environment [an idea called "effectance motivation," White (1959)] or freedom from excessive control. Cognitive evaluation theory underlines that such incentives should be accompanied by a sense of autonomy or a sense that the PFO can indeed influence his or her environment (is "in control"). Thus, PFOs must jointly experience feelings of competence and of autonomy. Self-determination theory (Ryan and Deci 2000b) further add the notion of relatedness, the need to feel connected to others, to belong to a community.

It is worth recalling that the original context of the self-determination theory of Ryan and Deci (2000a) appears to be set in an education environment in which intrinsic motivation is referred to learning. On the other hand, in forestry, the context is tertiarization, that is providing ecosystem services, through policy adoption, that have benefits for the whole of society. From an ethical point of view, intending to foster intrinsic motivation can be seen as a way to extract valuable nonmarket services from PFOs without due compensations. Although the issue does not seem to have been raised in the literature, such a posture is likely unsustainable, in the political sense that it may not receive much support. In this sense, payments for ecosystem services may be seen as more just since they not only retribute the service providers, but also constitute a means for the general public to express its value for these services.

However, as has been argued, the nonmarket nature of the ecosystem services, and their interconnections, makes estimating such value complex. Further, as discussed in the introduction to this chapter, for small or medium PFOs, PESs will likely be small and, therefore, little motivating, an issue that is compounded by the risk of crowding out. In the current process of tertiarization of the forestry sector, designing instruments that mix intrinsic values, complementarily to economic incentives, may help avoid these measurement and size issues while providing a just retribution for ecosystem services.

REFERENCES

Andreoni, J. 1990. Impure altruism and donations to public goods: A theory of warm-glow giving. *Economic Journal* 100(401):464–477.

Bénabou, R. and J. Tirole. 2006. Incentives and prosocial behavior. *American Economic Review* 96(5):1652–1678.

Bieling, C. 2004. Non-industrial private-forest owners: Possibilities for increasing adoption of close-to-nature forest management. *European Journal of Forest Research* 123(4):293–303.

Börner, J., Baylis, K., Corbera, S. et al. 2017. The effectiveness of payments for environmental services. *World Development* 96:359–374.

Cardenas, J.-C. 2004. Norms from outside and from inside: An experimental analysis on the governance of local ecosystems. *Forest Policy and Economics* 6(3–4):229–241.

Cetas, E. R. and M. Yasué. 2017. A systematic review of motivational values and conservation success in and around protected areas. *Conservation Biology* 31(1):203–212.

Chaudhuri, A. 2011. Sustaining cooperation in laboratory public goods experiments: A selective survey of the literature. *Experimental Economics* 14(1):47–83.

Côté, M. A., Gilbert, D. and S. Nadeau. 2015. Characterizing the profiles, motivations and behaviour of Québec's forest owners. *Forest Policy and Economics* 59:83–90.

Cowling, R. M. 2014. Let's get serious about human behavior and conservation. *Conservation Letters* 7(3):147–148.

Crutchfield, R. S. 1953. Correlates of individual behavior in a controlled group situation. *American Psychologist* 8(338):191–198.

d'Adda, G. 2011. Motivation crowding in environmental protection: Evidence from an artefactual field experiment. *Ecological Economics* 70(11):2083–2097.

De Young, R. 2000. New ways to promote proenvironmental behavior: Expanding and evaluating motives for environmentally responsible behavior. *Journal of Social Issues* 56(3):509–526.

DeCaro, D. and M. Stokes. 2013. Public participation and institutional fit: A social–psychological perspective. *Ecology and Society* 18(4):40.

Deci, E. L. and R. M. Ryan. 1985a. The general causality orientations scale: Self-determination in personality. *Journal of Research in Personality* 19(2):109–134.

Deci, E. L. and R. M. Ryan. 1985b. Cognitive evaluation theory. In *Intrinsic Motivation and Self-Determination in Human Behavior*, 43–85. Amsterdam: Springer.

Dedeurwaerdere, T., Admiraal, J., Beringer, A. et al. 2016. Combining internal and external motivations in multi-actor governance arrangements for biodiversity and ecosystem services. *Environmental Science and Policy* 58:1–10.

Disney, K., Le Grand, J. and G. Atkinson. 2013. From irresponsible knaves to responsible knights for just 5p: Behavioural public policy and the environment. In *Behavioural Public Policy*, ed. Oliver, A., 69–87. Cambridge, MA: Cambridge University Press.

Farcy, C., de Camino, R., Martinez de Arano, I. and E. Rojas Briales. 2016. External drivers of changes challenging forestry: Political and social issues at stake. *In Ecological Forest Management Handbook*, ed. Larocque, G., 87–105. Boca Raton, FL: Taylor & Francis Group/CRC Press.

Ferraro, P. J. 2009. Counterfactual thinking and impact evaluation in environmental policy. *New Directions for Evaluation* (122):75–84.

Festré, A. and P. Garrouste. 2015. Theory and evidence in psychology and economics about motivation crowding out: A possible convergence? *Journal of Economic Surveys* 29(2):339–356.

Food and Agriculture Organization of the United Nations. 2014. *State of the World's Forests 2014: Enhancing the Socioeconomic Benefits from Forests*. Rome: FAO.

Freeman, R. B. 1996. *Working for Nothing: The Supply of Volunteer Labor*. NBER Working Paper No. 5435. Cambridge, MA: National Bureau of Economic Research. http://www.nber.org/papers/w5435 (Accessed March 12, 2018).

Frey, B. S. 1992. Pricing and regulating affect environmental ethics. *Environmental and Resource Economics* 2(4):399–414.

Frey, B. S. and R. Jegen. 2001. Motivation crowding theory. *Journal of Economic Surveys* 15(5):589–611.

García-Amado, L. R., Ruiz Pérez, M. and S. Barrasa García. 2013. Motivation for conservation: Assessing integrated conservation and development projects and payments for environmental services in la Sepultura Biosphere Reserve, Chiapas, Mexico. *Ecological Economics* 89:92–100.

Gneezy, U. and A. Rustichini. 2000. A fine is a price. *Journal of Legal Studies* 29:1–17.

Heckman, J. J. 1979. Sample selection as a specification error. *Econometrica* 47:153–161.

Karppinen, H. 1998. Values and objectives of non-industrial private forest owners in Finland. *Silva Fennica* 32:43–59.

Kendra, A. and R. B. Hull. 2005. Motivations and behaviors of new forest owners in Virginia. *Forest Science* 51(2):142–154.

Kerr, J., Vardhan, M., and R. Jindal. 2012. Prosocial behavior and incentives: Evidence from field experiments in rural Mexico and Tanzania. *Ecological Economics* 73:220–227.

Mayer, A. L. and P. M. Tikka. 2006. Biodiversity conservation incentive programs for privately owned forests. *Environmental Science and Policy* 9(7–8):614–625.

Nair, C. T. S. 2004. What does the future hold for forestry education? *Unasylva* 216(55):3–9.

Oliver, A. 2013. *Behavioural Public Policy*. Cambridge, MA: Cambridge University Press.

Ostrom, E. 1990. Governing the Commons: The Evolution of Institutions for Collective Action. Cambridge University Press.

Pagiola, S., Honey-Rosés, J. and J. Freire-González. 2016. Evaluation of the permanence of land use change induced by payments for environmental services in Quindío, Colombia. *PLOS One* 11(3):e0147829

Polomé, P. 2016. Private forest owners motivations for adopting biodiversity-related protection programs. *Journal of Environmental Management* 183:212–219.

Rode, J., Gómez-Baggethun, G. and T. Krause. 2015. Motivation crowding by economic incentives in conservation policy: A review of the empirical evidence. *Ecological Economics* 117:270–282.

Ryan, R. M. and E. L. Deci. 2000a. Intrinsic and extrinsic motivations: Classic definitions and new directions. *Contemporary Educational Psychology* 25(1):54–67.

Ryan, R. M. and E. L. Deci. 2000b. Self-determination theory and the facilitation of intrinsic motivation, social development, and well-being. *American Psychologist* 55(1):68.

Souto, T., Deichmann, J. L., Núñez, C. and A. Alonso. 2014. Classifying conservation targets based on the origin of motivation: Implications over the success of community-based conservation projects. *Biodiversity and Conservation* 23(5):1331–1337.

Titmuss, R. M. 1970. *The Gift Relationship: From Human Blood to Social Policy.* London: Allen & Unwin.

Villalobos, R., Carrera, F., de Camino, R., Morales, J. P. and W. Flores. 2012. Construcción de cultura forestal para el desarrollo: tres historias de éxito en latinoamérica. Paper presented at the International Seminar on Scaling Up Rural Innovations.

White, R. W. 1959. Motivation reconsidered: The concept of competence. *Psychological Review* 66(5):297.

Wooldridge, J. M. 2012. *Introductory Econometrics.* Boston, MA: Cengage Learning.

ENDNOTES

1. Public goods: individuals cannot be excluded from consuming them, and consumption by one individual does not decrease that amount available to others.

2. Common pool resources: although they retain the nonexcludability quality of public goods, consumption by each agent limits the amounts others may consume.

3. It may be useful to distinguish motives and motivations. A motive is a reason for doing something, e.g., a person would (or not) do it for the money. Motivation is an incentive: the actual reward for doing something. In other words, a motive exists if the person cares about something, e.g., money, but the corresponding incentive/motivation might not be there. This distinction is useful in applied work in which forest owners may state motives that have no motivations, e.g., regret that there is no social recognition for some action as such recognition would be important for them.

4. d'Adda's (2011) experiment is actually an "artefactual field experiment" in which participants contributed money to a reforestation project.

5. Many surveys have been completed to produce typologies of PFOs, such as, e.g., Bieling (2004), but their goals did not include PFOs' motivations or how these may influence choices of provision of non-market services. Such surveys are not reviewed here.

Section IV

Globalization

16 Main Findings and Trends of Globalization

Yemi Adeyeye, Jiadong Ye, Sarah Sra,
and Louise Adam

CONTENTS

16.1 INTRODUCTION

For past 30 years, the globalization of social life has been accelerating and intensifying, constituting a societal process of primary importance, with large-scale and distant societies connected together in a variety of ways (Giddens and Sutton 2017). Scholars such as Held et al. (1999) and Scholte and Wallace (2001) identify globalization as a controversial or pluralistic word with many definitions and meanings since its appearance about 30 years ago, including some skepticism on its reality while agreeing on the importance of related changes. The commonly used definition focuses on economic perspectives and refers to the global process of market integration resulting from liberalization of trade, improvement, and expansion of transportation systems as well as the growth and progress of information and communication technologies. Other definitions are drawn to capture global issues including environmental protection, global politics, culture, and laws. Naghshpour (2008) characterizes globalization by the distribution of ideas, information, and people into a larger global network. Giddens and Sutton (2017) highlights "globalization as involving the fact that we all increasingly live in one world, as individuals, companies, groups and nations, and become ever more interdependent, interconnected and geographically mobile than ever before." As Robinson and Carson (2015) point out, in contrast to "internationalization," the term *globalization* implies a degree of purposive functional integration among geographically dispersed activities.

Potrafke (2015) asserts that defining and assessing globalization require the use of indices that encapsulate economic, social, and political aspects. Examples of such indices include the Kearney/Foreign Policy Magazine Global Index,[1] Centre for the Study of Globalization and Regionalization Globalization Inde,[2] Maastricht Globalization Index (Figge and Martens 2014), and the KOF[3] Index of Globalization developed by ETH Zurich (KOF and Dreher 2017). Potrafke (2015) further highlights that KOF index, developed for assessing a country's rate of globalization (KOF 2017), is often preferred due to its availability since 1970 for almost every country in the world.

For this very introductory chapter whose challenge is to briefly present such a broad and complex process, we are referring to the three interrelated dimensions of globalization used by KOF,[4] namely, economic, social, and political dimensions. Based on the study by Clark (2000) and the book by Norris (2000), KOF Index defines globalization as "the process of creating networks of connection among actors at the multi-continental distances, mediated through a variety of flows including people, information and ideas, capitals and goods. Globalization is a process that erodes national boundaries, integrates national economies, cultures, technologies and governance, and produces complex relations of mutual interdependence" Dreher (2006). The index measures globalization on a scale from 1 to 100. The underlying variables are divided into percentiles to smooth out outliers and reduce fluctuation over time (KOF 2017). The index benefits from a continuous improvement process and is often adapted in response to changing context.

Even though we acknowledge the relevance of the debates about the respective importance of economic, social, and political dimensions and their weight as drivers of globalizing processes (Giddens and Sutton 2017), the scope of this review is to provide a concise brief of globalization as a concept, and the dimensions and prevailing trends as effected by economic, social, and political factors, which coevolve and interact across spatial and temporal scales.

16.2 MAIN TRENDS OF GLOBALIZATION

Main trends in literature include debates about the effects and consequences of globalization. Examples include debates in land use changes (Meyfroidt et al. 2010; Lambin and Meyfroidt 2011; Li et al. 2017), human consumption habits (Hawkes 2006), gendered evolution of employment (Beneria et al. 2015), learning modalities (Fletcher 2015; Clark and Mayer 2016), and individual loss of the "sense of belonging" for a feeling of more freedom of choice (Giddens and Sutton 2017). Also, there is an emergence of debates about the potential coexistence of globalization with deglobalization, in an attempt to maintain some forms of stability across economic, social, and political lines (e.g., Lamy and Rudd 2016). Postelnicu et al. (2015) refers to deglobalization by the way of diminishing economic inter-dependence and integration between states as, for example, when the import share in gross product decreases (Van Bergeijk 2017). Example could be seen in the production sector, particularly with the United States introducing new border taxes (Carzana 2017).

As illustrated in Table 16.1, high-income countries and the member countries of the Organization for Economic Co-operation and Development, which are predomi-nantly in the global north, are higher on the globalization index (Grinin et al. 2012), European Union (EU) integration being also a key driver.

Table 16.1 demonstrates the assertion of Deese (2012) that social, economic, and political interactions of states, that is, positions of countries on the global-ization index, are primarily influenced by global connections, with implications for socioeconomic (in)equality and overall growth and development. Nonetheless, on a general scale, there has been an upward trend for globalization across the three dimensions since 1980 (Figure 16.1). The characterizations of the summa-tion of variables across the three dimensions of globalization are discussed in the following.

TABLE 16.1

Country Ranking in 2014 According to KOF Index of Globalization (N = 193)

S/No.	Country	KOF 2017
1	Netherlands	92.84
2	Ireland	92.15
3	Belgium	91.75
4	Austria	90.05
5	Switzerland	88.79
27	United States	79.73
48	Russian Federation	68.25
71	China	62.02
73	Brazil	61.40
189	Comoros	30.84
190	Micronesia	27.96
191	Equatorial Guinea	26.16
192	Eritrea	25.07
193	Solomon Islands	23.98

Source: Dreher, A., *Applied Economics*, 38, 1091–1110, 2006; With kind permission from *Springer Science+Business Media*: *Measuring Globalisation—Gauging Its Consequences*, 2008, Dreher, A. et al.; KOF, *KOF Index of Globalization*, KOF, 2017.

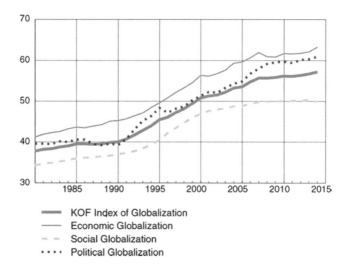

FIGURE 16.1   Evolution of KOF index of globalization since 1980. (From KOF, *KOF Index of Globalization*, KOF Swiss Economic Institute, Zurich, 2017.)

16.3   ECONOMIC GLOBALIZATION

16.3.1   INTRODUCTION

Economic globalization usually refers to a process that moves toward merging economies as a consequence of the rising worldwide movement of goods, services, and capital or the movement of labor and technology (International Monetary Fund 2008; Joshi 2009). Naghshpour (2008) clearly links globalization to economics and posits that the present-day rapid movement of goods, services, technology, and capital across borders is an outcome of the increasing internationalization of markets and declining national borders where national economies are continually adjusting as both capital and labor become progressively mobile. Economic globalization accelerated in the 1970s when countries increasingly began to exchange goods as well as the emergence of market liberalization and foreign investment by multinational companies (Greer and Hauptmeier 2016). Figure 16.2, which shows the preparation and assemblage of oak logs into containers in Belgium for export to China, exemplifies the movement of goods and services as a result of globalizations.

Economic globalization is understood by KOF as "long distance flows of goods, capital and services as well as information and perceptions that accompany market exchanges. [. . .] The economic dimension of globalization reflects the extent of cross-border trade and investment and revenue flows in relation to GDP as well as the impact of restrictions on trade and capital transactions" (KOF 2017). Economic indices are assessed using variables of actual flows (trade or portfolio investment) and restrictions (hidden import barriers or taxes on international trade) (KOF 2017). According to the KOF index, Singapore is in leading position followed by some EU countries such as Ireland (Table 16.2).

An assessment of the European economy conducted by Denis et al. (2006) demonstrates that despite the high position of the EU states in comparison to many other

FIGURE 16.2   Oak for export to China. (Courtesy of Ch. Farcy.)

TABLE 16.2

Country Ranking in 2014 According to KOF Index of Economic Globalization (*N* = 163)

S/No.	Country	KOF 2017
1	Singapore	97.77
2	Ireland	94.65
3	Luxembourg	94.06
4	Netherlands	93.06
5	Malta	91.74
54	United States	71.58
121	China	52.84
124	Brazil	52.30
125	Russian Federation	52.06
159	Iran	31.09
160	Burundi	28.45
161	Sudan	27.16
162	Ethiopia	26.9
163	Nepal	24.72

Source: Dreher, A., *Applied Economics*, 38, 1091–1110, 2006; With kind permission from *Springer Science+Business Media*: *Measuring Globalisation—Gauging Its Consequences*, 2008, Dreher, A. et al.; KOF, *KOF Index of Globalization*, KOF, 2017.

states, some states within the EU continually experience slower growth partly due to low technological advancement, in comparison to powerful economies such as Ireland and outside EU such as Singapore. More generally, as a result of greater interdependency with their neighbors, small countries will benefit from a higher ranking. Some larger economies around the world, such as China and Brazil, remain midranked. Similar to other dimensions of globalization, it is noticeable that countries in the global south are more represented at the lower end of the index (as shown in tables included in this chapter). Deese (2012) argues that the disenfranchisement of some states through the lack of inclusion in decision-making processes that have global economic consequences further entrenches the unequal distribution of the effects of economic globalization. That is, countries who are least involved or represented in global decision-making are more vulnerable to the negative effects of globalization, such as during the 1997–1999 international financial crisis which created more setbacks for a lot of emerging economies. Conversely, states that continually dominate the global decision-making platforms are able to disproportionately benefit from the positive effects of globalization. An illustration is the historic dominance of the G-7 countries, including the United States, France, Germany, and the United Kingdom.

16.3.2 TRENDS

Since the 1980s, capital flows on a global scale has been drastically increasing due to trade in equity and debt market. As foreign investors have access to local equity securities and domestic investors are able to transact foreign equity securities, it increased overseas investment and thus nurtured the integration of global financial markets (Bekaert et al. 2005). Reduced trade barriers such as import tariffs also trigger foreign investments. A more internationalized market creates a channel for diversified fund allocation across countries by which the funds can be effectively assigned to the most profitable investment for high returns. This, in turn, facilitates the expansion of domestic financial systems worldwide (De Gregorio 2006). Moreover, the emerging "electronic economy," in turn, underpins the broader economic globalization. "Banks, corporations, fund managers and individual investors are able to shift funds internationally with the click of a mouse" (Giddens and Sutton 2017).

Moreover, the World Bank (2017) states that the value of trade increased from 24% in 1960 to 58% in 2015. The "value of trade" epitomizes linkages between globalization and economics. The value of trade is the sum of exports and imports of goods and services, which is measured as a share of gross domestic product (GDP) and as a percentage of world GDP.

The increasing level of activities of transnational companies is another great example of trends of economic globalization. Transnational companies account for two-thirds of all world trade (Giddens and Sutton 2017). The combined sales of the world's largest 500 transnational companies totaled $14.1 trillion, nearly half of the value of goods and services. As a consequence, globalization continually coevolves with transnational economic interactions.

There is also an emerging trend in terms of changes in economic growth as indicated by recent developments in the GDP of many countries in the global south such

FIGURE 16.3 Annual GDP growth for 2016. (From International Monetary Fund, *World Economic Outlook (October 2017)—Real GDP Growth*, International Monetary Fund, Washington, DC, 2017.)

as India, China, Ethiopia, and Tanzania (Figure 16.3). GDP is a measure of the total value, in economic terms, of total goods and services produced by a country during a specific period. This is usually measured annually.

Figure 16.3 shows that the annual percentage change in the GDP for 2016, relative to the previous year. Although the figure does not provide information about changes in relative to the bigger economies, it provides some insight into potential trends and economic trajectory of emerging markets and developing economies—one of the derivatives of economic globalization.

16.4 SOCIAL GLOBALIZATION

16.4.1 INTRODUCTION

Social globalization can be thought of as a time–space compression, which has the ability to expand and intensify social relations on a global scale with tangible and visible effects (Steger 2014). Societies across the globe are continuously becoming connected in real time irrespective of language barriers (Gunter and Van der Hoeven 2004). Social globalization emerges based on the impacts of economic and political dimensions of globalization on society. Factors including transnational economic interactions, as addressed earlier, result in higher labour mobility and reduction in barriers for people to travel, study, and work in other countries (ILO 2003). In general, social globalization is attributed to various types of interpersonal contact that occurs on a daily basis.

Giddens and Sutton (2017) highlight the crucial role of the development of information and communications technologies "which have intensified the speed and scope of interactions between people all over the world, thereby creating genuinely global shared experiences—one prerequisite of a global society." Globalization and the Internet create channels for information flow across disciplines at local, national, and international scales, which has initiated social change and altered the way in which society receive, utilize, and perceive knowledge (Sheppard 2014). Social globalization allows increased opportunities for cultural contact among people worldwide while providing the easy flow of information from and about people and events in distant places, thereby resulting in the higher awareness of the interconnectedness of common purpose and responsibility in front of shared global concerns.

The KOF index of globalization measures the social dimension of globalization, defined as "the spread of ideas, information, images and people" (KOF 2017), in three categories. Firstly, it assesses cross-border personal contacts in the form of telephone calls, letters, and tourism flows as well as the size of the resident foreign population. Secondly, cross-border information flows are measured in terms of access to the Internet, TV, and foreign press products, and thirdly, the index attempts to measure the cultural proximity to the global mainstream by means of the number of McDonald's and Ikea branches as well as book imports and exports in relation to GDP (KOF 2017).

Table 16.3 on social globalization shows a similar pattern with economic globalization ranking; Singapore comes first, followed by some European countries, and the ranking ends with some African countries.

Consequentially, this pattern that accentuates the north–south divide has led to the rise of critics and skeptics of globalization. One of the implications is the

TABLE 16.3

Country Ranking in 2014 According to KOF Index of Social Globalization (N = 200)

S/No.	Country	KOF 2017
1	Singapore	91.61
2	Switzerland	91.13
3	Ireland	90.99
4	Netherlands	90.71
5	Austria	90.62
30	United States	78.82
52	Russian Federation	66.58
81	China	54.23
106	Brazil	45.58
196	Myanmar	17.87
197	Somalia	17.7
198	Central African Republic	17.54
199	Tanzania	17.52
200	Democratic Republic of the Congo	15.83

Source: Dreher, A., *Applied Economics*, 38, 1091–1110, 2006; With kind permission from *Springer Science+Business Media: Measuring Globalisation—Gauging Its Consequences*, 2008, Dreher, A. et al.; KOF, *KOF Index of Globalization*, KOF, 2017.

emergence of scholarship on globalization asymmetries. This include Yotopoulos (2006), Okome and Vaughan (2011), Deese (2012), and Aderibigbe et al. (2016). Deese (2012) highlights that processes of economic globalization such as financial integration tend to exacerbate social divisions, tensions, and disputes in the South. This is as a result of the continuous importation (some are forced and others invited) of neoliberal ideals in ways that unfavorably interact with existing social relations in societies and political structures in the south, which are not able to effectively manage the interference of external actors and connections that economic globalization introduces—as seen in the majority of African and Latin American countries. Furthermore, Aderibigbe et al. (2016) posit that the majority of the countries in the global south, particularly African states, have been alienated and exploited through the ideals of globalization. Similarly, Okome and Vaughan (2011) assert the need to systematically examine globalization processes from new fronts, including decolonization and postcolonial perspectives with a view of revealing erroneous assumptions of conventional theories, which in part does contribute to the prevailing status of colonized states.

16.4.2 TRENDS

The emergence of the Internet is central to the evolution of social globalization. A prevailing trend is the exponential growth of Internet usage. About 50% of the world population access and use the Internet (Table 16.4). Shortening space and time, the Internet increases the ease and speed at which information could be shared and influenced internationally. "Widespread use of the Internet and mobile phones is deepening and accelerating processes of globalization as more people are interconnected, including people in places previously isolated or poorly served by traditional communication" (Giddens and Sutton 2017).

TABLE 16.4
Spread of the Internet in the World in 2017

Region	Number of Inhabitants	Number of Internet Users	Percentage of Total Internet Users	Growth of Internet Users since 2000 (%)	Percentage of Inhabitants Being Users of Internet
Africa	1,246,504,865	388,376,491	10.0	8,503.1	31.2
Asia	4,148,177,672	1,938,075,631	49.7	1,595.5	46.7
Europe	822,710,362	659,634,487	17.0	527.6	80.2
Latin America/ Caribbean	647,604,645	404,269,163	10.4	2,137.4	62.4
Middle East	250,327,574	146,972,123	3.8	4,374.3	58.7
North America	363,224,006	320,059,368	8.2	196.1	88.1
Oceania/Australia	40,479,846	28,180,356	0.7	269.8	69.6
World total	7,519,028,970	3,885,567,619	100.0	976.4	51.7

Source: Internetworldstats, June 30, 2017. World Internet Usage and Population Statistics, Miniwatts Group.

Today, the World Wide Web has enabled a greater amount of information to be distributed to a wider audience with greater accessibility compared to printed documents (Nageswara Rao and Babu 2001). The Internet allows information to be accessed from a variety of mobile devices including laptops and smartphones, while social media and networking continue to have an impact on how information is obtained and subjectively interpreted. The innovation and improvements of communication technologies not only fasten the information exchange, but also diversify the means and media for information flow. With modern communication technologies such as cell phones, social media, and the Internet connection, the world is able to share a larger stock of information from various sources across geographical regions, social classes, and languages (Stromquist and Monkman 2014). Their power in forging, influencing, and strengthening social representations, for example, on environmental matters, is undeniable (see Chapters 7 and 10). Furthermore, as an adaptation to the trend as well as to make an efficient use of the massive information, big data analytics is increasingly popular in the analysis for social issues nowadays. The concept of "big data" refers to the information assets that need high volume, velocity, and variety in order to be analyzed by certain technologies and methods and then achieve its transformation into value (De Mauro et al. 2014). Big data not only exaggerate the flow, stock, and computation of information from the quantity perspective, but also generate qualitative changes for data analysis under globalization. Different from conventional data analysis, big data are able to incorporate disorganized sources especially for grasping and storing the real-time information that may only appear on the social media at a relatively short duration (Hilbert 2016).

16.5 POLITICAL GLOBALIZATION

16.5.1 INTRODUCTION

The political dimension of globalization is demonstrated through the emergence of new modes of governance at the global level that seek to reinforce corporations among nations or regions of the world on challenges and concerns having global scope (Global Policy Forum 2017). A revealing element is the present-day understanding of public goods and the general interests around it. This includes the way that natural resources such as forests and water are understood, contested, and governed by different actors that cut across local, national, and international levels.

Before the emergence of political globalization was an era of territorial sovereignty that dated back to the Treaty of Westphalia in 1648. The agreement was endorsed and recognized by various European leaders and respected the laws specific to each territory (Baylis et al. 2013). According to Finger and Tamiotti (1999), the challenges of industrial development question the effectiveness of territorial sovereignty, which results in the gradual emergence of the globalization of politics. The shared nature of environmental and social matters across spatial scale contributed to the globalization of politics. For example, environmental issues such as climate change are a challenge with global implications, as illustrated by the mobility of climate refugees across borders. This evolution toward globalization, which also implies a less sharp boundary between public and private rights and responsibilities (internationally and nationally), leads to huge changes in decision-making process, policy making, and policy implementation. Concurrently, the word *governance* becomes a cornerstone of a new conceptual umbrella. Arts and Visseren-hamakers (2012) refer to governance as arrangement schemes for provisioning of public goods and solving public problems as well as distinguishing different roles for the state including its absence.

According to KOF (2017) assessment, political globalization "characterizes the diffusion of government policies." Political globalization is measured in terms of the number of foreign embassies resident in a country, the number of international organizations of which the country is a member, the number of UN peace missions in which the country has been engaged, and the number of bilateral and multilateral agreements the country has concluded since 1945. As shown in Table 16.5, European countries are ranked at the top in 2014. This is due to high influence of European states in international organizations and their participation in international missions (Mr Globalization 2011; Statista 2017). France occupies the top rank in this subindex, relegating last year's front-runner Italy to second place. Belgium remains in third place and Brazil, very active in the international scene, at the ninth. The United States, China, and the Russian Federation record their best score. At the bottom of the field are small island states and archipelagos (Table 16.5). Compared to the previous year, the degree of political globalization rose in 2014 (KOF 2017).

TABLE 16.5

Country Ranking in 2014 According to KOF Index of Political Globalization (*N* = 207)

S/No.	Country	KOF 2017
1	France	97.29
2	Italy	97.25
3	Belgium	95.79
4	Sweden	95.56
5	Netherlands	95.41
9	Brazil	94.30
19	United States	91.43
20	Russian Federation	91.34
44	China	84.26
203	Guam	2.09
204	American Samoa	2.09
205	Isla of Man	1.27
206	Northern Mariana Islands	1.27
207	Channel Islands	1.00

Source: Dreher, A., *Applied Economics*, 38, 1091–1110, 2006; With kind permission from *Springer Science+Business Media: Measuring Globalisation—Gauging Its Consequences*, 2008, Dreher, A. et al.; KOF, *KOF Index of Globalization*, KOF, 2017.

16.5.2 TRENDS

Increasingly, politics and governance are becoming "multilevel" or "multiscalar" across substate (municipal and provincial) bodies, suprastate agencies (such as the EU), state organs, as well as international organizations such as the World Bank and the International Monetary Fund (Scholte and Wallace 2001). In essence, governance has been able to develop at a higher level through political integration schemes (Global Policy Forum 2017). Furthermore, various private sector and civil society actors have taken on regulatory roles. In this line of argument, states survive under globalization, but they are no longer the sole—and in some cases not even the principal—site of governance. This trend results in what Rosenau and Czempiel (1992) popularly describes as a transition from government to governance.

Another dominant trend is embodied by the systems of governance in powerful countries such the United States and the United Kingdom. Countries are gradually becoming self-centered with their governance approach—what scholars describe as protectionism (Nangle 2017). Prevailing protectionist political systems are exemplified by Brexit in the United Kingdom as well as the trade policies of the United States, through Trump's presidency that discourages free trade programs.

The Continued rise of militarism and international conflicts is also a trend which results from political globalization (Acemoglu and Yared 2010). Lutz and Lutz (2014) assert that as a result of political difficulties which are enabled by factors such as market capitalism that undermines the structure of local economies and the consequential violent reactions of marginalized groups, international conflicts have increased.

16.6 CONCLUDING REMARK

The KOF assessment gave us a broad and useful picture of the complex process of globalization, which has impacts in varying degrees on every corner of the planet. Nevertheless, we acknowledge the relevance of comments by authors such as Ebenthal (2007), questioning the use of single indices to assess globalization without taking into account the levels of development of countries and thus underestimating less visible social and technological progresses in less developed countries.

Globalization, in its economic, social, and political dimensions including their interactions, evolves in forms and meanings as the world changes. Jogdand and Michael (2009) accentuate this claim by referring to globalization as a double-edged sword, with positive and negative impacts and consequences at the same time. From the social dimension, for example, more and more voices are emerging to denounce the damage that the abusive use of "screens" or social networks is likely to have to children and young people (Bach et al. 2013). There are also debates about the real benefits of some kind of tourism in view of its counterpart in environmental or cultural terms (Giddens and Sutton 2017). Consequentially, society continually grapples with the challenge of how to keep pace with globalization-induced changes.

Across the three dimensions described earlier, globalization remains contested (Hosseini 2010). Historic occurrences such as the 2008 economic crisis partly impact the contestation. Casarões (2017) identifies that the consequences of the crisis include deep recession and massive unemployment, right-wing nationalism, populism, and protectionism. Nonetheless, Raube (2017) suggests that globalization has the potential to alleviate poverty and rising standards of living in Asia, Latin America, and Africa. Samimi and Jenatabadi (2014) also suggest that the advantages gained from trade, cross-national supply chains, and global capital flows result from economic globalization. Consequentially, even if globalization is contested and will place new demands on governance, Yin (2017) and Lai (2017) argue that international trade will continue among nations. Similarly, Dadush (2017) demonstrates that despite tensed political climates around the world, globalization will persist as a result of protectionist agendas.

Finally, from our review, we agree with the position that globalization is not a linear or one-way process leading to a clearly identified endpoint for the global economy, society, and policy (Giddens and Sutton 2017). Corrections and adjustments will undoubtedly take place depending on available leeway and according to the lessons learned from the experiences that globalization has made possible.

REFERENCES

Acemoglu, D. and P. Yared. 2010. Political limits to globalization. *American Economic Review* 100(2):83–88. http://www.jstor.org.ezproxy.library.ubc.ca/stable/27804968 (Accessed March 10, 2018).

Aderibigbe, G., Omotoye, R. W., and L. B. Akande. 2016. *Contextualizing Africans and Globalization: Expressions in Sociopolitical and Religious Contents and Discontents.* Maryland: Rowman & Littlefield. https://ebookcentral.proquest.com/lib/ubc/detail .action?docID=4729886 (Accessed March 10, 2018).

Arts, B. and I. Visseren-hamakers. 2012. Forest governance: A state of the art review. In *Forest-People Interfaces: Understanding Community Forestry and Biocultural Diversity*, 241–257. Wageningen: Wageningen Academic Publishers.

Bach, J., Houdé, O., Léna, P., and S. Tisseron. 2013. *L'enfant et les écrans, un avis de l'Académie des Sciences*, Paris: Le Pommier.

Baylis, J., Smith, S., and P. Owens. 2013. The globalization of world politics: An introduction to international relations. In *The Globalization of World Politics.* https://doi .org/10.2307/2604959

Bekaert, G., Harvey, C., and C. Lundblad. 2005. Does financial liberalization spur growth? *Journal of Financial Economics.* https://doi.org/10.1016/j.jfineco.2004.05.007

Beneria, L., Berik, G., and M. Floro. 2015. *Gender, Development and Globalization: Economics as if All People Mattered.* Abingdon: Routledge.

Carzana, A. 2017. *The Underestimated Perils of Deglobalisation.* Massachusetts: Columbia Threadneedle Investments. http://www.columbiathreadneedle.co.uk/media/11136986 /en_the_underestimated_perils_of_deglobalisation.pdf (Accessed November 21, 2017).

Casarões, G. 2017. *Future of Globalization: Strengthening Social Inclusion.* New Haven, CT: Yale School of Management: Global Network Perspectives. http://gnp.advancedmanage ment.net/article/2017/04/future-globalization-strengthening-social-inclusion (Accessed November 16, 2017).

Clark, W. C. 2000. Environmental globalization. In *Governance in a Globalizing World*, eds. Nye, J. and J. D. Donahue, 86–108. Washington, DC: Brookings Institution Press.

Clark, R. C. and R. E. Mayer. 2016. *E-Learning and the Science of Instruction: Proven Guidelines for Consumers and Designers of Multimedia Learning.* Hoboken, NJ: John Wiley & Sons.

Dadush, U. 2017. The future of globalization summary. *The Africa Portal.* http://www.ocppc .ma (Accessed February 5, 2018).

Deese, D. A. 2012. *Globalization: Causes and Effects.* Abingdon: Routledge.

De Gregorio, J. 2006. *Financial Integration, Financial Development and Economic Growth.* Chile: Centre for Economics. http://citeseerx.ist.psu.edu/viewdoc/download (Accessed November 1, 2017).

De Mauro, A., Greco, M., and M. Grimaldi. 2014. What is big data? A consensual definition and a review of key research topics. *AIP Proceedings* 1644:97–104.

Denis, C., Mc Morrow, K., and W. Röger. 2006. *Globalisation: Trends, Issues and Macro Implications for the EU Directorate-General for Economic and Financial Affairs.* EU Economic Papers, No 254, 97. Belgium: Directorate-General for Economic and Financial Affairs Publications. http://ec.europa.eu/economy_finance/publications/pages /publication668_en.pdf (Accessed February 19, 2018).

Dreher, A. 2006. Does globalization affect growth? Evidence from a new index of globalization. *Applied Economics* 38(10):1091–1110.

Dreher, A., Gaston, N., and P. Martens. 2008. *Measuring Globalisation—Gauging Its Consequences.* New York: Springer.

Ebenthal, S. 2007. Messung von Globalisierung in Entwicklungsländern: Zur Analyse der Globalisierung mit Globalisierungsindizes. *Berichte aus dem Weltwirtschaftlichen Colloquium der Universität Bremen* 104:1–36.

Figge, L. and P. Martens. 2014. "Globalisation continues: The Maastricht Globalisation Index revisited and updated." *Globalizations.* Vol. 11: Iss. 2, Article 5. Available at: http:// services.bepress.com/globalizations/vol11/iss2/art5 (Accessed February 19, 2018).

Finger, M. and L. Tamiotti. 1999. New global regulatory mechanisms and the environment— The emerging linkage between the WTO and the ISO. *IDS Bulletin, Institute of Development Studies* 30(3):8–15.

Fletcher, A. 2015. Youth and Globalization. https://freechild.org/youth-and-globalization/ (Accessed November 7, 2017).

Giddens, A. and P. W. Sutton. 2014. *Essential Concepts in Sociology.* Hoboken, NJ: John Wiley & Sons.

Global Policy Forum. 2017. *Globalization of Politics.* New York: Global Policy Forum https:// www.globalpolicy.org/globalization/globalization-of-politics.html (Accessed June 27, 2017).

Greer, I. and M. Hauptmeier. 2016. Management whipsawing: The staging of labor competition under globalization. *Industrial and Labor Relations Review* 69(1):29–52.

Grinin, L., Ilya, I., and K. Andrey. 2012. *Globalistics and Globalization Studies.* Moscow: Moscow State University, Publisher Teacher.

Gunter, B. G. and R. Van der Hoeven. 2004. The social dimension of globalization: A review of the literature. *International Labour Review* 143(1–2):7–43.

Hawkes, C. 2006. Uneven dietary development: Linking the policies and processes of globalization with the nutrition transition, obesity and diet-related chronic diseases. *Globalization and Health* 2:4. https://globalizationandhealth.biomedcentral.com/articles /10.1186/1744–8603-2-4 (Accessed February 2, 2018).

Held, D., Goldblatt, D., McGrew, A. and J. Perraton. 1999. *Global Transformations: Politics, Economics and Culture.* Cambridge, MA: Polity.

Hilbert, M. 2016. Big data for development: A review of promises and challenges. *Development Policy Review* 34(1):135–174.

Hosseini, S. A. 2010. *Contested Meanings of Globalization.* Australia: Global Alternatives. https://globalalternatives.wordpress.com/2010/03/19/contested-meanings-of -globalization/ (Accessed February 2, 2018).

International Labour Organization. 2003. *The Social Dimension of Globalization.* Geneva: International Labour Organization. http://www.ilo.org/public/english/wcsdg/globali /globali.htm (Accessed October 31, 2017).

International Monetary Fund. 2008. *Globalization: A Brief Overview.* Washington, DC: International Monetary Fund. https://www.imf.org/external/np/exr/ib/2008/053008 .htm (Accessed October 28, 2017).

International Monetary Fund. 2017. *World Economic Outlook (October 2017)—Real GDP Growth.* Washington, DC: International Monetary Fund. http://www.imf.org/external/data mapper/NGDP_RPCH@WEO/OEMDC/ADVEC/WEOWORLD (Accessed February 21, 2018).

Internetworldstats. 2017. World Internet Usage and Population Statistics. Miniwatts Marketing Group. https://www.internetworldstats.com/stats.htm (Accessed June 30, 2017).

Jogdand, P. G. and S. M. Michael. 2009. *Globalization and Social Movements: Struggle for a Humane Society.* New Delhi: Rawat Publications.

Joshi, R. M. 2009. *International Business.* New Delhi and New York: Oxford University Press.

KOF. 2017. *KOF Index of Globalization.* Zurich: KOF Swiss Economic Institute http:// globalization.kof.ethz.ch/media/filer_public/2017/04/19/variables_2017.pdf (Accessed June 29, 2017).

KOF and A. Dreher. 2017. *Does Globalization Affect Growth? Empirical Evidence from a New Index, 2017 Variables and Weight.* Zurich: KOF Swiss Economic Institute. http:// globalization.kof.ethz.ch/media/filer_public/2017/04/19/variables_2017.pdf (Accessed June 29, 2017).

Lai, E. L.-C. 2017. *Future of Globalization: The Global Value Chain.* New Haven, CT: Yale School of Management: Global Network Perspectives. http://gnp.advancedmanagement .net/article/2017/04/future-globalization-global-value-chain (Accessed February 8, 2017).

Lambin, E. F. and P. Meyfroidt. 2011. Global land use change, economic globalization, and the looming land scarcity. *Proceedings of the National Academy of Sciences of the United States of America* 108(9):3465–3472.

Lamy, P. and K. Rudd. 2016. *Interview: The Future of Globalization.* Cologny: World Economic Forum. http://reports.weforum.org/outlook-2013/the-future-of-globalization/#view/img-8 (Accessed September 4, 2017).

Leiner, B., Cerf, V., Clark, D. et al. 1997. Brief History of the Internet. Washington DC: Internet Society. Retrieved from https://www.internetsociety.org/internet/history-internet /brief-history-internet/%0Ahttp://www.isoc.org/internet/history/.

Li, L., Liu, J., Long, H., De Jong, W., and Y.-C. Youn. 2017. Economic globalization, trade and forest transition-the case of nine Asian countries. *Forest Policy and Economics* 76:7–13.

Lutz, B. J. and Lutz, J. M. 2014. Economic, social and political globalization and terrorism. *Journal of Social, Political, and Economic Studies* 39(2):186–218.

Meyfroidt, P., Rudel, T. K., and E. F. Lambin. 2010. Forest transitions, trade, and the global displacement of land use. *Proceedings of the National Academy of Sciences of the United States of America* 107(49):20917–20922.

Nageswara Rao, K. and K. H. Babu. 2001. Role of librarian in Internet and World Wide Web environment. *Informing Science* 4(1):25–34.

Naghshpour, S. 2008. *Globalization: Is It Good or Bad.* Hattiesburg, MS: University of Southern Mississippi. http://globalization.icaap.org/content/special/Naghshpour.html (Accessed June 20, 2017).

Nangle, T. 2017. The Underestimated Perils of Deglobalisation for Investment Professionals Only. Columbia Threadneedle Investments. Massachusetts: Columbia Threadneedle Investments. http://www.columbiathreadneedle.co.uk/media/11136986/en_the_under estimated_perils_of_deglobalisation.pdf (Accessed November 7, 2017).

Norris, P. 2000. *Global Governance and Cosmopolitan Citizens.* Washington, DC: Brookings Institution Press.

Okome, M. O. and O. Vaughan. 2011. Transnational Africa and globalization: Introduction. In *Transnational Africa and Globalization*, 1–15. New York: Palgrave Macmillan US.

Postelnicu, C., Dinu, V., and D.-C. Dabija. 2015. Economic deglobalization—From hypothesis to reality. *Economics and Management* 18:1–14.

Potrafke, N. 2015. The evidence on globalisation. *World Economy* 38(3):509–552.

Raube, K. 2017. *Future of Globalization: Embracing Opportunity*. New Haven, CT: Yale School of Management: Global Network Perspectives. http://gnp.advancedmanagement .net/article/2017/04/future-globalization-embracing-opportunity (Accessed February 9, 2018).

Robinson, G. M. and D. A. Carson. 2015. *Handbook on the Globalization of Agriculture*. Cheltenham; Northampton, MA: Edward Elgar.

Rosenau, J. and E. Czempiel. 1992. *Governance without Government: Order and Change in World Politics*. Cambridge, MA: Cambridge University Press.

Samimi, P. and H. S. Jenatabadi. 2014. Globalization and economic growth: Empirical evidence on the role of complementarities. *PLoS ONE* 9(4):e87824.

Scholte, J. A. and I. Wallace. 2001. Globalization: A critical introduction. *Environment and Planning C: Government and Policy* 19(6):928.

Statista. 2017. *Top 50 Countries in the Globalization Index 2017*. Hamburg: Statista. https:// www.statista.com/statistics/268168/globalization-index-by-country/ (Accessed April 8, 2017).

Steger, M. B. 2014. Globalization. In *The Encyclopedia of Political Thought*, 1488–1499. Indianapolis: Wiley-Blackwell.

Stromquist, N. P. and K. Monkman. 2014. Defining globalization and assessing its implications on knowledge and education. In *Globalization and Education: Integration and Contestation across Cultures*, 1–25. London: R&L Education.

Van Bergeijk, P. A. G. 2017. "One is not enough! An economic history perspective on world trade collapses and deglobalization" ISS Working Papers—General Series 98695, International Institute of Social Studies of Erasmus University Rotterdam (ISS), The Hague.

World Bank. 2017. World Trade (% of GDP) | Data. Washington, DC: World Bank https:// data.worldbank.org/indicator/NE.TRD.GNFS.ZS?end=2015&start=1960&view=chart (Accessed November 24, 2017).

Yin, Z. 2017. *Future of Globalization: Cross-Cultural Engagement*. New Haven, CT: Yale School of Management: Global Network Perspectives. http://gnp.advancedmanage ment.net/article/2017/04/future-globalization-cross-cultural-engagement (Accessed February 4, 2018).

Yotopoulos, P. A. 2006. *Asymmetric Globalization: Impact on the Third World (No. 270)*. Stanford, CA: Stanford Center on Global Poverty and Development. https://global poverty.stanford.edu/sites/default/files/publications/270wp.pdf (Accessed March 11, 2018).

ENDNOTES

1. http://foreignpolicy.com/.
2. https://warwick.ac.uk/fac/soc/pais/research/researchcentres/csgr/index/.
3. KOF coming from "Konjunkturforschungsstelle," which means business cycle research institute.
4. http://globalization.kof.ethz.ch/.

17 Financialization and the Forestry Sector

Patrick Meyfroidt

CONTENTS

17.1 INTRODUCTION

As seen throughout this book, the forestry sector at the global level and in many countries is currently influenced by several trends that affect all spheres of human activities. Urbanization, tertiarization, and globalization of societies and economies manifest themselves through multiple processes.

Over the recent decades, starting in the 1970s, three broad trends of transformations of capitalism have been identified (Epstein 2005): (1) neoliberalism, i.e., the decreasing influence of governments and public actors and the corresponding rise of influence of markets and private actors in economies and societies; (2) globalization, i.e., the increase in speed, reach, and amounts of flows of commodities, information, people, norms, and money across the World; and (3) financialization, which has been defined in various ways but relates to the increasing prominence and influence of finance and finance activities in capitalist economies.

Many works have focused on the trends and effects of neoliberalism and globalization in its trade component on the forestry sector. These three dimensions are strongly interdependent (Epstein 2005) and influence the forestry sector in conjunction. Although a few recent works have highlighted the importance of financial markets in understanding and governing changes in social–ecological systems (Galaz et al. 2015), the specific dynamics of financialization have received little attention in forestry and more broadly in agriculture and other land use sectors, as well as in the human–environmental change or social–ecological systems literature, compared to neoliberalization and globalization. This chapter presents an overview and discusses the effects of financialization in the forestry sector. This chapter starts by a general background on what financialization means and how it has been unfolding over the recent decades in the general organization of capitalist economies, distinguishing three major dimensions of financialization. Then, this chapter discusses the dynamics and specificities of financialization in the forestry sector, as well as more broadly in the land-based commodity sectors including agriculture, where this process has been stronger and studied more. This chapter then shows that these three dimensions of financialization occur in forestry, through three main processes: (1) its effects on commodity markets and in particular on prices and prices volatility; (2) its relation with changes in land ownership and the related effects; and (3) changes in behaviors of forestry companies, with an increasing emphasis on financial activities, including through carbon forestry.

17.2 FINANCIALIZATION: TRANSFORMATIONS OF FINANCE AND THEIR ROLE IN CAPITALISM

In the loosest sense, financialization refers to "the increasing role of financial motives, financial markets, financial actors and financial institutions in the operation of the domestic and international economies" (Epstein 2005). "Financial" here refers to "activities relating to the provision (or transfer) of liquid capital in expectation of future interests, dividends, or capital gains" (Krippner 2005). A more analytical approach identifies several definitions, which correspond to a set of distinct although strongly interrelated dimensions of the phenomenon, i.e., financialization as a mode of capital accumulation, a change in the value orientation of corporations, and a growth and complexification of financial activities (Krippner 2005; Dore 2008; van der Zwan 2014).

First, financialization is defined as "a pattern of accumulation in which profits accrue primarily through financial channels rather than through trade and commodity production" (Krippner 2005). In this process, the financial industry has increased its share of GDP in advanced economies, but in addition, nonfinancial corporations have also increasingly derived profits from financial activities rather than from productive investment (van den Zwan 2014). Financialization has been argued to result in slower aggregate economic growth, by leading to shorter planning horizons, a declining allegiance of stakeholders to long-term corporate goals, and a large increase in the percentage of cash flow paid to financial market agents (Crotty 2003). Three forms of responses to the latter process from nonfinancial corporations were cutting wages and benefits to workers, engaging in fraud and deception to increase apparent profits, and moving into financial operations to increase profits (Epstein 2005).

Second, financialization corresponds to the rise of "shareholder value" as a mode of corporate governance. This refers to the increasingly strong assertion of the property rights of owners and the monetary returns to these shareholders as transcending all other forms of social accountability for business corporations (Froud et al. 2000; van den Zwan 2014). This trend has been supported by governments promoting the idea of an "equity culture," i.e., a society where every citizen would hold shares in market activities, for different internal and geopolitical motives (Dore 2008). This has also been called the "financialization of the everyday" (van den Zwan 2014).

Several legislative changes toward deregulation in the 1970s and 1980s, particularly in the United States, also encouraged public investment funds, such as pension funds, to diversify their investments away from their traditional reliance on low risk, fixed-income securities such as government and corporate bonds to more risky investments such as stocks (Gunnoe and Gellert 2011; Gunnoe 2014). This shift bears important consequences on power and control of economic decisions in modern corporations (Krippner 2005). In the early twentieth century, managers increasingly held power within corporations because of the rise of stockownership, which detached stakeholders from the daily engagement with business operations. Over the recent decades, through financialization, finance and nonfinance activities are increasingly tied in joint operations, and as discussed earlier, finance activities are increasingly important sources of profits for nonfinancial firms. With the rise of the "shareholder value" model, the autonomy of managers has strongly reduced. Managers find themselves under close supervision of a board that represents and exclusively promotes the interests of shareholders. Various considerations that may have represented important factors in the decision-making of managers, such as social responsibility, service to customers, responsibility toward employees, personal reputation, and ethic, or the sense of contributing to the development of a region, city, country, or an economic sector, now have to be considered only in terms of their potential effects on shareholder value (Dore 2008; Khurana 2010). Shareholders are typically expecting a steadily rising, rather than a stable, return on their equity. Note that this change in managers' roles does not operate through plain coercion, but through a change in the culture of the managers; rise in salaries' inequalities; as well as carefully designed systems of carrots and sticks of stock options, bonus systems; and the overhanging threat of instant dismissal.

Third, the growth and increasing complexity in financial trading through multiple new financial instruments and intermediating activities between savers and the users of capital in the real economy. These activities very largely take a speculative character (Dore 2008). A great variety of instruments has been developed to support the broader trend of financialization, such as commodity index funds (Irwin and Sanders 2011), trading houses (Gibbon 2014), and derivatives markets (Galaz et al. 2015). This increasing complexity of financial products make it increasingly complicated to trace the decision-making responsibilities and to govern supply chains by providing the appropriate incentive to the appropriate actor.

Empirical evidence of the different dimensions of financialization is provided, discussed, and reviewed in Krippner (2005), Epstein (2005), Dore (2008), and van der Zwan (2014), among others. These different processes result in a set of consequences, which are also sometimes used as indicators or definitions of financialization, such as the emergence and increasing political and economic power of a particular social class of "rentier," i.e., a "functionless investor" (van der Zwan 2014), and the growing dominance of capital market financial systems over bank-based financial systems (Krippner 2005). Financialization has been sometimes considered as the major force sustaining liberalization and globalization, and the scientific literature has widely discussed many of its consequences, including on inequality in the distribution of income and wealth, through lessening the dependence of nonfinancial firms on productive activities and thus on labor force, and on the social welfare policies (e.g., Dore 2008). Financialization has profound effects on management, including because the time horizon of finance actors can be extremely rapid, as massive amounts of almost real-time information makes it possible to reevaluate investment decisions continuously to obtain the optimal return (see Chapter 14). Financialization also supports an increasing global interconnection and fluidity of capital markets.

17.3 FINANCIALIZATION IN THE FORESTRY AND AGRICULTURAL COMMODITY SECTORS

Following this general introduction of financialization, this section analyzes in more details how financialization has been occurring in the land-based commodity sectors, i.e., agriculture and forestry, and what have been its major impacts. Although the book focuses on forestry, this section also reaches out to other commodity markets where the financialization process has been stronger.

This section focuses on three main manifestations of financialization: (1) its effects on commodity markets and in particular on prices and prices volatility; (2) its relation with changes in land ownership and the related effects; and (3) changes in behaviors of forestry companies, with an increasing emphasis on financial activities, including through carbon forestry.

17.3.1 FINANCIALIZATION IN COMMODITY MARKETS AND EFFECTS ON PRICE, INCLUDING VOLATILITY

Financialization in commodity markets takes several forms, among which one of the most important is the establishment of commodity futures markets. A futures contract is an instrument through which traders buy (take a long position) or sell (take a short position) a volume of a specified commodity for a decided price, in a certain period in the future (Mehrotra and Carter 2017). Unlike forward contracts, futures contracts are marketable, allowing futures market participants to exit a contract before maturity by taking an offsetting position. Commodity futures markets have been rapidly developing over the recent decades (Galaz et al. 2015). Futures market trading is generally considered to impact commodity prices through several economic mechanisms (Cheng and Xiong 2014): (1) the decision to store that commodity instead of consuming it, with such delay depending on the difference between the current value and the expected future value; (2) the distribution of the risks among producers, consumers, and intermediaries in the supply chain; and (3) the aggregation of dispersed information held by market participants around the world about global supply and demand of commodities. Through the latter effect, prices of timber futures markets can serve as a forecast of the future spot price (Mehrotra and Carter 2017). Financialization has been shown to contribute to the rise in comovement between prices of the major commodities that have futures contracts, including a set of agricultural products such as corn, soybean, wheat, rice, coffee, sugar, cocoa, cattle, and pork meat, as well as timber and a set of mineral commodities such as oil, natural gas, gold, silver, and copper (Tang and Xiong 2010; Pradhananga 2016).

Timber futures market remain rather inefficient compared to other commodities, i.e., the information on changes in supply and demand is not readily incorporated into prices changes (Kristoufek and Vosvrda 2014).

Futures markets affect the economic context into which timber producers operate. Indeed, housing starts announcements have been shown to affect timber prices, but the effects of a housing starts shock is buffered for longer-term futures contracts compared to short-term contracts (Karali and Thurman 2009). Commodities prices are typically more volatile than for manufactured products, but historically, sawn wood is among the commodities with the lowest price volatility (Ecorys 2012). The volatility of timber prices has notably increased since the 1990s (Ecorys 2012). Although it remains undemonstrated that this increased volatility is directly linked to the cointegration of timber with other commodities markets, this would be consistent with the cointegration trend given the higher volatility of these other commodities prices compared to timber. The volatility of timber prices, by creating uncertainty, impacts the ability of foresters to plan for their harvests in an economically sound way (Rinaldi and Jonsson 2013). Another aspect linked to volatility is the speed at which investments decisions are made, such as through algorithmic trade (Galaz et al. 2015). The effects of this speed are still unclear, but they may conflict with the very long-term horizon at which forestry management decisions have to be made.

17.3.2 CHANGES IN OWNERSHIP IN US TIMBERLANDS

Financialization also manifests itself through changes in forestland ownership. This process has been mainly studied in the United States, where institutional and financial investors, mainly organized through timber investment management organizations (TIMOs) and Real Estate Investment Trusts have become large landowners and active players in the timber markets (Healey et al. 2005; Newell and Eves 2009; Bliss et al. 2010). Increases in concentrated stock ownership among institutional investors and the use of incentive-based compensation have been indeed considered as manifestations of the adoption of shareholder value strategies in the US forest products sector (Gunnoe 2016). The value of US timberland properties owned by institutional investors increased from $1.5 billion in 1990 to over $30 billion in 2009 (Newell and Eves 2009) and corresponds to more than half of the nation's private industrial timberland (Gunnoe and Gellert 2011). "Institutional landownership" encompasses the ownership and control of land by a broad array of financial actors, including pension funds, endowments, sovereign wealth funds, hedge funds, and private equity firms, operating through various organizational structures from real estate investment trusts to newly created land management entities such as TIMOs (Gunnoe 2014). These financial investors are often referred to as "passive" investors, in the sense that they do not actively engage in the forestland management (Zinkhan and Cubbage 2003).

Healey et al. (2005) explain this interest of institutional investors by a set of characteristics of timber investments, i.e., primarily its historical low price volatility and its low correlation with prices of other financial assets. For a similar expected return, timberland is shown to reduce the volatility of a financial portfolio (Zinkhan and Cubbage 2003). Beyond these fundamentals, the rise of financial funds dedicated to forestry in the 1980s helped institutional investors to get hold on timberland in the United States (Zinkhan and Cubbage 2003). Within a general context of financialization, including the deregulation of investments rules, institutional investors such as pension funds, banks, mutual funds, and insurance companies began to invest in real estate (Gunnoe and Gellert 2011). Meanwhile, in a context of deindustrialization and perceived undervaluation of timberlands in the stock market, timber companies were looking for investors willing to acquire their timberlands.

This transformation of timberland ownership in the United States was influenced by and at the same time reinforcing a reshaping of managerial strategies, toward an increasing prominence of the pursuit of "shareholder value" (Gunnoe and Gellert 2011; Gunnoe 2016). Selling timberland was indeed seen, by timber companies, as one way to create value for shareholders under this new managerial mode (Gunnoe 2016). The rising prominence of the shareholder value model had profound impacts on forestry as an activity and on forest landscapes. Institutional investments focus on a time horizon of typically 10–15 years, which is very short in forestry. The goal is not to produce timber but to maximize the market value of any investment. In search of the highest return, institutional investors are ready to convert pieces of timberland to other uses, thereby inducing fragmentation of forests. Being absentee landowners, institutional investors can hardly interact with local communities and care little about the impacts of their investments and management style on these communities (Gunnoe and Gellert 2011). On the other hand, many of these investors recognized the services values of the forest in addition to timber that they intend to cash on in the future as these services become internalize in markets.

17.3.3 CHANGES IN BEHAVIORS OF FORESTRY COMPANIES

As reflected in the selling of timberland by timber companies in order to maximize shareholder value, financialization modified the behavior and decisions of forestry companies. One aspect of financialization is the increase in the share of profit that enterprises derive from financial activities rather than productive activities. Carbon forestry corresponds to one way through which forestry companies can derive profit through a financial activity (Knox-Hayes 2013; Sullivan 2013). Carbon forestry is itself one manifestation of the broader "carbon economy" that is made of a set of interconnected compliance and voluntary carbon markets (Boyd et al. 2011). Carbon forestry includes several types of activities through which forestry can generate income through its relation with carbon storage and sequestration. These activities include reforestation that can generate credits such as under the clean development mechanism or voluntary carbon offset markets (Reeson et al. 2015) as well as public and private donors funding for forest conservation channeled through the international reducing emissions from deforestation and forest degradation scheme (Chow 2015). Discussing in depth the details and architecture of carbon forestry is way beyond the scope of this chapter, but it can be noted that this participates to the broader dynamic of financialization, and it has been shown to modify the management style of forestry companies. Carbon is a particularly uncertain commodity, as both its supply and demand are strongly influenced by public policies (Reeson et al. 2015). Carbon forestry is thus increasing the level of uncertainties that land

BOX 17.1   INVESTMENTS OF GREEN RESOURCES IN UGANDA

Green Resources is a Norwegian-owned forestry company active in multiple countries. A set of papers examined the activities of this company in activities in Uganda, where the company holds licenses for timber production as well as the sale of carbon credit (Lyons and Westoby 2014; Richards and Lyons 2016). These studies showed that financial flows channeled through private forestry companies such as Green Resources were a major way through which the government and public institutions were expecting to achieve forest conservation and reforestation in the country. Beyond environmental goals, international investments were seen as a major factor in development such as through the building of infrastructures and the payment of taxes. In essence, the national development agenda was thus embedded within the financialization of the forest sector. Yet investments by private companies in land have also been described as "neoliberal enclosure of land for plantation forestry" and a form of "state enabled land grab." The displacement and loss of access to key resources negatively affected the livelihoods of local communities.

managers have to face. Financialization through carbon forestry has contributed to a surge in land investments in tropical, developing countries, contributing to what has been called "new corporate enclosures" inducing dispossession, marginalization, and impoverishment of local communities in forest areas (see Box 17.1).

Mainly driven by the "shareholder value" model, these forestry activities are hardly concerned by the fate of local communities. Yet, under certain conditions, private investments in forestry can indeed create benefits for local communities, but these conditions remain insufficiently clear (Zoomers and Otsuki 2017). Carbon forestry, when taking the form of intensive plantations with rapid growth, also creates environmental trade-offs in terms of potential soil degradation, water and pesticide uses, and habitat for biodiversity. However, in some contexts, carbon pricing, and the internalization of forest services externalities could favor continuous cover forestry and adoption of more sustainable forestry practices instead of intensive plantations.

17.4   CONCLUSION

We have started by identifying three main dimensions of financialization as a general process of transformation in the world economy, particularly in high-income countries. These main dimensions are (1) a change in the pattern of accumulation toward an increasing share of profits derived from financial activities, (2) a change of value of corporations toward an increasing prominence of "shareholder value" as the benchmark against which any firm activity has to be evaluated, and (3) an increasing complexity in financial trading through multiple new financial instruments and intermediating activities.

We have then shown that these three trends are manifested in the forestry sector, e.g., through (1) a shift in accumulation pattern toward an increasing share of profit derived from financial activities such as selling timberland to institutional investors and engaging in carbon forestry, (2) the rise of the "shareholder value" model in forestry management in both developed and developing countries, and (3) the increasing weight of commodity futures markets in the forestry sector, directly both through the development of timber futures markets and through the increasing integration between timber futures markets and other commodities futures markets.

These trends may pose significant risks for the resilience of the forestry sector, as demonstrated by several examples where large investors or banks have failed to consider large-scale ecological risks of their managerial approaches (Galaz et al. 2015). Yet the rising influence of financial actors also opens opportunities to influence forestry companies to improve the sustainability of their practices. The probable increased demand for forest services, which could enter financial markets, could encourage forestry operators to manage forests for these services in addition to timber in order to maximize revenues. Financialization could help speed this transition. By raising awareness of financial actors on the sustainability risks of their investments, it may be possible to leverage their influence on forestry operators, although "green" finance currently remains a tiny niche market within the finance sector (Scholtens 2017). But the increasing complexity of financial products make it increasingly complicated to trace the decision-making responsibilities. The appropriate governance of financialized supply chains requires tools to trace the different actors involved in the supply and value chains of a given commodity, in order to trace these to the social and environmental impacts at the production place, and thereby providing the appropriate incentives to the appropriate actors (Godar et al. 2016; Gardner et al. under revision). This chapter is only a short exploration of the trends and effects of financialization in the forestry sector, which should be followed up by more in depth studies.

REFERENCES

Bliss, J. C., Kelly, E. C., Abrams, J., Bailey, C., and J. Dyer. 2010. Disintegration of the US industrial forest estate: Dynamics, trajectories, and questions. *Small-Scale Forestry* 9(1):53–66.
Boyd, E., Boykoff, M., and P. Newell. 2011. The "new" carbon economy: What's new? *Antipode* 43(3):601–611.
Cheng, I. H. and W. Xiong. 2014. Financialization of commodity markets. *Annual Review of Financial Economics* 6(1):419–441.
Chow, J. 2015. Forests as Capital: Financial Mechanisms for Tropical Forest Conservation. *Journal of Sustainable Forestry* 34:517–533.
Crotty, J. 2003. The neoliberal paradox: The impact of destructive product market competition and impatient finance on nonfinancial corporations in the neoliberal era. *Review of Radical Political Economics* 35(3):271–279.
Dore, R. 2008. Financialization of the global economy. *Industrial and Corporate Change* 17(6):1097–1112.
Ecorys. 2012. *Mapping Resource Prices: The Past and the Future: Summary Report—Final Report.* Rotterdam: European Commission–DG Environment.
Epstein, G. A. 2005. *Financialization and the World Economy.* Cheltenham: Edward Elgar Publishing.

Froud, J., Haslam, C., Johal, S., and K. Williams. 2000. Shareholder value and financialization: Consultancy promises, management moves. *Economy and Society* 29(1):80–110.

Galaz, V., Gars, J., Moberg, F., Nykvist, B., and C. Repinski. 2015. Why ecologists should care about financial markets. *Trends in Ecology and Evolution* 30(10):571–580.

Gardner, T. A., Benzie, M., Börner, J., Dawkins, E., Fick, S., Garrett, R. et al. 2018. Transparency and sustainability in global commodity supply chains. *World Development*, in press, https://doi.org/10.1016/j.worlddev.2018.05.025.

Gibbon, P. 2014. *Trading Houses during and since the Great Commodity Boom: Financialization, Productivization or . . .?* DIIS Working Paper. Copenhagen: Danish Institute for International Studies.

Godar, J., Suavet, C., Gardner, T. A., Dawkins, E., and P. Meyfroidt. 2016. Balancing detail and scale in assessing transparency to improve the governance of agricultural commodity supply chains. *Environmental Research Letters* 11(3):035015.

Gunnoe, A. 2014. The political economy of institutional landownership: Neorentier society and the financialization of land. *Rural Sociology* 79(4):478–504.

Gunnoe, A. 2016. The financialization of the US forest products industry: Socio-economic relations, shareholder value, and the restructuring of an industry. *Social Forces* 94(3):1075–1101.

Gunnoe, A. and P. K. Gellert. 2011. Financialization, shareholder value, and the transformation of timberland ownership in the US. *Critical Sociology* 37(3):265–284.

Healey, T., Corriero, T., and R. Rozenov. 2005. Timber as an institutional investment. *Journal of Alternative Investments* 8(3):60–74.

Irwin, S. H. and D. R. Sanders. 2011. Index funds, financialization, and commodity futures markets. *Applied Economic Perspectives and Policy* 33(1):1–31.

Karali, B. and W. N. Thurman. 2009. Announcement effects and the theory of storage: An empirical study of lumber futures. *Agricultural Economics* 40(4):421–436.

Khurana, R. 2010. *From Higher Aims to Hired Hands: The Social Transformation of American Business Schools and the Unfulfilled Promise of Management as a Profession.* Princeton, NJ: Princeton University Press.

Knox-Hayes, J. 2013. The spatial and temporal dynamics of value in financialization: Analysis of the infrastructure of carbon markets. *Geoforum* 50:117–128.

Krippner, G. R. 2005. The financialization of the American economy. *Socio-economic Review* 3(2):173–208.

Kristoufek, L. and M. Vosvrda. 2014. Commodity futures and market efficiency. *Energy Economics* 42:50–57.

Lyons, K. and P. Westoby. 2014. Carbon colonialism and the new land grab: Plantation forestry in Uganda and its livelihood impacts. *Journal of Rural Studies* 36:13–21.

Mehrotra, S. N. and D. R. Carter. 2017. Forecasting performance of lumber futures prices. *Economics Research International* 2017:165036.

Newell, G. and C. Eves. 2009. The role of US timberland in real estate portfolios. *Journal of Real Estate Portfolio Management* 15(1):95–106.

Pradhananga, M. 2016. Financialization and the rise in co-movement of commodity prices. *International Review of Applied Economics* 30(5):547–566.

Reeson, A., Rudd, L., and Z. Zhu. 2015. Management flexibility, price uncertainty and the adoption of carbon forestry. *Land Use Policy* 46:267–272.

Richards, C. and K. Lyons. 2016. The new corporate enclosures: Plantation forestry, carbon markets and the limits of financialised solutions to the climate crisis. *Land Use Policy* 56:209–216.

Rinaldi, F. and R. Jonsson. 2013. Risks, information and short-run timber supply. *Forests* 4(4):1158–1170.

Scholtens, B. 2017. Why finance should care about ecology. *Trends in Ecology and Evolution* 32(7):500–505.

Sullivan, S. 2013. Banking nature? The spectacular financialisation of environmental conservation. *Antipode* 45(1):198–217.

Tang, K. and W. Xiong. 2010. *Index Investment and Financialization of Commodities.* NBER Working Paper w16385. Cambridge, MA: National Bureau of Economic Research.

Van der Zwan, N. 2014. Making sense of financialization. *Socio-Economic Review* 12(1): 99–129.

Zinkhan, F. C. and F. W. Cubbage. 2003. Financial analysis of timber investments. In *Forests in a Market Economy*, eds. Sills, E. O. and K. L. Abt, 77–95. Dordrecht: Springer.

Zoomers, E. A. and K. Otsuki. 2017. Addressing the impacts of large-scale land investments: Re-engaging with livelihood research. *Geoforum* 83:164–171.

18 Social and Technological Innovations in Forestry

*Laura Secco, Elena Pisani, Mauro Masiero,
and Davide Pettenella*

CONTENTS

18.1 INTRODUCTION

In Europe, when referring to innovation in forestry, the dominant discourses mostly deal with technological innovation based on large-scale industrial investments (EC 2003). "Innovation is rather often used synonymously with technological innovation" (Kubeczko et al. 2006, p. 706). This is supported by a biased (limited) approach to the bioeconomy[1] in current strategies, where attention is almost completely focused on the development of biorefineries, i.e., on innovative plants that produce power, heat, a potentially large set of biochemicals, and, in some cases, pulp, normally using huge amounts of low-value biomasses from forestry, agriculture, or organic wastes (McCormick and Kautto 2013; Scarlat et al. 2015; Fund et al. 2015). The needs for industrial-scale economies that characterize large forest-based biorefineries are creating a demand for woody biomass that is frequently not covered by the potential local supply, so industrial plants are located in proximity to port facilities with a process of internationalization of not only the investment capital, but also wood procurement (Pülzl et al. 2017). Moreover, although it has been pointed out "the need to focus on innovation as a socially embedded phenomenon that should stretch across all economic sectors, [this concept] has mostly been applied in policy practice in high-tech fields, often with a technological focus or bias" (EC 2003;

von Tunzelmann and Acha 2003), rather than in forestry (Rametsteiner and Weiss 2003, p. 692). This is confirmed by recent studies: current forest bioeconomy innovations in Europe are technologically oriented (Lovrić et al. 2018).

Other emerging and innovative initiatives, such as the creation of nature-based businesses connected with the establishment of payment schemes for ecosystem (or environmental) services that try to obtain value from the management of public goods such as water, biodiversity, and human well-being (e.g., Wunder 2005), are often not considered as strategic choices to be invested in for the development of national economies,[2] despite their potential in rural development (e.g., by means of income generation and employment creation) and innovation[3] (Matilainen et al. 2011; Slee 2011; O'Driscoll et al. 2017; Tyrväinen et al. 2017). However, it was recently stressed that a new policy narrative is needed that "should emphasise a sustainable and socially inclusive forest-based bioeconomy" (Winkel 2017, p. 153), i.e., a holistic bioeconomy "[…] that recognises and mobilises the entire spectrum of ecosystem services that Europe's forests can provide for the benefit of Europe's societies" (Winkel 2017, p. 153).

This chapter introduces and discusses the various implications of social and technological innovations on the forestry sector, especially in Europe. In Section 18.2, links are made with the various components of globalization. In Section 18.3, both technological and social approaches are presented based on commonly used definitions. In Section 18.4, the two approaches are illustrated by means of concrete examples, while their pros and cons (in terms of positive and negative consequences) are pointed out and briefly compared. In Section 18.5, insights into how to integrate the two approaches are proposed and discussed in relation to the current perspectives of globalization and future development. The special role that information technologies can play in the two cases is highlighted.

18.2 INNOVATION AND GLOBALIZATION

Innovation is understood in this chapter in its common definition, i.e., "the implementation of a new or significantly improved product (good or service), or process, a new marketing method, or a new organizational method in business practices, workplace organization or external relations" (OECD 2005, p. 46). The concept has drastically evolved over the last 50 years. In the 1960s, it was considered as the product of a discrete event uprising from isolated individuals that developed technical solutions to identified problems. Nowadays, it is considered as a process, involving different social actors, and based on a combination of tangible (physical, technological, financial) and intangible forms of capital (human and social). As suggested by Landry et al. (2002), this evolution depicts different features of the knowledge based innovation process: innovation is specifically a problem-solving process that occurs primarily in firms and is based on the interactions of the organizations with the different actors of their environment. These interactions are based on formal and informal networks where different learning processes are taking places (learning by doing, learning by sharing, and learning by using). Additionally, the learning processes involve the exchange of tacit and codified knowledge, and the interactive process among actors generate a system labeled in different ways (innovation system, *milieu innovateur*, and innovation cluster). These general concepts—mainly deriving from economic and institutional theories—have been explored in relation to forestry by a specialized literature on innovation and policy-related issues [e.g., the studies by Rametsteiner and Weiss (2006) and Ollonqvist et al. (2011) and the book by Weiss et al. (2011)].

18.2.1 GLOBALIZATION OF MARKETS, FINANCE, AND ECONOMY

The dominance of technological innovation in the forestry sector, pushed by industry-oriented forest countries such as the Nordic ones, has traditionally been driven by globalization of markets, finance, and economy. On the one hand, concentrating production geographically, investing in technologies that increase the efficiency of wood harvesting and processing, and improving the dimensions of companies—eventually creating clusters or networks of enterprises (both horizontally and vertically)—is considered the best way[4]—not only in a capitalistic neoliberal-oriented economic world—to improve the efficiency of the forest-biomass value chain and production process, reduce wastes, increase profits and profitability of investments [e.g., the book by Weiss et al. (2011)], and, more recently, contribute to European Union (EU) member countries' 2020 energy goals.

Moreover, transnational corporations are assuming a key role in dominating global finance and the economy, not always acting in a socially responsible and environmentally sustainable way (Chomsky 2017).[5] Corporations and large companies investing in biomass and paper production, energy, and forest plantations are no exception. Rather, they follow the same internationalization trend that is criticized by social and environmental movements, blaming transnational large-scale forestry corporations for causing environmental degradation, natural resources depletion and overexploitation, land grabbing, social conflicts, and social exclusion (Fenton 2017).

18.2.2 GLOBALIZATION OF SOCIAL AND ENVIRONMENTAL PROBLEMS

Indeed, although there can be various drivers or determining factors, in parallel with the globalization of markets, finance, and the economy, there is also a globalization of social and environmental problems. Together with the depletion and overexploitation, climate change pressures, social and political instability, unbalanced distribution of resources, new and larger migration flows, and conflicts will affect the forestry sector in the immediate and long-term future.

In this context social inclusion, social capital, and social innovation are increasingly considered key intangible factors to guarantee successful and effective policy implementation and business development not only in the field of rural development and agriculture (Pisani et al. 2017), but also in forestry. In the EU 2020 strategy, social innovation is identified as a core element to promote smart, inclusive, and sustainable growth in the region.

18.2.3 GLOBALIZATION OF INFORMATION

A crucial role in supporting this development path can be provided by information technologies applied to forestry. Indeed, in parallel with the globalization of the economy, forestry today is also influenced by the globalization of information, taking advantage of the advances in both forest-specific and not forest-specific information technologies. Forest information technologies are increasingly recognized as useful tools for remote sensing and monitoring at the forest management unit level (Watson and Dal Bosco 2014). Collecting data and sharing information can assist in monitoring phenomena that are globally relevant, such as forest fires, illegal logging, or forest degradation. Depending on the target users and goals, data collected and information provided are used to support internal management decisions or communicated worldwide, as a marketing tool to increase the reputation of the company or country with respect to its commitments to sustainable forest management and timber supply from legal sources. The timely sharing of data and information through communication technologies can result in a reduction of information asymmetries. The use of these technologies is connected to an increasing demand[6] expressed by society for more responsible forest management, greater visibility of harvesting operations (via satellite images and geographic information system) and forest degradation, and improved tracking of raw materials associated to the need to monitor and stop illegalities. Information handling and spreading is also a political matter connected to the advocacy for responsible forest management and to getting the consensus and support by politicians. The empowerment of environmental and social nongovernmental organizations (NGOs)[7] based on just-in-time knowledge sharing is one of the consequences of the globalization of information, while companies can use information sharing and reporting to reduce risks of boycotts or conflicts.

It is also worth noting that information technologies, in general, provide new options for social networking and civil society involvement in forest policy making, as well as in citizen science initiatives. A number of new apps have recently been created and launched to allow the pro-active participation of citizens in scientific data collection or field monitoring activities in forestry and related fields (e.g., biodiversity, urban forestry, pests and disease monitoring, and monumental trees identification[8]). Social networks have proven to be effective tools in spreading information worldwide, thus raising the attention of the global community on a specific site or issue ("shame mobilization") and the support of international public opinion, and related coalitions, to protect specific forests, as proved by the recent case of the Białowieża forest in Poland (*The Guardian* 2017; various articles in 2017). Several international organizations are based on strong networking for creating or consolidating their coalitions and/or lobbying capacities.[9]

The globalization of information and related information technologies in forestry contribute nowadays and will continue in the coming years, to shape the future development of the forestry sector. In addition to their traditional applications in forestry, e.g., to facilitate the collection of data on large and remote forest areas, these instruments can play a key role in supporting innovative solutions for the development of forest-based local economies in rural areas, grounded on the creation of new small-scale social relationships, networks, and civil society engagement rather than on large-scale industrial technological investments. In other cases, data collected by researchers through sensor and positioning technologies associated to the use of social media from an emerging big data perspective are used to measure the use of urban green infrastructures and the time–spatial distribution of urban park users, thus providing valuable information to support decision-making (e.g., Chen et al. 2018). However, while recently, it has been observed that the most important sources for forestry sector specific information are websites and blogs, together with professional publications and specialized media, conventional face-to-face contacts have been found to be the most important communication and marketing channel to promote services and products (Watson and Dal Bosco 2014). While public forest administrations play a strong role as sources of information, especially in Eastern European countries, it was reported that less than 10% of innovators considered information from government or private nonprofit research institutes and from universities or other higher educational establishments as very important (Rametsteiner and Weiss 2006). Lastly, it is important to remember the extremely high power of social media and dominant discourses in driving public opinion and politicians and the potential negative consequences (e.g., dramatic oversimplifications, misinterpretations, and fuzzy topics of discussion) of any debate that might be related to important forestry issues (e.g., a national forest reform) mainly based on an improper and/or violent use of Facebook posts, Twitter tweets, and other media that can host "hate speeches." Social media are recognized as not always being effective tools for promoting constructive dialogue and building reciprocal trust (Hakansson and Witmer 2015). This is even more relevant in the current era of "post-truth" discourses—i.e., according to the Oxford Dictionary—circumstances in which objective facts are less influential in shaping public opinion than appeals to emotion and personal belief. These considerations suggest the idea not only that technological tools alone cannot completely replace social processes, but also that social processes can become drivers of a new (or rediscovered) role of forests for the benefit of the whole society.

18.3 TECHNOLOGICAL AND SOCIAL INNOVATIONS: WHAT ARE THEY?

18.3.1 TECHNOLOGICAL INNOVATION

The concept of technological innovation is primarily grounded on industrial management and business consulting contexts. It was defined as a nontrivial change in products and processes where there are no previous experiences [Nelson and Winter 1977; as cited by Rametsteiner and Weiss (2006)], and it is commonly applied at an enterprise or company level. This definition seems to encompass both product innovation and process innovation (OECD 2005).[10] According to the interpretation by Kubeczko et al. (2006), technological innovation is a subcategory of process innovation. Indeed, in forestry, technological advancements are traditionally connected to the mechanization of wood harvesting and wood processing (e.g., the use of new technical equipment and machines in manufacturing/treating wood). However, technological advancements regard products too, with the manufacturing and commercialization of engineered new wooden-based products (e.g., nanocelluloses from wood waste). As noted by previous research,[11] innovation policies mainly supported the diffusion of new technologies in timber production and processing.

As mentioned, one of the dominant areas for technological innovation investments is currently the biobased economy, in particular biorefineries. Biorefineries are "increasingly at the core of the bioeconomy vision at the EU level and worldwide" (Sauvée and Viaggi 2016), while the development of a biorefinery system is "a key factor in the transition to a bio-based economy" (Scarlat et al. 2015). According to 2017 data collected by the Nova Institute on behalf of the Bio-based Industries Consortium (BIC), 224 biorefineries[12] have been identified and mapped across Europe (21 countries). However, several other biorefineries are currently planned and/or under construction, and the list of those existing is probably not exhaustive. Within this framework, biorefineries are defined as integrated production plants using biomass or biomass-derived feedstocks to produce a range of value-added products and energy. Wood-based biorefineries (not including facilities for production of pulp just for paper) correspond to 25 plants, i.e., about 12% of the total. Roughly 60% of these are concentrated in Finland (33%) and Sweden (25%), while the contribution of southern European countries is limited: Italy, France, and Portugal together total just four plants. Countries with the larger investments in biorefineries are also among those where the forest sector provides a high contribution to the national gross domestic production (Figure 18.1).

Wood-based biorefineries mainly produce pulp, tall oil, specialty cellulose, bioethanol, and energy. Figure 18.2 reports the geographical location of wood-based biorefineries in Europe in 2017.

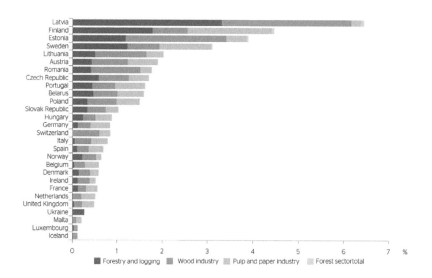

FIGURE 18.1 Contribution of the forest sector to the GDP in selected countries—percentage of the GDP 2010. (From Forest Europe, *State of Europe's Forests 2015*, Forest Europe, Madrid, 2015.)

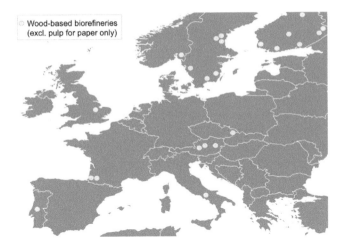

FIGURE 18.2   Map of wood-based biorefineries in Europe in 2017. (Adapted from Nova-Institut GmbH and Bio-based Industries Consortium, *Biorefineries in Europe 2017*, 2017.)

18.3.2   SOCIAL INNOVATION

The concept of social innovation is an emerging one, especially in its application to the field of forestry. Its main initial focus was to address social disadvantage and exclusion in a wide range of contexts, more often urban than rural (Moulaert et al. 2005; MacCallum et al. 2009). Social innovation provides a renovated role to society, being considered—at a time of major budgetary constraints—an effective way of responding to social challenges by mobilizing people's creativity, promoting an innovative and learning society, and creating the social dynamics behind technological innovations (BEPA 2011). So far, a few scholars have proposed how to interpret the concept in the rural arena (Neumeier 2012). Bosworth et al. (2016) identified the key elements of social innovation in the case of the EU Leader programme by using the Schumpeterian (1934) approach and framework to analyze innovation.[13] Bock et al. (2016) worked more on theoretical conceptualization.

One recent proposal that draws from a wealth of research and work in a variety of fields, including economics, sociology, ecology, and political sciences, and that tries to integrate the previously existing approaches while focusing on rural areas, is suggested by Polman et al. (2017, p. 12)[14]: "the reconfiguring of social practices, in response to societal challenges, which seeks to enhance outcomes on societal well-being and necessarily includes the engagement of civil society actors." A catalogue of 50 examples of social innovation, which have been identified according to this definition within the fields of agriculture, forestry, and rural development in marginalized rural areas in EU and extra-EU Mediterranean countries (Bryce et al. 2017; Price et al. 2017), has been compiled and published online by the Social Innovation in Marginalised Rural Areas (SIMRA) project.[15] The catalogue is neither fixed nor comprehensive. Rather, it provides an initial overview on how wide the variety of already implemented social innovation cases can be, although a commonly accepted definition and theoretical conceptualization are still under construction and specific policy instruments are still lacking.

Social innovation in forestry is probably more widespread than reported so far by the scientific literature, as the concept refers *de facto* to a wide range of initiatives dealing with different societal challenges: from the new social uses of forests (e.g., "forest bathing" for the disabled, elderly people, or children) to the creation of new public–private partnerships to produce, transform, and commercialize new types of wild forest products (e.g., insects), the inclusion of migrants/refugees in forest management activities as a means for social, and multicultural integration, and others. While a number of social innovation examples are likely to exist in Europe and other regions, it seems that data and information are so far available only as spots or case studies, not having yet been systematized or collected in a structured way. Recently, Rogelja et al. (2018) noted that EU policies have emphasized market economic features of social innovation, such as efficiency and effectiveness of social investment and budgeting, consequently prioritizing social business over social movements (EC 2013; Jenson 2017; Moulaert et al. 2017) and undermining the relevance of the broader sociopolitical context for the development of bottom-up initiatives (Demming 2016; Moulaert et al. 2017).

18.4   TECHNOLOGICAL VS. SOCIAL APPROACH: PROS AND CONS

The technological approach is typically based on a one-way, top-down process of innovation, where the knowledge is created by one actor (and intellectual property is strictly protected by means of patents). The social approach is likely based on inter-sectoral network-based interactions, where the knowledge is shared and emerges from a more collaborative learning process. While the first approach is linked to the "linear concepts of innovation, […] [which] continue to be widely applied in research, business and business consulting contexts, especially in a firm level context" (Rametsteiner and Weiss 2006, p. 692), the latter is connected with "innovation systems," conceptualized as "a complex nonlinear process" involving a range of actors and institutions, which do not necessarily belong to the same sector, interact each other and contribute to the development and diffusion of innovations in forestry (Rametsteiner and Weiss 2006, p. 693). Figure 18.3 visually represents the two approaches. Both have pros and cons and positive and negative consequences on various aspects.

While it is clear that there are positive consequences of investing in industrial plants technology advancements from an economic point of view in terms of corporations' efficiency and profitability, the adoption of a strategy of development based only or predominantly on industrial technological innovation has several limits and often an unbalanced distribution of costs and benefits for the local rural communities with a high number of small forest ownerships. Table 18.1 compares the two approaches, obviously simplifying and taking the issues under discussion to the extreme.

First of all, any concentration of industries and corporations leads to the concentration of power (on not only the market, but also—through powerful industrial lobbies—on politicians and thus on decisions taken). The increasing international

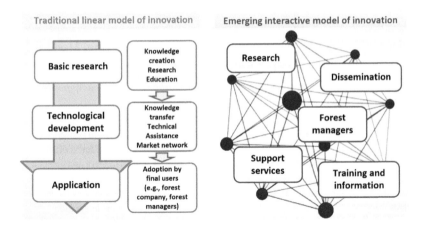

FIGURE 18.3 Two models of innovation. (Adapted from Illuminati, R., Innovazione in agricoltura e capital sociale. Impatto della Misura 1.2.4 del PSR 2007–2013 in Umbria, MSc Thesis, University of Perugia, Perugia, 2015.)

flows of raw materials and growing competitive advantages by highly efficient large-scale forest industries determine a progressive marginalization or exclusion from the international timber market of less specialized countries such as southern EU countries, with the marginalization of the forestry sector in the national economy and disconnection of national timber industries from domestic sources. One consequence, also having other drivers such as urbanization (see Chapter 8), is forestland abandonment (e.g., in Italy). An increasing level of mechanization/automatic wood-processing processes can lead to a reduction of labor, sometimes having as a consequence less demand for low qualified forestry workers. The focus of national bioeconomy strategies mainly on timber and paper/pulp production is an indicator of the limited interest or inappropriate recognition of the importance of other forest functions, products, and services, such as nonmarket ecosystem services (Pülzl et al. 2017). Large-scale investments are typically needed (see the two examples in Box 18.1) but only large-scale investors and transnational corporations have the financial resources to afford and support them, often with a high share of contribution by public funds. It has been found that larger forest holdings have a higher level of innovative activity with respect to smaller forest holdings[16] (Rametsteiner and Weiss 2006). Governments often participate by cofunding large private investments, with the justification that they will have positive impacts on the national GDP and will contribute to reaching the national commitments on renewable energy and emissions reduction. These are excellent justifications from an industrial point of view at national level and might be proper drivers of policy making and investments in those countries where forests significantly contribute to the national GDPs with their products and processes.

However, public cofinancing large scale, single plant investments raises the problem of equity in cost-benefit distribution, these industrial investments having less positive impacts on local development of rural communities than those that would derive from more widespread small-scale investments involving larger numbers of forest managers and small-scale enterprises. According to the forest-based sector technology platform,

TABLE 18.1

Dominant, Traditional Technological Approach vs. Emerging, Modern Social Approach: A Nutshell Comparison Related to Forestry

	Technological Approach	Social Approach
Focus on	• Technological innovations (toward a low carbon emissions economy) • Large-scale investments (capital intensive) • Industry-based forest economy	• Social innovations • Small-scale enterprises (labor-intensive) and networks • Rural-based forest economy • Urban forestry and greening
Vertical vs. horizontal relations	• Value chain perspective • Sectorial development • Vertical integration	• Network economy • Intersectorial development • Horizontal integration
Inputs and outputs diversification	• Low-quality woody biomass as the only, cheap input material • Specialization in high added value outputs (→ weaker resilience to financial/economic global crises and other unpredictable events)	• Diversification in inputs (industrial wood, biomass, nontimber or wild forest products, other ecosystem services) • Diversification in high added value outputs (→ stronger resilience to financial/economic global crises and other unpredictable events)
Market power	Increased market power of the industrial companies controlling the advanced technologies (→ higher risks connected to the companies consolidation trends)	Balanced market power among the various diversified operators (→ lower risks due to higher diversification)
Measure of performance (examples)	Ecoinnovation scoreboard (national level assessment approach), by the EU Eco Innovation Observatory (http://www.eco-innovation.eu/)	Spot, site-specific (e.g., ongoing pilot cases by Horizon 2020 SIMRA project: http://www.simra-h2020.eu; those carried out by some local action groups within the EU Leader approach: http://enrd.ec.europa.eu/enrd-static/leader/en/leader_en.html)
Model regions	North and central European countries and coastal areas	South European and Mediterranean countries and mountain regions
Stakeholders and public involvement	Risks of lack of public consensus around the industrial investment (not in my back yard effect) and need for addressing the process of social inclusion	Potential for social inclusiveness, both in R&D processes (citizen science and network-innovation) and cofunding that can increase the stakeholders and public empowerment in forestry
Drivers	• Patented (private) R&D initiatives, with public support/funds • Top-down, linear transfer and creation of innovation	• Public–private initiatives in education, training, and nonpatented innovations • Network-based transfer and creation of innovation

**BOX 18.1   TECHNOLOGICAL APPROACH TO THE
BIOECONOMY IN FORESTRY: TWO EXAMPLES**

1. The *Tees Renewable Energy Plant* (REP), a planned power plant in the
   United Kingdom, will be located in the Port of Teesside, Middlesbrough,
   and will have a capacity of 299 MW, thus becoming the largest biomass
   power plant in the world. The project's engineering and construction is
   expected to cost more than €600 million and create around 1100 jobs
   during the construction phase. The renewable energy generated is anticipated to be equivalent to the power consumed by 600,000 households
   in the United Kingdom. The company (MGT Teesside Ltd) website
   says the plant will help meet the United Kingdom's nationwide renewable energy goal of 15% of all energy consumed by 2020. The company
   also projects that the plant will save approximately 1.2 million t of CO_2
   each year. MGT Teesside Ltd reports that the project is expected to
   break ground as soon as funding is secured by early 2016, and the plant
   will be operational by 2019—just in time to help offsetting the coal and
   gas usage and contribute to the United Kingdom's 2020 energy goals.
   Wood pellets and chips from sustainable forestry sources will fuel the
   Tees REP. A tentative forecast of the wood biomass consumption is
   1.2 M t chips/year, which will be imported by ship mainly from the
   United States. In terms of environmental statement, the website reports
   that "the wood pellets are produced from the co-products of the saw-
   timber industry and are sourced entirely from commercial forestry,
   which does not contribute to deforestation because forestry is always
   re-establishing after removals." The company also states that its suppliers of pellets and chips "will be subject to regular third party audits
   to ensure the ongoing sustainability" of the supply chain.

   *Tees REP: http://www.power-technology.com
   /projects/tees-renewable-plant-teesside/; MGT Teesside
   Ltd.: http://www.mgtteesside.co.uk/#tees-rep/; Tees REP
   general info: https://tecnicasreunidas-sct-teesrep.com*

2. The *Metsä Group bioplant*: The Metsä Group is planning the biggest
   investment in the forest industry in Finland, about €1.1 billion to convert and expand a traditional large pulp mill into a bioproduct mill. The
   project plans to refine wood into biomaterials, bioenergy, biochemicals, and fertilizers sustainably and with great resource efficiency. The
   planned annual pulp production is 1.3 million t, with an annual wood
   consumption of 6.5 million m^3. The consumption of wood will approximately triple, as current consumption is 2.4 million t. This will contribute to wood mobilization. According to the project, over 2500 jobs will
   be created throughout the whole value chain in Finland, including new

jobs in harvesting and wood transport, and there will be the need for a competent workforce. Internal financing is approximately 40%. The project is expected to help Finland reach its targets for the use of renewable energy, as it contributes 1400 GWh/annum electricity generation, 7000 GWh/annum district heating and steam and 1200 GWh/annum wood energy. The necessary technological innovations will allow the use of raw materials and 100% side streams as products and bioenergy, without using fossil fuels, and the choice of equipment and machinery will emphasize the criterion of energy efficiency. The stated advantages are "efficient production of high-quality pulp," "integrated production of new bio-products," and "resource-efficient way of using all production side streams." However, organizational innovation is also needed. According to the project, "the operating model will be based on an efficient partner network," where "new products will be created in collaboration with various experts joining the network," and "create opportunities especially for small and medium-sized enterprises to produce innovative bioproducts with high added value." These last elements are coherent with cluster-based strategies and regionalization processes, where the linearity of the technological innovation model remains internal to each industry/corporation or their clusters.

Metsä Group: https://www.metsagroup.com

the current research and development (R&D) investments in Europe reach an amount of €2.5 billion in total, with the total public funding contributing with 1.7 billion (68% of total). However, R&D is mainly focused on technical problems, creation of new patents, and a linear top-down approach to innovation, not always able to grasp the social aspects (e.g., potentially excluding workers who are not highly qualified, latent social conflicts, and protests against the industrial plant that can create potential risks for the reputation of companies and investors). The implementation of the Strategic Research and Innovation Agenda 2020 (SRA), released in 2006 and revised in 2013, resulted in the launch of more than 230 research projects relevant for the European forest-based sector and an amount of over €1 billion of EU funding (FTP 2017). The SRA introduces 19 research and innovation areas (RIAs) identified as key to unlock the potential of the forest-based sector and ensure its future competitiveness. However, looking at the list of RIA titles, the orientation appears clear: 12 out of 19 RIAs are mainly technologically oriented (e.g., enhanced biomass production, secured wood supply, forest operations and logistics, cascade use, reuse and recycling systems, resource efficiency in manufacturing, biorefinery concepts, new biobased products, and intelligent packaging solutions), only 3 out of 19 are mainly socially oriented (e.g., citizen's perception of the sector, policies and good governance, new business models, and service concepts), while 4 out of 19 can be considered as mixed (e.g., multipurpose management of forests, forest ecology, and ecosystem services). This unbalance is even more significant in the strategic innovation and research agenda published by the public–private partnership BIC (2013). The document is almost totally focused on biomass technology and market developments, with no social or environmental contents.

The traditional technological innovation obviously includes investments in information and communication technologies, which play a fundamental role in collecting, processing, and analyzing large amounts of technical data to support and monitor the internal industrial processes; tracking (Tzoulis and Andreopoulou 2013) and organizing the distribution of products and in general solving logistic issues; marketing and managing relations with satellite activities, suppliers, and clients; and managing internal and external communication. But innovative information technologies (e.g., global positioning system devices and drones) (see Figure 18.4) and software are also increasingly needed for remote sensing control of large-scale forest and plantation areas, to create large datasets or improve the quality of data for internal uses and to update forest inventory by limiting costs and other applications. Data collected by means of these technologies are often sensitive, privately owned by the company, and used for internal managerial purposes. However, they can also be (and are) used for periodic reporting and marketing, providing evidence on the achievements of the company in terms of sustainable forest management, increasing transparency[17] and contributing to raise public awareness about forestry and forest-resources management issues.

In our opinion, while the traditional technological approach seems to have really good opportunities in well-connected and industrially developed areas, e.g., coastal and flat areas in north and central European countries, there are limited chances for the remote mountain regions, especially those located in the southern (Mediterranean) countries, to be competitive in the mass product market based on the large-scale use of wood for industrial purposes; this is the case, for example, of the biofuel production supported by the bioeconomy strategies that have recently been launched in the EU (Pülzl et al. 2017).

The social approach, whose efficiency should be highlighted by its integration in a strategic land use planning and development scheme at the regional level, might be more effective than the vertical approach in supporting job expansion and in taking advantage of the diversified forest resources available at small scale in remote rural areas. These areas (e.g., Alentejo, Catalonia, Provence, Trentino, Tuscany, and Istria) are often characterized by small-scale multifunctional forest activities, considered essential elements of a diversified rural development, timber being just one of the several territorial ecosystem services that can be delivered by forest management [e.g., the studies by Vuletić et al. (2010), Slee (2011), Gatto et al. (2014), and Tyrväinen et al. (2017)] and not always (or anymore) the most valuable one (Maso et al. 2007). Moreover, the social innovation approach is frequently adopted by green movements and local citizens' groups involved in new social uses of urban green areas (e.g., green care initiatives, urban gardening, and urban social horticulture) (Schicklinski 2017) (see examples in Box 18.2).

Our arguments here are mainly based on considerations about harvesting costs and logistics. On the one hand, the productivity is higher if technologically advanced machineries (e.g., large harvesters) can be used; timber harvesting and transporting costs are lower in flat areas with good connections (e.g., a high density of road networks); the logistic for the land transport and for the long-distance shipping of wood products can remarkably reduce the procurement costs for the

(a)

(b)

FIGURE 18.4 Drones used to collect forest data. (a) (Courtesy of OpenForests, Krefeld, 2018.) (b) (Courtesy of Etifor, Legnaro, 2017.)

industrial plants located in coastal areas. On the other hand, in fragmented forests located in remote mountainous regions, where large clear cutting areas are limited by environmental constraints, with limited logistics facilities (e.g., low density of road networks and few railways lines), the costs of wood products harvesting and transport can be significantly high (Spinelli et al. 2017). Small-scale forestry in such areas is hardly competitive with respect to large-scale industrial wood production

**BOX 18.2   SOCIAL APPROACH TO THE BIOECONOMY
IN FORESTRY: TWO EXAMPLES**

1. For more than 20 years, *the International Model Forest Network* (IMFN) (see Chapter 3) has been implementing a participatory-based approach at landscape level to the sustainable management of natural resources, included forests. The approach was not pioneering in its international networking goals, but it was innovative at that time in proposing and adopting principles and governance mechanisms able to promote a voluntarily based partnership and collaborative work among local stakeholders. Although the principles and attributes required for becoming a model forest recognized at the international level are quite general and aspire to sustainable forest management and good governance concepts, they include aspects that are also characteristic of social innovation. For example, the options set in principle 1 (partnership) for a neutral forum where both private organizations (often businesses), public administration (typically, local municipalities), and civic society representatives (e.g., NGOs) of interests and values are welcome to participate. Moreover, according to principle 5 (program of activities), the activities undertaken have to reflect the landscape vision and stakeholder needs and challenges. In short, we can argue that even if in the IMFN, the single model forests were not conceptualized as social innovations, it is likely that some of them are *de facto* social innovations (e.g., in terms of innovative partnerships and governance procedures, voluntary engagement of stakeholders and forest-based activities that are designed to solve socio-economic needs and societal challenges). Nowadays, the IMFN includes more than 60 large-scale landscapes in six regional networks, covering a total of ca. 84 million ha in 31 countries. One of the regional networks active in Europe, established in 2008, is the Mediterranean model forest network, which includes 12 landscapes in eight countries (Spain, France, Italy, Croatia, Greece, Turkey, Tunisia, and Morocco). Each single landscape is a local network, so that the regional one is a larger network of local networks, where ideas, best practices, knowledge, and information are exchanged.

http://imfn.net/mediterranean-model-forest-network

2. *Associazione Forestale di Pianura* (Italy): Urban and periurban forests in lowland areas of the Po Valley in the North of Italy are often crucial for recreational activities. In addition, they can be catalysts for social aggregation. Forests located near densely populated or intensively visited areas, if planned and managed for being accessible to a broad range of visitor categories (e.g., families with children;

disabled persons; elderly people with mobility limitations; sportspersons passionate about outdoor activities such as biking or running, and birdwatchers), may be relevant resources to invest in. They can attract visitors and initiatives, thus contributing to the growth of the local economy. The areas and patterns can be set up in a way that is functional for different social uses of the forest, providing support to various recreational services, give options for employment opportunities, and contribute to the well-being of local communities. If this implies voluntary engagement of the civil society, new types of relationships between private and public actors, and/or new governance procedures, it can be a social innovation in forestry (Figure 18.5).

One example is a lowland forest area in Veneto (northeast Italy), located close to Venice and well-known beaches along the Adriatic Sea (e.g., Jesolo and Eraclea). The area is visited by about 3 million tourists every summer and, starting from 35 years ago, has been subject to a large afforestation program.

There are currently 24 forests that are owned by eight local municipalities. The areas are managed for use by different target groups, including disabled persons and families. The management activities are carried out through various forms: direct management by municipalities, management agreements with private companies or not-for-profit entities, private rentals, etc. Since 2002, forest owners and managers are aggregated in and supported by the Lowland Forest Association (*Associazione Forestale di Pianura*, AFP),

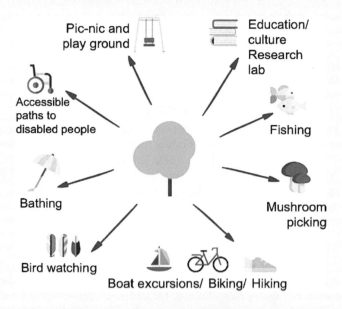

FIGURE 18.5 Activities in peri-urban lowland forests. (Courtesy of Associazione Forestale di Pianura, 2017.)

which collaborates with various environmental and social organizations (e.g., Legambiente, World Wide Fund for Nature Italy, and the Italian Association of Forest Sciences Students) and plays an active role within the Local Action Group Venezia Orientale (a local development agency) of the EU Leader program. AFP and its network are a unique case of this type of private–public cooperation in lowland coastal forest management in Italy (Figure 18.6).

Management operations are financed not just through municipal budgets, but also through funds raised by the AFP via other sources, such as the EU Rural Development Programme, private investors (including "AzzeroCO$_2$,"—a broker of carbon credits; "E-on"—a renewable energy agency; and Alí—a supermarket chain) and crowdfunding. In the period 2009 to May 2017, the area attracted in total ca. €1.7 million (€200,000/year): it has been estimated that €1.0 invested by each AFP member resulted in €7.1 of resources available for management. With the innovative approach of attracting donors and actively involving the local civic society (for example, by means of the crowdfunding project), the AFP determined a key change: while in the past, resources were mainly represented by regional and public funds, in the last few years, funds have started to be mainly international and private. In 2017 the forests got forest management certification according to the Forest Stewardship Council (FSC) standards, joining the WaldPlus group. This, together with an improvement in the investments made in communication, has significantly contributed to increasing the visibility of the area. In 2016, as a preparatory activity to the forest certification process and audit, a new forest management plan was developed, focused on interventions to increase the capacity of the forests to deliver

FIGURE 18.6 *Eraclea Mare* pine forest. (Courtesy of Eraclea Mare at Park Hotel Pineta, 2017.)

ecosystem services. They include products and services that are already sold to the investors (e.g., pine nuts for ca. €200–300/ha/year; carbon sequestration, ca. 5.99 tCO_2eq/ha/year, at a price of €17–24/tCO_2eq) and products and services that are potentially relevant, such as biodiversity enhancement and wild truffle production. Recreation is one of the cultural services that are planned to be enhanced with investments in the coming years: while in the period 2009–2011, 255 ha (70% of the total forest area) have been restored and made accessible for nature-based recreation, the 2025 target is to have 100% of forest areas restored and accessible, also with a diversification of the services (for example, it is planned to test the establishment of a kindergarten or school in the forest and various types of green care programs).

Own elaboration based on Secco et al. 2017

and processing in more accessible and better infrastructured contexts (Esteban and Carrasco 2011).

However, it is not only a matter of competitive advantages based on geomorphological, infrastructural, or forest management characteristics, which might be more or less favourable to the adoption of technologically advanced harvesting or processing solutions. Other territorial-based development models are possible, where the economic development is more linked to specific local forest resources; high labor-intensive and socially inclusive economic activities represent both a value and a strategic objective, and benefits are more widely and equitably distributed among actors and not concentrated in the hands of a few number of large companies. As already mentioned, nonwood forest products, such as chestnuts, mushrooms, truffles, pine seeds, medicinal, and aromatic herbs, might represent high added value products and an annual source of income based on organizational or institutional innovations, rather than technological ones (Pettenella and Secco 2006).

Unfortunately, the social approach has had a too limited political visibility for many reasons:

- This sector of the economy is a constellation of niche markets: Diversification is the key element, but it is often difficult to reach a critical mass of products and services to satisfy the potential consumers; the market organization is complex and fragmented (cross-sectorial and interlinked products and services).
- Also, for these reasons, only a few statistical data are available.
- Social capital (i.e., trust, relations, and other typical elements of social innovation) is far from being the main component of the dominant R&D culture.
- Products and services should be promoted with strong investments in technical assistance and communication innovation services, exactly the opposite of what is happening in many Mediterranean countries where these services are the first to be exposed to budget cuts or where the initiatives are too small for enough funds and resources being allocated to special forest information and communication technologies (despite recognition of their potential usefulness).

18.5 POSSIBLE INTERACTIONS BETWEEN TECHNOLOGICAL AND SOCIAL INNOVATIONS

Even if we argue that the technological approach is predominant in the north and central European countries and the social one is promising and emerging in the south and Mediterranean areas, the two approaches are obviously interlinked. They do not necessarily exclude one another, and they coexist *de facto* in many countries. For example, in a Scandinavian wood industry-oriented country such as Finland there are examples of social forestry; in the United Kingdom both the approaches are very well developed; the large number of examples of social innovations in Italy does not mean that investments in technological innovations with the creation of medium-scale industrial plants producing innovative biochemicals are not possible[18] or increasing the efficiency of wood-harvesting and processing is not useful. If properly coordinated, both approaches can contribute to support an "inclusive, smart and sustainable growth" as required by the Europe 2020 strategy and the United Nations Sustainable Development Goals. The two approaches, and their reciprocal interactions in dealing with regional or global socioeconomic challenges, are outlined in Figure 18.7.

On the one hand, technological innovations (in their Schumpeterian meaning of product innovation, market innovation, etc.) are typically designed to be applied in a single company, or group of companies, to obtain profits directly benefitting investors and industrial owners (and only indirectly benefitting the local communities by means of employment opportunities or provision of funds to support social

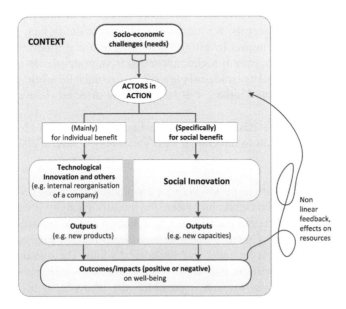

FIGURE 18.7   Possible links between technological and social innovation (Authors' own elaboration inspired by the Social Innovation in Marginalised Rural Areas Research Team.)

events—as a compensation for their environmental impacts). Technological innovations are typically oriented toward tangible outputs, such as new products (e.g., nanocellulose-based fibers and hydrogels used to rebuild human body parts—see, for example, the study by Syverud (2017)—that when used by company's clients and/or final consumers bring positive outcomes on the whole society (e.g., medical applications). On the other hand, social innovations are specifically designed to seek to determine positive social benefits, i.e., broader benefits on human well-being that influences the quality of life of not only various members of the local community but also other people and networks. As outcomes, social innovation might have an increased capacity of collaborating or the improvement of other human or social capacities that, in a long-term perspective, bring positive impacts on the community. While we observe that technological and other innovations are qualitatively different from social innovations, given the socially intended goals of the latter and the nonmaterial nature of the innovation, we also recognize that technological innovations can lead to social innovation (Neumeier 2012; Cajaiba-Santana 2014). The overlapping areas in Figure 18.7, both in the process and in the outcome boxes, represent the potential reciprocity in supporting each other. Outputs from technological and social innovations can both contribute to the impacts on the society. However, this is an oversimplification of the various possible interactions: we can also find examples of social innovations internally to a firm, designed for solving personal attitudes and behaviors of firms' employees and coworkers when they impede the firm's innovation project implementation (Rametsteiner and Weiss 2006). Or vice versa, we can find examples of technological innovations (e.g., creation of high-tech products and applications) used in social-oriented and network-based innovation projects.

More recently, there has also been increasing interest in social innovations from the point of view of investors, not necessarily at local scale. In fact, the financial sector of the so-called impact investing[19] is looking at social innovation as a core field of action. However, even if social innovation is, in principle, designed to have positive impacts on society, trade-offs are often unavoidable: while some people will be positively affected, others will be negatively affected (Kluvánková et al. 2017).

Such integration is not easily realized in practice. If we start from a technological-based approach and want to integrate it with social issues, several challenges have to be considered:

- Ethical values, with respect to both the community where the raw material is exploited and the community where the industrial activity is carried out.
- Efficiency vs. participation dilemma, well-known and old but still valid.
- Social inclusion of more vulnerable and disadvantaged groups, such as newcomers, disabled people, and unqualified youth. The question here is whether this perspective is possible or should rather—more realistically— be considered as a "mission impossible," the involvement of disadvantaged groups being a cost and an organizational challenge for large-scale industry-oriented investments that seek to increase efficiency and profitability.

If we start from a social-oriented approach and want to integrate it with techno-logical advancements, different challenges have to be considered, none less relevant:

- Investments in social-oriented R&D are obviously needed, probably with the involvement of private sponsorships.
- Scaling up, outscaling, and replicability are an issue, as social innovation is often local specific and happens at local level (Secco et al. 2017).
- There is undoubtedly a potential for an improved role of citizen science, but this needs to be regulated and have more investments to make it possible and workable in practice (by means of coordination, open platforms for collecting, cleaning and interpreting data, and apps).
- Social exclusion issues risk arising when innovation projects refer to groups of people and/or areas that are less developed or advanced in technologies and, in particular, information technologies.

One option is to try to support their interconnection and integration by spreading or reinforcing the use of information and communication technologies, both specific to forestry and not. On the one hand, involving people in a technological innovation process will increase the capacity of industrial-oriented investments in the more advanced industrial countries and businesses to be legitimated and supported. On the other, linking small-scale and fragmented socially oriented initiatives in larger networks, for example, through a smart use of social networks and the media, will give them higher visibility and recognition by policy makers, as well as more capac-ity to create a critical mass able to influence future development paths. As previously mentioned, the integration of the two approaches can be useful for supporting the growth of both the European economy and society and a more inclusive and sustain-able development of forestry.

Large-scale technological investments often involve stakeholders only for con-sultation on very general issues (e.g., environmental impacts of industrial activities), and stakeholders do not always have a real capacity to influence decision-making at the higher jurisdictional levels. In the technological approach, science expert-based knowledge is predominant over nonexpert-based and local knowledge (thus miss-ing a lot of potentially valuable information). The potential role of citizen science is underestimated, and there are risks of new asymmetry of information. However, several examples exist of very successful small-scale (social) innovations in forestry that do or can have positive impacts. They are progressively increasing, with a grow-ing involvement of the public, not just of stakeholders (Kleinschmit et al. 2018), but also in funding innovative solutions to support local investments (e.g., crowdfund-ing). Some examples found in Italy are given in Box 18.3.

To summarize, we think that the main reason for technological innovation domi-nating the discourses and policy choices so far is that they are able to obtain high vis-ibility, as large-scale projects often with significant cofunding from public resources, and derive from power concentrations, so are becoming well-known examples world-wide of technological progress—even if social and environmental costs are often disregarded. On the one hand, in most cases, confidential information (e.g., business-finance models, licenses for special technological solutions of product processing,

BOX 18.3   PUBLIC PARTICIPATORY PLATFORMS AND INITIATIVES REGARDING FORESTRY IN ITALY

Public participatory initiatives regarding forestry in Italy, such as mass science initiatives, are still limited and can be summarized in five main groups:

- Online forums aim to inform people about ongoing initiatives and collect comments/feedback or facilitate interactions among users. Relevant examples are those provided by the forums for the discussion of national forest certification standards, but they are applied in many other circumstances.
- Storytelling initiatives are more recent and try to deliver scientific research to nonspecialists by adopting a simplified language and appropriate communication channels. Although not specifically intended for forestry, the recently launched start-up Learnscapes[20] is a good example.
- Information capturing from social networks is becoming increasingly frequent, also by scientists, and forestry is no exception. Examples include the use/statistical elaboration of georeferenced photos gathered from social networks such as Panoramio and Flickr to assess recreational services provided by natural resources, including forests.
- Social networks and blogs also play a role in forestry and related fields, functioning as platforms for sharing information, ideas, contacts, experiences among media, groups of stakeholders, and individuals at various levels interested in forest resources and their management. An example in Italy is the social networks and related online communities managed by the technical/professional national journal *Sherwood—La Compagnia delle Foreste* (more than 2300 Facebook and 1200 Twitter followers as of February 2018). Professional social networks might also be relevant (e.g., ResearchGate, Academia, and Linkedin).
- Citizen science initiatives, with the active cooperation of citizens and scientists in collecting, delivering, validating, and sharing information, are still marginally implemented in Italy. Examples mostly refer to environmental/biodiversity monitoring initiatives that in some cases also took place in forest areas. The list includes specific projects (e.g., Citizen Science Monitoring—CSMON Life[21]; Monitoring of insects with public participation—MIPP Life[22]; U-SAVEREDS[23] for the conservation of red squirrels; and GESTIRE Life[24] for the collection of fauna and flora data regarding Natura 2000 sites) and other (mostly voluntary) initiatives, such as I-naturalist Italia groups (alien species,[25] butterflies,[26] and amphibians and reptiles[27]) and Bioblitz[28] experiences and activities promoted within the Italian Long-Term Ecological Research Nework[29] as a member of the European Citizen Science Association.

- Crowdfunding is used to collect funds from private organizations and individuals. It is particularly effective in collecting funds for socially oriented initiatives (e.g., support for disabled people and arts development), but some examples exist also for forest or forest-based initiatives. There are nowadays several platforms managing crowdfunding in Italy, but *Produzioni Dal Basso* (https://www.produzionidalbasso .com/) (in English, the meaning would be Bottom-up Productions) was one of the first launched, currently having 350 projects funded, with a total of €170,600 collected so far. Out of 2600 projects currently open for funds collection, about 30 are directly linked to forest and forest resources (e.g., green care initiatives in forest sites, such as forest schools, and creation of protected areas with a bottom-up approach), while ca. 120 are indirectly linked to them (e.g., they refer to documentary films on forests or artistic installations inspired by forests). Crowdfunding is an effective way of engaging people, who feel they are being part and driver of change.
- Online petition platforms are other forms of citizens' self-engagement in topics that might have forestry relevance. As an example, from 2007 until March 2018, the well-known Italian petition platforms *Firmiamo.it* and *Change.org* have promoted, respectively, 68 and 28 petitions related to forestry issues (including petitions for the protection of Nordic forests and wild animals, such as wolf and bear; for the reduction of the number of forestry workers in the south Italian regions; for the protection of forests against fires; or against the use of forest biomasses to produce energy). While some petitions have only a few number of signatures (less than 50), others were able to collect more than 12,000 signatures in a few weeks.

profits, and their distribution along the value chains) are not open and publicly accessible, i.e., shared with public opinion. On the other, examples of social innovation have a very limited visibility (if any), and small-scale projects are often acting in isolation/ individually (apart from a few examples like the model forest networks), they are *de facto* a large number of microscale or medium-scale examples barely interconnected to each other. In the EU, industrial interests are much more politically relevant than the interests and political strength of private forest owners–managers (small, weak, and poorly represented), and when forest owners' interests are represented, as in the case of the Confederation of European Forest Owners, the large and industrial-oriented landowners from the Nordic and central European countries play a major role, because of their critical mass and recognized key role in national economies.

Clearly, there is a need to increase the capacity to use the information and communication technology to consolidate, enlarge networks, and let the fragmented small-scale examples become more visible to the public and more influential. However, in this case, it must be clear that technology is not enough, if intangible factors such as reciprocal trust, willingness to collaborate and share information and/or organizational innovations such as new types of public-private agreements, and renegotiation of forest ownership rights are lacking.

18.6 CONCLUSIONS

So far, the technological approach is largely predominant in implementing a bioeconomy development strategy in Europe, while very limited attention and investments in R&D are linked with the social dimension of future alternative models of economic development. However, several examples of the social innovation approach do already exist in forestry in those countries with fewer industrial investments. These investments are more relevant in terms of provision of ecosystem services such as well-being, recreation, and health rather than biomass production. Quite obviously, a possible reasonable and feasible path is to pursue the integration of these two approaches, rather than think of them as alternative and mutually exclusive solutions.

However, such an integration is not easy to realize in practice. Two different development paths and innovation models should probably be chosen, depending on the area, cluster, and region's prevailing characteristics. Indeed, this is what can be observed in practice.

In those areas rich in forest resources, high industrial investment capacity and interest, and good logistical connections, the path should be to continue to pursue technological innovation models. Technologies (new materials and new industrial processes, forest information technologies, etc.) will remain the most important instruments for industrial advancement, but transnational and large-scale industrial corporations should integrate them with social issues, finding a way to introduce mechanisms of equity, social inclusion, stakeholders' involvement, and social and environmental responsibility reporting. This will help wood-based companies to create more consensus, increase their reputation, reduce the risk of conflicts and potential international campaigns against their business goals, and thus increase their attractiveness to new investors searching for "impact investments."

In those areas rich in forest resources and limited industrial investment capacity and interest, and/or with limited connections (i.e., with higher logistical costs), or in those areas with poor forest resources, it would seem better to pursue social innovation models based on local endogenous resources, small-medium scale enterprises and networks, social capital (trust, shared social values and norms, and traditional knowledge), civic society engagement, and willingness to be part of the change. They are likely to have more positive and significant impacts on the resilience and productive capacity of forestry based on local communities and economies. Here technology can be seen as an operational instrument for consolidating social relationships (e.g., social networks for social capital building, social cooperatives based on a short value chain able to commercialize their high-quality products worldwide via the web/online shops, social science to increase the knowledge and awareness of nonexperts, and public opinion about the importance of forest resources for human well-being). This approach also seems to be valid in supporting the path toward increasing interest in urban forestry and urban greening.

In both approaches, information and communication technologies in their current trend of globalization, can be powerful and useful instruments. However, the social processes and direct social relations are (and will remain or will return to be) the glue of forest societies at local level in both rural and urban areas.

REFERENCES

Aggestam, F., Pulzl, H., Sotirov, M., and G. Winkel. 2017. The EU policy framework. In *What Science Can Tell Us Series 8, towards a Sustainable European Forest-Based Bioeconomy—Assessment and the Way Forward*, ed. G. Winkel, 19–35. Joensuu: European Forest Institute.

BEPA (Bureau of European Policy Advisers). 2011. *Empowering People, Driving Change: Social Innovation in the European Union*. Brussels: European Commission. https://ec.europa.eu/migrant-integration/index.cfm?action=media.download&uuid=2A18225B-A4EF-443D-9D074439D071447D (Accessed April 14, 2018).

BIC (Bio-based Industries Consortium). 2013. *Strategic Innovation and Research Agenda (SIRA). Bio-based and Renewable Industries for the Development and Growth in Europe*. BIC Brussels. https://ec.europa.eu/research/participants/data/ref/h2020/other/legal/jtis/bbi-sira_en.pdf (Accessed April 14, 2018).

Bock, B. 2016. Rural marginalisation and the role of social innovation: A turn toward nexogenous development and rural reconnection. *Sociologia Ruralis* 56(4):552–573.

Bosworth, G., Rizzo, F., Marquardt, D., Strijker, D., Haartsen, T., and A. Aagaard Thuesen. 2016. Identifying social innovations in European local rural development initiatives. *European Journal of Social Science Research* 29(4):442–461.

Bryce, R., Valero, D., and M. Price. 2017. Creation of Interactive Database of Examples of Social Innovation, Deliverable 3.2, Social Innovation in Marginalised Rural Areas (SIMRA). p. 24.

Cajaiba-Santana, G. 2014. Social innovation: Moving the field forward: A conceptual framework. *Technological Forecasting and Social Change* 82:42–51.

Chen, Y., Liu, X., Gao, W., Yu Wang, R., Li, Y., and W. Tu. 2018. Emerging social media data on measuring urban park use. *Urban Forestry and Urban Greening* 31:130–141.

Chomsky, N. 2017. Who Rules the World? Penguin Books Ltd.

Demming, K. 2016. *Making Space for Social Innovation: What We Can Learn from the Midwifery Movement*. Toronto: Ontario Institute for Studies in Education of the University of Toronto.

EC (European Commission). 2003. *Innovation Policy: Updating the Union's Approach in the Context of the Lisbon Strategy*. COM (2003) 112 final. Brussels: EC. https://web.archive.org/web/20080412213430/http://ec.europa.eu/enterprise/innovation/communication.htm (Accessed April 14, 2018).

EC. 2012. *Innovating for Sustainable Growth: A Bioeconomy for Europe*. COM (2012) 60 final. Brussels: EC. http://ec.europa.eu/research/bioeconomy/pdf/201202_innovating_sustainable_growth_en.pdf (Accessed April 14, 2018).

EC. 2013. Guide to Social Innovation. Brussels: European Commission. https://ec.europa.eu/eip/ageing/library/guide-social-innovation_en (Accessed April 14, 2018).

Esteban, L. S. and J. E. Carrasco. 2011. Biomass resources and costs: Assessment in different EU countries. *Biomass and Bioenergy* 35(1):S21–S30.

Feliciano, D., Slee, B., Weiss, G., Matilainen, A., and T. Rimmler. 2011. The contribution of Leader+ to the implementation of innovative forest-related projects. In *Innovation in Forestry: Territorial and Value Chains Approaches*, eds. Weiss, G., Pettenella, D., Ollonqvist, P., and B. Slee, 87–100. London: CAB International.

Fenton, E. 2017. *Snake Oil or Climate Cure: The Effect of Public Funding on European Bioenergy*. FERB Bioenergy and Forests Briefing Note 2. http://fern.org/sites/default/files/news-pdf/snake%20oil%20or%20climate%20cure.pdf (Accessed March 26, 2018).

Forest Europe. 2015. *State of Europe's Forests 2015*. Madrid: Forest Europe.

FTP (Forest-Based Sector Technology Platform). 2017. *Database and Strategic Research and Innovation Agenda 2020*. Brussels. FTP http://new-www.forestplatform.org/#!/pages/6 (Accessed March 26, 2018).

Fund, C., El-Chichakli, B., Patermann, C., and P. Diechoff. 2015. *Bioeconomy Policy (Part II). Synopsis of National Strategies around the World*. Berlin: Office of the Bioeconomy Council.

Gatto, P., Vidale, E., Secco, L., and D. Pettenella. 2014. Exploring the willingness to pay for forest ecosystem services by residents of the Veneto Region. *Bio-based and Applied Economics* 3(1):21–43.

Hakansson, P. and H. Witmer. 2015. Social media and trust—A systematic literature review. *Journal of Business and Economics* 6(3):517–524.

Illuminati, R. 2015. Innovazione in agricoltura e capital sociale. Impatto della Misura 1.2.4 del PSR 2007–2013 in Umbria. MSc Thesis, Perugia: University of Perugia.

Jenson, J. 2017. Modernising the European Social Paradigm: Social Investments and Social Entrepreneurs. *Journal of Social Policy* 46:31–47.

Kleinschmit, D., Pülzl, H., Secco, L., Arnaud, S., and I. Wallin. 2018. Orchestrating in political processes: Involvement of experts, citizens, and participatory professionals in forest policy making. *Forest Policy and Economics* 89:4–15.

Kluvánková, T., Gežik, V., Špaček, M. et al. 2017. *Transdisciplinary Understanding of SI in MRAs*, Deliverable 2.2, Social Innovation in Marginalised Rural Areas (SIMRA).

Kubeczko, K., Ramesteiner, E. and G. Weiss. 2006. The role of sectoral and regional innovation systems in supporting innovations in forestry. *Forest Policy and Economics* 8:704–715.

Landry, R., Amara, N. and M. Lamari. 2002. Does social capital determine innovation? To what extent? *Technological Forecasting and Social Change* 69:681–701.

Lovrić, M., Lovrić, N., and R. Mavsar. 2018. *Synthesis on Forest Bioeconomy Research and Innovation in Europe*. European Forest Institute, SCAR SWG FOREST. Funded through CASA (Common Agricultural and wider bioeconomy reSearch Agenda) project. https://scar-europe.org/images/FOREST/Documents/SWG_forestry_study.pdf (Accessed April 12, 2018): Joensuu, Finland.

MacCallum, D., Moulaert, F., Hillier, J., and S. V. Haddock. 2009. *Social Innovation and Territorial Development*. Farnham: Ashgate.

Maso, D., Secco, L., and D. Pettenella. 2007. NWFP&S marketing: Lessons learned and new development paths from case studies in some European countries. *Small Scale Forestry* 6(4):373–390.

Matilainen, A., Weiss, G., Sarvašová, Z., Feliciano, D., Nastase, C., and M. Prede. 2011. The role of cooperation in enhancing innovation in nature-based tourism services. In *Innovation in Forestry: Territorial and Value Chains Approaches*, eds. Weiss, G., Pettenella, D., Ollonqvist, P., and B. Slee, 169–188. London: CAB International.

McCormick, K. and N. Kautto. 2013. The bioeconomy in Europe: An overview. *Sustainability* 5:2589–2608.

Moulaert, F., Martinelli, F., Swyngedouw, E., and S. González. 2005. Toward alternative model(s) of local innovation. *Urban Studies* 42(11):1969–1990.

Moulaert, F., Mehmood, A., MacCallum, D., and B. Leubolt. 2017. *Social Innovation as a Trigger for Transformations—The Role of Research*. Luxembourg: Publications Office of the European Union.

Nelson, R. R. and S. G. Winter. 1977. In search of useful theory of innovation. *Research Policy* 6(1):36–76.

Neumeier, S. 2012. Why do social innovations in rural development matter and should they be considered more seriously in rural development research? Proposal for a stronger focus on social innovations in rural development research. *Sociologia Ruralis* 52:148–169.

Nova-Institut GmbH and Bio-based Industries Consortium. 2017. *Biorefineries in Europe 2017*. http://news.bio-based.eu/map-of-224-european-biorefineries-published-by-bic-and

-nova-institute/ and http://biconsortium.eu/sites/biconsortium.eu/files/downloads/Mapping BiorefineriesAppendix_171219.pdf (Accessed March 26, 2018).

O'Driscoll, C., Leonardi, A., and M. Masiero. 2017. *Are Nature-Based Businesses Really Innovative? An Assessment of European Entrepreneurial Initiatives.* Report by ECOSTAR Natural Talents, EU Erasmus+ project. http://www.ecostarhub.com/wp-content/uploads/2017/05/ECOSTAR_WP3_Innovation-Report-2017–1.pdf (Accessed March 26, 2018).

OECD (Organisation for Economic Co-operation and Development). 2005. *OSLO Manual: Guidelines for Collecting and Interpreting Innovation Data.* 3rd Edition. Paris: OECD. http://dx.doi.org/10.1787/9789264013100-en (Accessed April 14, 2018).

Ollonqvist, P., Nord, T., Pirc, A. et al. 2011. Networks and local milieu as a furniture industry innovation platform. In *Innovation in Forestry: Territorial and Value Chains Approaches*, eds. Weiss, G., Pettenella, D., Ollonqvist, P. and B. Slee, 233–253. London: CAB International.

Pettenella, D. and L. Secco. 2006. Small-scale forestry in the Italian Alps: From mass market to territorial marketing. In *Small-Scale Forestry and Rural Development: The Intersection of Ecosystems, Economics and Society*, ed. S. Wall, 398–408. Proceedings of IUFRO 3.08 Conference. Galway: Galway-Mayo Institute of Technology.

Pisani, E., Franceschetti, G., Secco, L., and A. Christoforou. 2017. *Social Capital and Local Development: From Theory to Empirics.* London: Palgrave Macmillan.

Polman, N., Slee, W., Kluvánková, T. et al. 2017. *Classification of Social Innovations for Marginalized Rural Areas*, Deliverable 2.1, Social Innovation in Marginalised Rural Areas (SIMRA).

Price, M., Miller, D., McKeen, M., Slee, W., and M. Nijnik. 2017. Categorisation of Marginalised Rural Areas (MRAs), Deliverable 3.1, Social Innovation in Marginalised Rural Areas (SIMRA). p. 57.

Pülzl, H., Giurca, A., Kleinshmit, D. et al. 2017. The role of forests in bioeconomy strategies at the domestic and EU level. In *Toward a Sustainable European Forest-Based Bioeconomy—Assessment and the Way Forward*, ed. G. Winkel, 36–51. Joensuu: European Forest Institute.

Rametsteiner, E. and G. Weiss. 2006. Innovation and innovation policy in forestry: Linking innovation process with systems models. *Forest Policy and Economics* 8:691–703.

Rimmler, T., Coppock, R., Oberwimmer, R., Pirc, A., Posavec, S., and G. Weiss. 2011. How to Support Firm Competitiveness in Timber Industries? Clusters as Policy Means in Four European Countries. In *Innovation in Forestry: Territorial and Value Chains Approaches*, eds. Weiss, G., Pettenella, D., Ollonqvist, P. and B. Slee, 101–113. London: CAB International.

Rogelja, T., Ludvig, A., Weiss, G., and L. Secco. 2018. Implications of policy framework conditions for the development of forestry-based social innovation initiatives in Slovenia. Paper presented at the 2nd International Forest Policy Meeting.

Sauvée, L. and D. Viaggi. 2016. Biorefineries in the bio-based economy: Opportunities and challenges for economic research. *Bio-based and Applied Economics* 5(1):1–4.

Scarlat, N., Dallemand, J., Monforti-Ferrario, F., and V. Nita. 2015. The role of biomass and bioenergy in a future bioeconomy: Policies and facts. *Environmental Development* 15:3–34.

Schicklinski, J. 2017. *The Governance of Urban Green Spaces in the EU: Social Innovation and Civil Society.* Oxford, UK: Routledge.

Schumpeter, J. 1934. *The Theory of Economic Development: An Inquiry into Profits, Capital, Credit, Interest and the Business Cycle.* Harvard Economic Studies, vol. 46. Cambridge: Harvard College.

Secco, L., Pisani, E., Burlando, C. et al. 2017. *Set of Methods to Assess SI Implications at Different Levels: Instructions for WPs 5 and 6.* Deliverable D4.2, Social Innovation in Marginalized Rural Areas Project (SIMRA), Demonstrator to the European Commission.

Slee, B. 2011. Innovation in forest-related territorial goods and services: An introduction. In *Innovation in Forestry: Territorial and Value Chains Approaches*, eds. Weiss, G., Pettenella, D., Ollonqvist and B. Slee, 118–130. London: CAB International.

Spinelli, R., Magagnotti, N., Jessup, E., and M. Soucy. 2017. Perspectives and challenges of logging enterprises in the Italian Alps. *Forest Policy and Economics* 80:44–51.

Syverud, K. 2017. Building human body parts from wood: Innovative use of nanocellulose. Paper presented at the EFI 2017 Annual Conference Seminar.

The Guardian. 2017. *EU Court Orders Poland to Stop Logging in Białowieża Forest.* Various articles: https://www.theguardian.com/environment/2017/jul/28/eu-court-orders-poland-to-stop-logging-in-bialowieza-forest (Accessed March 26, 2018).

Tyrväinen, L., Plieninger, T., and G. Sanesi. 2017. How does the forest-based bioeconomy relate to amenity values? In *Toward a Sustainable European Forest-Based Bioeconomy—Assessment and the Way Forward*, ed. G. Winkel, 92–100. Joensuu: European Forest Institute.

Tzoulis, I. and Z. Andreopoulou. 2013. Emerging traceability technologies as a tool for quality wood trade. *Procedia Technology* 8:606–611.

Verhegghen, A., Hugh, E., Desclée, B., and F. Achard. 2016. Review and combination of recent remote sensing based products for forest cover change assessments in Cameroon. *International Forestry Review* 18(2):2016–2017.

von Tunzelmann, N. and V. Acha. 2003. *Innovation in "Low-Tech" Industries.* TEARI Working Paper no. 15. Oslo: University of Oslo.

Vuletić, D., Posavec, S., Krajter, S., and E. Paladinić. 2010. Payments for environmental services (PES) in Croatia—Public and professional perception and needs for adaptation. *South-East European forestry* 1(2):61–66.

Watson, A. and G. Dal Bosco. 2014. *Use of Forest Information Technologies and Marketing of Forestry Services and Products.* Bonn: OpenForests.

Weiss, G., Ollonqvist, P., and B. Slee. 2011. *Innovation in Forestry: Territorial and Value Chains Approaches.* London: CAB International.

Winkel, G. 2017. *Toward a Sustainable European Forest-Based Bioeconomy—Assessment and the Way Forward.* What Science Can Tell Us Series 8. Joensuu: European Forest Institute.

Wunder, S. 2005. *Payments for Environmental Services: Some Nuts and Bolts.* CIFOR Occasional Paper 42. Bogor: CIFOR.

ENDNOTES

1. Bioeconomy encompasses the production of renewable biological resources and their conversion into food, feed, biobased products, and bioenergy. It includes agriculture, forestry, fisheries, food, and pulp and paper production, as well as parts of chemical, biotechnological, and energy industries (EC 2012).
2. Insights into these issues are provided in Chapter 15 (motivation of service providers) of this book.
3. Fostering knowledge transfer and innovation in agriculture, forestry, and rural areas has been established as one of the six priorities for the rural development policy 2014–2020 [e.g., the study by Aggestam et al. (2017)].
4. Even if recent trends show rather the emergence of regionalization processes (Winkel 2017) and cluster strategies (Rimmler et al. 2011).
5. Insights into these issues are provided in Chapter 17.
6. It was found that despite increasing demand, the supply of these technologies and related software is still insufficient (Watson and Dal Bosco 2014).
7. Examples of NGOs using reporting, mass media information, and campaigning are the Environment Investigation Agency, Forests Monitor, Forests and the European

Union Resource Network (FERN), Global Witness, World Wildlife Fund, Greenpeace, Friends of the Earth and many others.

8. An example can be seen at http://www.treezilla.org.
9. Examples are the World Conservation Union (IUCN), Forests and the European Union Resource Network, World Rainforest Movement, Taiga Rescue Network, and others.
10. Other types of innovation, i.e., marketing innovation, organizational innovation (OECD 2005), and institutional innovation (Weiss et al. 2010), are not in the scope of this chapter.
11. Particularly relevant have been the European Forest Institute (EFI) Project Center INNOFORCE "Toward a Sustainable Forest Sector in Europe: Fostering Innovation and Entrepreneurship" (2004–2008); the COST Action E51 "Integrating Innovation and Development Policies for the Forest Sector" (2006–2010); and a recent a study realized by the EFI on the behest of SCAR Strategic Working Group on forests and forestry research and innovation (Lovrić et al. 2018).
12. Most biorefineries in Europe are oil-/fat-based (53% of total), mainly producing biodiesel or oleochemicals. Sugar-/starch-based biorefineries are also relevant (28%) and mainly produce bioethanol but also products for use in food, feed or biochemicals (Nova Institute 2017).
13. The contribution of LEADER+ to the implementation of innovative forest-related projects was explored by Feliciano et al. (2011).
14. Specifically, innovation theory, endogenous and neoendogenous development, social capital, socioecological systems, regional development, social enterprises, and entrepreneurship are considered prominent precursors to social innovation in marginalised rural areas [see the studies by Kluvánková et al. (2017), Polman et al. (2017).
15. More information about the EU-funded Horizon 2020 project SIMRA that is at the basis of this definition is available at http://www.simra-h2020.
16. In particular, in the central European countries, "the percentage of innovative forest holdings larger than 500 ha is at least 4 times higher than that of forest holdings with properties smaller than 500 ha" (Rametsteiner and Weiss 2006, p. 695).
17. Outside Europe, an interesting case is the use of remote sensing control instruments by monitoring organizations in charge of keeping very large forest areas and a high number of companies under control, such as in the case of the Cameroonian government initiative to contrast illegal logging (Verhegghen et al. 2016).
18. See, for example, the Chimica Verde-associated members at the http://www.chimica verde.it.
19. Impact investing is a type of investing (that can be made by companies, organizations, and funds) that aims to generate measurable, beneficial social or environmental impacts alongside a financial return. This emerging finance sector was initially developed by the intervention and pioneer applications of some institutional investors (e.g., European development finance institutions).
20. http://www.learnscapes.co.
21. http://www.csmon-life.eu.
22. http://lifemipp.eu/mipp/new/index.jsp?language=en_US.
23. http://usavereds.eu/en_GB/.
24. http://www.naturachevale.it/en/.
25. http://www.inaturalist.org/projects/osservatorio-italiano-specie-aliene.
26. http://www.inaturalist.org/projects/farfalle-d-italia.
27. http://www.inaturalist.org/projects/italian-herps-betha.
28. http://www.bioblitzitalia.it/index.html.
29. http://www.lteritalia.it/it/content/citizenscience.

19 Globalization and Employment in Forests and Tree Product Value Chains
Are Women Losing Out?

*Jennie Dey de Pryck, Marlène Elias,
and Bimbika Sijapati Basnett*

CONTENTS

19.1   INTRODUCTION

Globalization is transforming employment in forestry and tree product value chains, including related services, in ways that are bringing substantial benefits for some people but losses for others.[1] However, such changes are not linear. Today's winners could be tomorrow's losers, and vice versa. Few rural areas are untouched by global processes of market integration fostered by trade liberalization; deregulation; and improved transportation, information, and communication technologies, so most rural dwellers are increasingly engaging in the global market economy as both producers and consumers. Through technical changes that allow different parts of a production process to be split and geographically dispersed (Ghosh 2009), and the worldwide spread of modern production technologies (Robinson and Carson 2015)[2] forest- and tree-product producers are being linked to consumers along increasingly complex and lengthening value chains that stretch across countries and continents. At the same time, the composition of the labor force engaged in forestry and tree product value chains keeps changing as new opportunities in globalizing labor markets stimulate the growth of both agricultural and nonagricultural rural jobs, while also promoting rural outmigration to urban areas. This is particularly evident in the Global South where, despite significant regional variations, these migration flows tend to be male-dominated (FAO 2017), resulting in increasingly feminized forestry and agroforestry landscapes in many countries.

While growing demand for industrial roundwood and fuelwood (HLPE 2017) is maintaining employment in these traditional forest subsectors (Agrawal et al. 2013), some of the most striking employment impacts of globalization are visible in the more rapidly globalizing value chains of exotic or early season foods (fruits, nuts, coffee, and cocoa) and tree products for the food industry (palm oil) or nonfood industrial use (biofuels, argan oil, shea butter, and palm oil) (Robinson and Carson 2015). With labor-intensive production and processing concentrated in developing countries, global firms are benefiting from fiscal advantages, better access to specialist skills, lower labor costs, and/or fewer labor and environmental regulations. While employment in such value chains is opening up possibilities for many of the poor to escape poverty, there is increasing concern that global firms are using a business model that exploits low-cost, low-skilled labor under insecure casual contracts without social benefits, often in physically poor or risky conditions.[3] Women tend to predominate in these jobs.

An additional concern is that these largely monoculture plantation and agricultural tree crops are often transforming forested and agroforestry landscapes in ways that undermine existing local employment and livelihoods through deforestation, and loss of biodiversity and the risk-mitigation capacity of mixed land use systems. This entails displacement of local people without compensation for their lost land and with little regard for the importance of forests in their culture and spiritual rituals. At a more global level, these processes are often destroying valuable global public goods provided by forests and trees, especially ecosystem services that are vital to combat climate change, biological and cultural diversity, and recreation and tourism. These all have implications for employment.

A corollary to these multifaceted threats, globalization processes are provoking a broader "internationalization of concerns" that embrace a range of interrelated economic, social, cultural, ecological, political, and rights concerns (Farcy et al. 2015). These are reflected in a growing number of international agreements, treaties, and conventions, for example, on climate change, biological diversity, desertification, and labor standards. The overarching sustainable development goals (SDGs) are eliciting multiscale responses to address such environmental, employment, poverty, social equity and gender equality, and human rights issues. These responses are complex, often involving policy and ethical dilemmas that call for politically and socially sensitive trade-offs. They also call for new global governance arrangements that involve multiple state, private, and civil society stakeholders—working together across and within countries and regions, from local to international levels.

The forestry and agroforestry sector is also grappling with broader concerns that transcend its long-standing focus on environmental issues (deforestation, desertification, reforestation, environmental degradation, and biodiversity). These include global concerns about the people who inhabit and manage forests and agroforests. Influenced by the discourse on and responses to the broader international concerns mentioned earlier, societal responses include legislation and other measures to protect the rights of local and indigenous peoples to forest and tree resources; promote more effective community governance systems; and improve implementation of labor laws and regulations, certification of forest and tree products, and schemes such as payments for ecosystem services and reduction in emissions from deforestation and degradation (REDD+) schemes.[4] These responses have implications for the ways forests are managed and the employment of the millions of women and men who inhabit forested and agroforestry landscapes worldwide. Hence, any evaluation of the effects of globalization on forests and agroforestry and their peoples must consider both the costs and benefits that it has brought to the lives and employment of a wide range of people who are affected in different ways as well as the health of our planet.

In what follows, we undertake this task, with a specific focus on the gender-differentiated impacts of globalization on employment in forests and tree value chains and in services supporting forest ecosystems and forest and tree products. We focus on the Global South, where employment conditions in globalizing industries, especially for women, tend to be more exploitative than in the North. After reviewing available information on employment in the sector and flagging some critical data gaps, we identify different types of employment within forestry and some major tree product value chains, and some gendered impacts of globalization on the terms and conditions of such employment. Although women's employment status in the sector varies by factors such as socioeconomic class, ethnicity, religion, education, marital status, and stage in the lifecycle, we show that broader trends along gender lines are still discernible. Following Benerìa et al. (2000), Kabeer (2012), and Razavi et al. (2012), we argue that these globalization processes operate through institutions[5] and structures of the economy that are intrinsically *gendered*. These affect the ways in which women and men engage in global value chains and the terms and conditions on which they do so. Likewise, gendered institutions mediate women's and men's opportunities to participate in and benefit from policies and programs to protect global forest public goods, such as REDD+ schemes, or to certify forest and tree products. These institutions, and the outcomes of globalization, are context specific, varying across and within countries, by sector, culture, gender, and class (among other factors). Drawing on case studies, we illustrate how broad globalizing processes interact with context-specific economic institutions and sociocultural norms that govern gender relations at the household and community level. These mediate women's access to and control over productive resources (land, trees, labor, and capital) and income and what are considered acceptable roles and responsibilities for women in forestry and tree value chains, including wage work outside the household (Barrientos 2001; Dunaway 2014). As employment outcomes are also affected by public and private policies regulating the chains at global, national, and subnational levels, we conclude by flagging some entry points and dilemmas that governments, global firms, and the international community need to address.

19.2 EMPLOYMENT IN FORESTS AND AGROFORESTRY: TRENDS AND DATA GAPS

Some 1.6 billion people worldwide (22% of the world's population) are estimated to derive part of their livelihoods from forests/tree-based systems (Vira et al. 2015).[6] However, it is difficult to untangle the contributions of forests to employment as much of the work is informal and often interspersed with subsistence activities.[7] The most authoritative estimates of employment are those of the Food and Agriculture Organization of the United Nations (FAO), which estimated that in 2011, the formal forest sector employed some 13.2 million people across the world (about 0.4% of the global workforce) while at least another 41 million were employed in the informal sector (FAO 2014). These are likely to be underestimates as the formal sector figures only covered forestry and logging activities, sawnwood and wood-based panel production, and pulp and paper production and excluded other major forest-based activities such as rubber and oil palm plantations and related manufacturing as well as nature-based tourism. The calculations for informal employment were heavily biased to wood fuel and charcoal production for which more data were available and most likely underestimated the share of employment in nonwood forest products.

The absolute size of the workforce employed in the formal forest sector was the highest in Asia and Oceania (6.9 million), followed by Europe (3.2 million), Latin America and the Caribbean (1.3 million), North America (1.1 million), and Africa (0.6 million). However, because of the larger population in Asia and Oceania, the share of the total workforce employed in the formal forest sector was only 0.3% compared with 0.9% in Europe, which had the largest share of the total workforce employed in the sector (FAO 2014).

Global employment trends in forestry and logging decreased slightly over the period 2000–2010, for reasons that varied by region. The decline in Europe and North America was triggered, at least in part, by the global economic downturn in 2008–2009 when the construction industry entered into a period of deep recession. The employment in China fell by 21% due to a decline in export markets for processed forest products (such as furniture) and in house building, while in Australia and New Zealand, rising productivity was accompanied by a fall in employment. Other regions (Latin America, west and central Africa, and south and southeast Asia) experienced increased employment in the sector as a result of factors such as increasing exports of forest products, afforestation, reforestation, and industrial development. Brazil and

Uruguay also benefited from legislation to promote the development of the forestry sector that led to long-term employment benefits (Whiteman et al. 2015). Despite the declines in employment, value-added contributions and exports increased in most regions in the period of 2000–2006 (Agrawal et al. 2013).

Relatively few countries collect sex-disaggregated data on employment in forestry. For the 29 countries, representing 17% of the global forest area, that reported time series data for the five-yearly global forest resources assessments (FRAs), the share of female employment in forestry and logging rose from 20% in 1990 to 32% in 2010 (FAO 2016). Countries with the largest share of female employment in 2010 were Mali (90%), Mongolia and Namibia (45%), and Bangladesh (40%). The FRA 2015's commentary suggests that the increased share of female employment was due to the use of different definitions and concepts, better data collection methods, and a real increase in female labor in the sector.

FAO's *State of the World's Forests 2014* (FAO 2014) provides complementary estimates based on International Labour Organization (ILO) data for the formal forest sector. This dataset contained sex-disaggregated statistics on employment in the wood-processing industry for many countries, although much less information was available on employment in forestry (FAO 2014). Overall, women accounted for 24% of formal employment in the sector, with a relatively low proportion of total employment in forestry at 18% but higher in solid wood processing and in pulp and paper, at 23% and 27%, respectively. The FRA and FAO 2014 data are not strictly comparable as the FRA included values for both formal and informal employment in forestry and logging, while the ILO data covered only formal employment and in a broader subsector that included forestry, solid wood processing, and pulp and paper. Since these are the only available global estimates of female employment in the sector—and they do not cover all subsectors—there is a clear need to collect more comprehensive and comparable sex-disaggregated data.

While women account for about a quarter of the global workforce in the formal forestry sector and rather more in the informal sector, little is known about the type of work women (and men) perform in these areas. Rather more is known from case studies about men's and women's work in informal forestry and agroforestry activities, whether for subsistence or income. Although such data are not generalizable as gender roles vary considerably between and within regions and countries (Sunderland et al. 2014), they provide valuable insights for policy at a national or subnational level. Macrolevel data on employment in agricultural tree value chains, including on the growing feminization of labor-intensive tree commodity processing both in factories and in less visible homeworking, are also sparse despite the importance of many such commodities in international (and domestic) trade and gross domestic product. Again, case studies provide evidence of the complex gender division of labor within such value chains, as, for example, in the shea butter value chain (Box 19.1; Figure 19.1). Better data are needed to identify areas where women and men are concentrated, opportunities to increase women's involvement and returns, and gender-specific training needs (Whiteman et al. 2015).

A prerequisite to analyzing the gendered impacts of globalization on employment, and addressing negative impacts such as on the *quality* of employment, two data gaps need urgently filling. These concern the common failure to (1) disaggregate

BOX 19.1   GENDERED WORK IN GLOBAL
SHEA BUTTER VALUE CHAINS

Shea kernels and butter derive from the *Vitellaria paradoxa* tree that grows in Africa's Sudano–Sahelian region. Women across the region, but primarily in west African countries, are responsible for nut collection and processing into kernels or butter. In Burkina Faso, rural women processors sell their shea kernels to male and female vendors who market these in progressively larger kernel quantities moving downstream the value chain (Elias and Arora-Jonsson 2016). Via wholesalers (mainly men) who work on behalf of large-scale Ghanaian importers (Chalfin 2004) or urban-based Burkinabè exporters, shea kernels make their way to west African coastal countries and overseas to European, Indian, Malaysian, and Japanese agrofood industries and refineries (Rousseau et al. 2016). Shea butter is then exported from one European country to another for incorporation in foods, cosmetics, and pharmaceuticals, which are consumed worldwide.

FIGURE 19.1   Burkinabé woman washing shea kernels during the shea butter extraction process. (Courtesy of Barbara Vinceti, Bioversity International, Rome.)

labor data by employment status and gender and (2) collect systematic and comparable data (disaggregated by gender) on the terms of employment (types of contracts, occupational segregation, wage rates,[8] and benefits) and working conditions including health and safety provisions. The former is important as women's labor is usually overrepresented in the categories of unpaid family labor and unskilled wage labor and the latter as women often predominate in casual, insecure jobs with low wages. A complementary data gap concerns women's ubiquitous unpaid care work that enables men to focus on remunerated work while limiting women's opportunities to engage in productive own-account or wage labor. While case studies offer valuable insights into these issues, such macrolevel data are essential to increase women's visibility among policy makers to ensure more empirically informed, socially just, and equitable decision-making in the sector.

19.3 GENDER-DIFFERENTIATED IMPACTS OF GLOBALIZATION PROCESSES ON EMPLOYMENT IN FORESTRY AND TREE VALUE CHAINS

Each forest enterprise or tree product value chain is likely to involve a variety of labor regimes with diverse outcomes under globalization for different categories of workers, depending on class, gender, and generation (Hall et al. 2017). We make a further distinction, arguing that employment and livelihood outcomes for different worker categories also depend on the following:

1. The type of enterprise, further differentiated by the following:
   - Production or collection model [small- and medium-scale commercial production by independent farmers (individuals, families), collective enterprises, contract farming, plantations, and estates (owned by multinational/national companies, often with on-farm processing, with or without contract farmers)]
   - Processing model (artisanal, multinational/national companies, and outsourced homeworking)
   - Type of market (local, national, or export market)
   - Type of trade model (conventional markets versus alternative/ethical trade, often third-party certified, or market niches that offer a price premium)
2. Their employment status within these enterprises (own-account producers, salaried managers/supervisors, and paid or unpaid workers)[9]

Rural dwellers engage in several ways in one or more of these enterprises or value chains, working at different moments, on different days or in different seasons as independent producers (sometimes hiring in labor), unpaid family workers, cooperative members, and wage laborers. They might sell products in both export and local markets, for example, oranges in Morocco (Sippel 2015) and in both alternative and conventional markets, for example, organic coffee in Uganda (Bolwig and Odeke 2007). These family and individual risk-reducing and income-smoothing strategies are flexible, integrating new opportunities brought by globalization while helping ride out negative impacts. Nevertheless, integration as producers or workers into global value chains creates new vulnerabilities for the rural poor that are often gendered precisely because access to and the conditions of these jobs are determined by local institutions that are intrinsically gendered.

19.3.1 GENDERED EMPLOYMENT OUTCOMES OF GLOBALIZATION IN DIFFERENT PRODUCTION MODELS

19.3.1.1 Small- and Medium-Scale Production by Families and Individuals

19.3.1.1.1 Collecting and Processing Forest Products

There is considerable evidence of marked gender differentiation in rights to and practices in collecting and processing forest products that vary by culture and context within and across countries and regions (Agrawal et al. 2013; Sunderland et al. 2014; Vira et al. 2015; Colfer et al. 2016). A comparative study in 24 countries in Africa, Asia, and Latin America using household level data from the Poverty Environment Network (PEN) project, found that globally, men and women contribute almost equally to the value of household income from unprocessed forest products (Sunderland et al. 2014). There were marked regional differences with men bringing considerably more income in Latin America. The opposite trend occurred in Africa and their contribution was similar in Asia. Globally, men brought a much higher share of processed forest product income (61%) than women (25%), with male income dominance uniform across the three regions. Overall, most of the products were used for household consumption. Although the specific gender specialization by product varied by geographical region and culture, the PEN study confirmed observations in the gender-focused literature that some forest products tend to be ascribed to women (especially among plant foods—Figure 19.2) or men (structural and fiber and for timber), while women and men worked jointly or in complementary

FIGURE 19.2 Women collectors of wild foods, Indonesia. (Courtesy of Icaro Cooke Vieira, Center for International Forestry Research, Bogor, Indonesia.)

ways on other products, for example, on Brazil nuts in Latin America and bush mango in Central Africa.

New globalizing markets for forest products such as shea butter, argan, gum arabic, locust bean, neem, Brazil and other nuts, and mushrooms (Vira et al. 2015) may solidify or change these customary labor roles, with different gender outcomes. Where women and men work jointly, the outcomes may be equitable. In other cases, new outlets or improved remuneration for female products formerly produced for subsistence or local trade with low returns have often increased men's interests in these "female" resources. This trend is typified by the example of shea nut collection (Box 19.2). The male takeover of "female" production and trade activities as these gain economic value has been reported in various contexts and for multiple agricultural (Dey 1981; Njuki et al. 2011), horticultural (Dolan 2001), and tree crops (oil palm: Martin 1984; shea: Elias and Carney 2007; argan: Biermayr-Jenzano et al. 2014). This is particularly relevant in sub-Saharan Africa, where women and men have distinct "purses" and financial responsibilities within the household, and where women's income-earning opportunities are traditionally more limited than men's. Since there is a wealth of evidence that women's income is more often spent on food and children's health and education than income controlled by men (FAO 2011; Sunderland et al. 2014), changes in gender-specific production and trade roles have implications for women's independent control over forest product income and their families' welfare.

Another concern is that although new global markets for forest products can offer both men and women new employment opportunities, the main benefits often go to downstream actors rather than rural collectors and processors, particularly women. This is typified by shea butter and argan, whose value addition is largely captured by intermediaries, exporters, importers, refiners, and retailers (Carr et al. 2000; Biermayr-Jenzano et al. 2014; Elias 2016). Efforts to organize women shea nut and argan nut collectors and processors to use new productivity- and quality-enhancing processing technologies and improve their access to international markets remain

BOX 19.2   MALE ENCROACHMENT ON WOMEN'S SHEA NUT COLLECTION

As international shea markets grow and shea market prices increase, (young) men in Burkina Faso who are not traditionally involved in shea nut collection are entering this activity. These men collect shea nuts on a bicycle, motorcycle, or donkey cart, which helps them canvass greater and more distant areas than women. This gives them an advantage over women, particularly as rising prices coupled with years of poor yields require collectors to explore more distant areas to find untapped trees. Men's involvement in this activity threatens women's monopoly over the production node of the value chain. The more remunerative echelons of the chain continue to be dominated by men, whose expansive trading networks and access to capital permits them to exploit emerging opportunities (Chalfin 2004; Elias and Arora-Jonsson 2016).

small scale, with most women still largely dependent on intermediaries for market access (Carr et al. 2000; Biermayr-Jenzano et al. 2014; Elias 2016).

19.3.1.1.2 Employment in Agroforestry, Orchards, and Plantations

The gender division of labor as well as gender-specific asset endowments often result in different types or species of (tree) crops being primarily associated with women or men. For instance, women's rights to land and trees are often mediated through their relationships with men, such as through marriage, divorce, and widowhood, and are thus often less secure than men's rights (Hecht 2007; Mwangi et al. 2011; Sunderland et al. 2014). The customary division of labor often precludes women from clearing land, which would give them usufructory rights, permitting them to grow tree crops (Dey 1981; Dancer and Tsikata 2015). When land belongs to men, women are frequently prohibited from planting trees for themselves as this can be considered a land claim. Women may also have less incentive to invest in land that they do not have secure claims over. These weaker land claims partly explain women's greater reliance on common property resources in many countries particularly in Latin America and Asia (Agrawal 2001; Agarwal 2002; Sunderland et al. 2014). Differences are thus manifested across genders in relation to the extraction and sale of products from planted versus spontaneously growing (or "wild") trees and to the physical spaces where trees are located (Fortmann and Bruce 1988; Howard and Nabanoga 2007).

Nonetheless, both customary practices and more recent legislation in some countries mean that women can and do own and farm tree crops in some areas. For instance, in the oil palm plantation area in Ghana's western region, where inheritance is matrilineal, some women inherited oil palm farms. In contrast, the mango production area in the eastern region has a patrilineal inheritance system that prevents women from inheriting land, so only a few women are owners/managers of mango farms (Yaro et al. 2017). In western Ghana, where cocoa farms are usually owned by men, women received gifts of land planted with cocoa trees from their husbands in compensation for their labor on their husband's cocoa farms (Quisumbing et al. 2003). Agarwal (1994) found that in Sri Lanka, women sometimes received coconut trees as dowry, and their brothers would periodically send them a share of the harvest. In some countries, women are able to enter into sharecropping arrangements and buy or access land collectively, often with the help of nongovernmental organizations (NGOs) (Agarwal 2003; Lastarria-Cornhiel et al. 2014). For example, leases and share contracts (in oil palm and citrus) are becoming more common in commoditized areas of northern Ghana, which facilitates women's access to sharecropping (Dancer and Tsikata 2015). However, in other cases, as globalization spurs the expansion of rental, lease, and land markets for plantations and commercial farming for global markets, women's customary weaker access to land has sometimes been further eroded (Yaro et al. 2017). Many Indonesian palm oil plantations were developed without consultation or compensation for the people who live on the land and were dispossessed of food and cash crops, medicines, and other forest materials. Because women were seldom included in the decision-making processes regarding oil palm development, their interests were widely ignored with the result that they lost access to large tracks of swidden land that they had informally owned and/or managed for household food provisioning (Li 2015; Elmhirst et al. 2017). The result is increasing resort to wage labor among poor women as well as men (Figure 19.3), although there is no clear evidence to establish whether the benefits outweigh the losses and burdens.

Considerable evidence, reviewed by FAO (2011), shows that women farmers are as productive as men if they have the same level of inputs. For example, Kavoi et al. (2003) found that in Tanzanian tea, where access to inputs was relatively gender equitable, female-managed farms were, in general, more efficient and profitable than male-managed farms (Figure 19.4).

FIGURE 19.3   Oil palm laborers going to work, Indonesia. (Courtesy of Icaro Cooke Vieira, Center for International Forestry Research. Bogor, Indonesia.)

FIGURE 19.4   Tea grower, Tanzania. (Courtesy of Nkumi Mtimgwa, Center for International Forestry Research, Bogor, Indonesia.)

However, most evidence points to widespread gender asymmetries in access to inputs and correspondingly lower productivity among women. The challenge is thus to close the gender gap in access to assets, knowledge, and inputs. Access to labor can be subject to significant social constraints as Hill and Vigneri (2014) found in Ghana cocoa where poor men could draw on mutual labor exchange (i.e., working on each other's farms in rotation without pay) for physically demanding (male) tasks such as tree felling. Since labor exchange groups were separate for women and men, poor women were obliged to hire male labor, if they could afford it. While improving access, particularly to land and trees, is likely to involve changes to laws, customary rules, and social norms, development and other actors are contributing to closing these gaps in creative ways. For example, the world's largest chocolate company, Mondelēz, is working directly with cocoa farmers through its Cocoa Life program in Côte D'Ivoire, Ghana, Indonesia, India, Dominican Republic, and Brazil to provide training and community planning skills. The activities also aim to close gender gaps, for example, through capacity building in farming, business management and financial skills, establishing village savings and loan associations to increase women's access to finance, and promoting women's membership in cooperatives (IFC 2016). Mondelēz partnered with the International Finance Corporation and CARE International in Indonesia to evaluate gender gaps and opportunities and define very specific interventions for women cocoa farmers. The United Nations Industrial Development Organization (UNIDO) trained women entrepreneurs in Chefchaouen, Morocco, in better methods for harvesting olives, controlling the oil quality, and organizing a marketing system (packaging, labeling, storage, and promotion). Productivity increased by up to 40% and selling through kiosks in town, rather than "at the farm gate," helped increase sales by at least 85%. Overall earnings have doubled (FAO et al. 2010).

There is some evidence of globalization triggering changes in the gender division of labor and increasing the monetization of family labor, especially male labor, as well as the use of hired labor. This is particularly notable among richer families integrating into global markets, where, in some cases, social norms are legitimizing the exclusion of women from remunerated production for these markets. Hall et al. (2017) found that large- and medium-scale commercial mango farmers in Ghana and coffee producers in Kenya were increasingly resorting to paying family members to work on their farms, particularly adult sons, as a way of encouraging them to remain or return from work elsewhere. As these commercial farms became more established, women were often excluded from farming or moved into trading. Some wealthy mango farmers in Ghana did not want their wives to work with them so they would not know how much their husbands earned and demand a greater share of family income (Yaro et al. 2017). In Uganda, men gave their wives a small part of their organic pineapple income for their personal needs, although women provided limited labor for pineapples, whereas in the case of organic coffee, for which women did the majority of the work, men directly managed all the coffee income. A possible reason could be that, as a relatively new crop, pineapple is less subject to customary male control as in the case of the long-established coffee crop (Bolwig and Odeke 2007).

In the Amazon estuary in Brazil, where the açaí palm economy has boomed since the 1970s in response to growing demand for palm hearts and, more recently, the açaí fruit, local families have intensified forest management and agroforestry techniques (Brondizio et al. 2014). As production has expanded, it has become increasingly gendered, with men tending to control and work on management, production, harvesting, and marketing and women participating only in cases where they share land ownership with their husband or children, or at best, in processing stalls. The gendered nature of the açaí economy has contributed to pushing women out of agriculture and to increasing women's migration to urban areas for work and study.

In the case of poorer farm families, integration in global value chains can increase the workload of women who, if unpaid with little or no say over the use of the income, may have limited incentives to work in these new value chains. These increased workloads can also reduce the time available for their own independent production, with consequences for household well-being and women's autonomy, particularly in African countries where men and women often have separate purses and responsibilities (Dancer and Tsikata 2015). However, such views need to be nuanced. For example, in Uganda, women felt that their increased workloads in organic coffee controlled by their husbands were worthwhile as the increased income benefitted their families (Bolwig and Odeke 2007).

With male outmigration, spurred by new opportunities in a globalizing economy, women often take over larger labor roles, although the extent to which they gain in autonomy depends on local social norms. Yabiku et al. (2010) reviewed studies in Mexico, Morocco, Armenia, Guatemala, and Bangladesh that pointed to rural women gaining decision-making and management responsibilities in their husbands' absence. Sijapati Basnett's (2011) research in Nepal in contexts where the vast majority of men migrated for foreign employment and/or for seasonal work in neighboring cities found that women took over the majority of the roles except for certain traditional male tasks, particularly "ploughing," that were attached with sexual connotations. This caused acute labor shortages during peak agricultural seasons. Migration had an ambiguous effect on decision-making at the household level. Women in nuclear households, with husbands migrating for 2–3 years at a time, tended to experience a greater increase in overall decision-making than those in joint or multigeneration households with husbands migrating for seasonal work only. Even in nuclear households, in communities with high levels of gender inequality, the women left behind were merely responsible for executing orders that their husbands relayed from afar and/or were only able to exercise greater voice and autonomy in household decision-making while their husbands were away.

19.3.1.2 Employment in Contract Farming

As global agribusinesses expand to meet the growing demand for exotic or early season foods, contract farming for tree crops, such as cocoa, coffee, tea, fruit, olives, oil palm, as well as sugar cane, cotton and horticultural crops, is often seen as a win–win strategy to integrate smallholder family farms into national and transnational value chains (World Bank 2007; FAO 2009). There are various different contractual arrangements (Eaton and Shepherd 2001; UNCTAD 2009), some of which are linked to nucleus estates and processing plants as, for example, in the nucleus estate-outgrower model.[10] These schemes help reduce management and supervisory costs and secure investment, while minimizing land dispossession; enhancing smallholder access to markets, credit, and technology; and improving local incomes (von Braun and Meinzen-Dick 2009; Hall et al. 2017). Although numerous governments have supported such schemes (UNCTAD 2009; Vermeulen and Cotula 2010), the large volumes and the rigorous quality standards required by global companies mean that the transaction costs involved in dealing with large numbers of smallholders can create incentives for the companies to deal with large-scale (male) producers (Vermeulen and Cotula 2010). Dolan (2005) observed that the stringent "farm to fork" traceability systems of supermarket chains in Europe, most notably in the United Kingdom, had resulted in a substantial decline in smallholder sourcing of fruit and vegetables in Kenya by the end of the 1990s. Furthermore, in smallholder schemes, there is often a risk of local elite capture, and some contractual arrangements can result in smallholders being caught in risky and exploitative contracts (Reardon and Barrett 2000; Oya 2012; Smalley 2013; Hall et al. 2017), although there is evidence that farmers also cheat. For example, they may use the company's fertilizer on noncontracted crops or side sell to alternative markets (Smalley 2013).

Few agribusiness companies have explicit corporate gender policies and strategies, so women's involvement is generally governed by prevailing sociocultural norms and practices that tend to penalize them in relation to the costs and benefits (Wonani et al. 2013). In other cases, companies have imposed their norms on local communities, especially by dealing only with men on the implicit assumption that they are household heads with full decision-making power. These practices in the Indonesian oil palm area, where ethnic groups were traditionally more gender-egalitarian, have imbued men with a new authority by giving them the management of smallholder palm oil plots (called "plasma") linked to nucleus estates, while weakening women's rights and autonomy (Elmhirst et al. 2017). In most countries, women are rarely the contract holders, although they often do a large part of the work as unpaid family labor. Men generally obtain the contracts because they have more secure access to and control over land, as well as greater access to family or wage labor, capital, and technology (Dolan 2001, 2005; Maertens and Swinnen 2009; Razavi et al. 2012; Dancer and Tsikata 2015). There are nuances, as Rocca (2016) found when Zambia Sugar Plc. expanded operations in response to changes in the EU sugar trade regime in 2007 and the production companies supplying Zambia Sugar took on a small number of smallholder outgrowers. Although the original contractees were almost entirely male, after some years, between 16 and 27% were women who had inherited the land or lease (Wonani et al. 2013; Rocca 2016). This was not due to company policies but reflected changing attitudes to women inheriting land, reinforced by the 1989 Zambia Intestate Succession Act that stipulated widows' rights to at least 20% of the inheritance if there was no will. In northern Ghana, women represented 12% of the 1200 organic mango outgrowers of the Integrated Tamale Fruit Company, a significant finding as women represented between 2 and 10% of members of mango producer groups in the country as a whole. The reason can be partly, at least, explained by the fact that the company restricted outgrowers to 0.4 ha per individual, and since mango was far more profitable than other crops, the incentive for a family to register multiple members made it easier for women to join on their own account. Also, chiefs made considerable areas of community land available for community members wishing to grow mangoes. The women, who participated in this outgrowing scheme, used their earnings for household expenses (King and Bugri 2013). However, if contract farming brings low returns, men are less likely to engage. For example, in Meru County, Kenya, women outgrowers produced French beans on small plots but as their earnings were limited, better-off smallholders, especially men, chose not to participate in the scheme (Hakizimana et al. 2017; Hall et al. 2017).

Nonetheless, women and children often become "shadow workers" since they contribute invisibly by providing largely unpaid labor for family farms controlled by their husbands (Dolan 2001; Dolan and Sorby 2003).[11] This has been observed for mangoes in Ghana (King and Bugri 2013) and oil palm in Indonesia (Sijapati Basnett et al. 2016). Yet the lack of control over one's production can be a disincentive for women to participate. A palm oil processing company in Papua New Guinea successfully addressed this problem by paying men and women separately, with women workers receiving payment directly into their own bank accounts, and by hiring more female extension agents to provide technical assistance [Koczberski 2007, cited by Rubin and Manfre (2014)]. However, the result can be intrahousehold conflicts and struggles to secure cash payments or other rewards for family labor, depending on women's bargaining power and alternative income earning opportunities (Oya 2012; Smalley 2013; von Bülow and Sørensen, cited by Rubin and Manfre (2014) for tea in Kenya). Other unintended consequences can also arise from such changes in gender relations. Bolwig and Odeke (2007) found that women accepted a greater workload after their households converted to contract farming for organic export-oriented coffee production in Uganda, despite their lack of control of the income, as the increased income was beneficial to their households.

Where outgrower and contract farming schemes are linked to a plantation or estate, with or without processing facilities, there may be significant opportunities for casual wage employment as well as permanent and seasonal work in the processing mills and on the nucleus estate. These opportunities vary by crop: for example, Zambia sugar cane is seen as a man's crop as men do all the strenuous cane cutting. Although women do about 15–30% of the work, they are confined to planting, weeding, and some disease control and irrigation, tasks that are largely casual, while men are more likely to have fixed term or seasonal work with higher remuneration (Wonani et al. 2013; Rocca 2016). In a community in northern Ghana, women performed 58% of the work in mango production on a nucleus estate and processing plant, perhaps compensating for their reduced access to land for their own farming (Tsikata and Yaro 2014). Women accounted for 40% of employees in the Integrated Tamale Fruit Company across the plantation, nursery, pack house, beekeeping, and oversight of the outgrowers. They represented 72% of the seasonal workers (working 3–4 months a year) but only formed 15% of the permanent employees who also enjoyed additional benefits such as paid holidays, employer contributions to social security and pension schemes, accommodation, and medical care (King and Bugri 2013).

A family's involvement in contract farming, or work on nucleus estates, or in processing plants, can have negative consequences by displacing food or cash crops that were often under female control (Dolan 2001). Tanzania's sugar industry displaced food crops commonly controlled and farmed by women [Sulle and Smalley 2015 in the study by Hall et al. (2017)], while in Zambia, the expansion of sugarcane contract farming was often at the expense of traditionally female-marketed crops such as groundnuts (Rocca 2016). In Kalimantan, Indonesia, the loss of forest to oil palm production has affected diets, with less pork (hunted by men) and fish that were commonly trapped by women in small streams. Women's opportunities to earn income from handicrafts using rattan and bamboo have also declined (Elmhirst et al. 2016).

19.3.1.3 Cooperatives

Cooperatives and other group farming enterprises can provide a safe mechanism for women to engage in agriculture and tree crops in cultures that consider supporting individual women entrepreneurs as threatening to male dominance within households (Dancer and Tsikata 2015). They can help women access land, other assets, and services in countries where these are generally male controlled. The membership of cooperatives and other groups can also help women retain their earnings, to spend on what they regard as priorities (Biermayr-Jenzano et al. 2014).

Being organized in cooperatives is often a precondition for producers to access alternative markets, such as certified organic and fair trade channels entailing agreements with socially motivated overseas companies. Cooperatives reduce transaction costs in supplying large quantities of quality products to international clients and can more easily meet certification procedures and product standardization. Their members can also benefit from external financial and technical support and services.

Women's benefits from cooperative membership partly depend on whether they belong to women-only or mixed cooperatives. Sometimes, cooperatives are women-only if the product (for example, argan oil) is a traditional female product, while in other cases, they may be promoted as women's cooperatives by an international buyer committed to corporate social responsibility as, for example, the fair trade coffee cooperatives, Café Femenino in Peru and Las Hermanas in Honduras (Rubin and Manfre 2014). A deliberate "gender-smart" activity, Café Femenino charges a premium that is passed on to the women farmers (IFC 2016). Women-only cooperatives are also able to provide more specific services for women, as, for example, in some of the argan cooperatives in Morocco whose services included literacy, financial management, home economics courses (cooking, sewing, nutrition, and health), daycare facilities, and shops with lower prices as the cooperatives bulk buy (Biermayr-Jenzano et al. 2014). Social norms often restrict women from exercising leadership roles and voice in mixed cooperatives and associations, which mean their gender-specific interests are not represented. As Nestlé noted, "women do more than two-thirds of the work involved in coffee farming in Kenya. However, fewer than 5% of leadership roles in coffee cooperatives in the country are currently held by women. We are encouraging them to move into leadership roles, so they can be adequately represented in decision-making" [cited in the study by IFC (2016)].

Cooperatives and producer associations can also strengthen social bonds and a sense of identity, empowerment, common culture, and community among producers (Le Mare 2008; Arora-Jonsson 2013). For example, the associative structure of a Burkinabè women's shea butter producer association increased the sharing of market information and knowledge; promoted innovation, joint production, and economies of scale; and improved product quality (Elias 2010). Female argan producers in Morocco reported that cooperative membership gave them a sense of empowerment by strengthening their ability to retain their own income and increasing the respect they received from their husbands and other men (Biermayr-Jenzano et al. 2014). Coffee cooperatives in Rwanda have contributed not only to increased productivity but also to

FIGURE 19.5   Smallholder coffee farmer in Colombia's Cauca region. (Courtesy of Neil Palmer, International Center for Tropical Agriculture, Cali, Colombia.)

women's empowerment; the economic opportunities they provide helped those women to increase their say in household decision-making [Ya-Bititi et al. (2015) cited by ILO (2016)]. Aware that they lacked decision-making power within the family and in the coffee trade, women in Colombia's Cauca region came together to found the Association of Rural Women Almaguereñas, with the support of UN Women (2017). The women received training in marketing (Figure 19.5) and business planning and management and were helped to obtain new coffee roasting and grinding machines for their processing factory, while the UN Women-sponsored trainings also engaged male local leaders and family members who became more supportive of women's rights. Although this may not be true of all collectives, such a space outside the home, sanctioned by external authority, where women can come together around issues that concern them, can enable them to forward their own agendas (Arora-Jonsson 2013).

Associations should not be idealized, however, as not all collectives are empowering or emancipatory nor are they egalitarian places of harmony. There can be risk of elite capture as in argan (Biermayr-Jenzano et al. 2014). Fair trade premiums for coffee in Guatemala were often used to pay for nicer facilities or more staff rather than for social services (Haight 2011). As Elias and Arora-Jonsson (2016) demonstrate in the case of a shea producers association, collectives can also bring exclusions based on gender, ethnicity, socioeconomic status, or other factors of social differentiation in their wake.

19.3.2 Labor Markets

Labor market institutions interact with the business model adopted by most globalized firms operating in forestry and tree value chains. This model exploits low-cost, largely unskilled, and increasingly casual and flexible labor in low-wage countries, rather than privileging technology that would increase productivity and wages (Benerìa et al. 2000). At the same time, organizational changes within global production chains are increasingly characterized by the concentration of ownership and control, greater dispersion and more layers of outsourcing and subcontracting (Ghosh 2009; Razavi et al. 2012).

The increasing complexity of these vertical value chains, particularly when combined with the practice of subcontracting through agents, can mask the responsibility for the exploitative conditions sometimes endured by small-scale producers and workers, especially those engaged in informal unregulated cottage industries or homeworking. Employment conditions in globalized industries are subject to the hosting countries' labor laws and the global "parent" companies' policies (if any) on ethical and responsible international labor standards (often under pressure from consumers). Yet these laws and policies are not always implemented. Moreover, they typically exclude workers on flexible contracts (seasonal, casual, and temporary) as well as outworkers such as smallholders (Dolan 2005). As global value chains expand, the lead firms tend to negotiate only with the first tier of subcontracting and have little, if any, control over the subcontractors and laborers further down the chain (Carr and Chen 2004). Indeed, many governments and unions acquiesce in the resulting decline in labor standards, including the growth of employment relationships not covered by labor legislation or social protection, as the price to attract foreign investment and new jobs or to preserve existing jobs (Carr et al. 2000; Carr and Chen 2004). These trends run in the face of the international community's commitment to meet the SDG targets of *decent* work.[12]

We explore in the following section some of these issues with regard to wage labor in plantations and estates and in homeworking, reviewing evidence that women tend to fare worse than men due to systemic institutional gender inequalities. These inequalities operate at three interrelated levels—the household/community, labor market organizations, and the public sphere (Dey de Pryck and Termine 2014). At the household/community level, social norms determine gender stereotypes about "suitable" work for men and women while restricting women's access to productive resources. Industries and labor market organizations build on these stereotypes to assign women to inferior jobs and keep them out of leadership roles, while public institutions often fail to provide or enforce policies, laws, or other measures to ensure gender-equitable rights at work, largely because of male dominance in leadership positions and stereotypes that men are the main breadwinners with women in secondary, supporting roles.

19.3.2.1 Wage Labor

Globalization has spurred the growth of plantation and estate production, and related processing and packing plants, creating considerable wage labor opportunities for both men and women. However, traditional social norms that determine what is regarded as "suitable" work for women are being perpetuated in globalizing industries to justify occupational segregation that often confines women to lower skilled, manual work (UNDAW 2009; Fontana and Paciello 2010). Women are mainly engaged in "low-skilled," poorly paid work where manual dexterity is prized (Dolan and Sorby 2003), reinforcing the perception of these female jobs as low-pay and low-status work (Dey de Pryck and Termine 2014). While men may also undertake low-skilled work, they monopolize most managerial and skilled technical jobs or undertake the physically heavier work. Interestingly, Deere (2005) found that occupational segregation by gender in Nicaraguan cotton, coffee, and tobacco plantations broke down under conditions of (male) labor shortage. Nonetheless, occupational segregation represents a barrier for women to move to better jobs but does not prevent them from losing existing work if it is mechanized or becomes more remunerative and is taken over by men (Fontana and Paciello 2010).

As in other sectors, plantations and processing/packing plants are also increasingly subject to employers shifting more of the workforce into informal temporary (short term) or casual (daily) jobs or piecework (FAO 2011; Smalley 2013; Dey de Pryck and Termine 2014) with lack of job security, benefits, or adherence to minimum wages. Such informal, flexible labor arrangements enable global firms to better withstand production risks (due to climatic variations, pests, and diseases) and commercial risks (strict standards, changing demand, and "just-in-time" production methods) (Dey de Pryck and Termine 2014)]. Women tend to comprise a large and sometimes predominant share of such work (e.g., Chilean fruit: Jarvis and Vera-Toscano 2004; South African fruit: Barrientos 2007; Brazilian grapes: Selwyn 2010; mangoes and oil palm in Ghana: King and Bugri 2013; Yaro et al. 2017; Kenyan coffee: Hakizimana et al. 2017). While one reason relates to the type of work (delicate, repetitive) that is considered as part of women's "innate" abilities, another reason is that firms seek to reduce the nonwage social costs of formal female labor such as maternity benefits and more absences to deal with family illnesses. This was striking in the Brazilian São Francisco Valley export grape farms where women joined the trade union's campaign in the 1990s and early 2000s that secured improved workers' rights and conditions, including the provision of crèche facilities, a paid day a month for women workers to visit doctors, the right for women with babies in the crèche to breastfeed for an hour per day, a 2-month period of paid maternity leave, and the right to return to employment following such leave (Selwyn 2010).

A number of studies reviewed by Fontana and Paciello (2010), FAO (2011), and Dey de Pryck and Termine (2014) reported that women rural wage workers were paid less than men. Hakizimana et al. (2017) found that Kenyan coffee farms paid casual male laborers more, while Sjiapati Basnett et al (2016) noted that male jobs were better remunerated in Indonesian oil palm. Women earned less than men in wage employment in Chilean grape (Jarvis and Vera-Toscano 2004). However, generalizations can be misleading. Women doing piece work in South African fruit often earned more than men (Dolan and Sorby 2003) and women working piece rates in Chilean grape packing sheds during the peak season had higher average daily earnings than wage workers who tended to be male (Jarvis and Vera-Toscano 2004).[13] Selwyn (2010) reported that permanently employed women in the Brazilian grape farms enjoyed the same pay scales as men. King and Bugri (2013) also found that there were no obvious differences in pay in Ghanaian mangoes. Gender wage gaps appeared smaller in some nontraditional agroexport industries (Fontana and Paciello 2010) that tended to provide higher wages and better working conditions than in traditional agricultural employment (FAO 2011). While these industries (that include high-value fruit) are opening up more equitable work opportunities for women, the impact of these trends has so far received little attention (FAO 2011).

Many studies suggest that women are pushed into low-paid temporary or casual work through poverty (sometimes compounded by being divorced, separated, or widowed), tenure insecurity, and/or lack of other options as in the case of female workers in Indonesian oil palm (Box 19.3—Figure 19.6). Nonetheless, this income can make a critical difference, enabling them to combine casual work with own-account farming, nonfarm employment, and reproductive activities (Kabeer 2004; Smalley 2013; Hakizimana et al. 2017).

BOX 19.3   FEMALE WORKERS IN OIL PALM

In Indonesia, the world's biggest producer and consumer of palm oil, palm oil accounted for 11% of total export earnings in 2013, second only to oil and gas. The industry employs approximately 3 million people and has contributed to poverty reduction and economic expansion in a country where more than 28 million people live below the poverty line (*The Guardian* 2014). While figures of how many women are employed in palm oil plantations remain uncertain, the industry produces considerable employment opportunities for women with maintenance jobs of fertilizing, pesticide spraying, and weeding reserved for women while harvesting is reserved for men (Li 2015; Sijapati Basnett et al. 2016). Although these jobs provide regular employment, the vast majority of women are employed as casual workers, with difficult working conditions, high exposure to chemicals, low wages, and few or no entitlements to benefits such as sick leave. Critics argue that the palm oil industry has deliberately created a fictitious labor market of abundant supply and finite demand, thereby assisting corporate producers to drive down wages, benefits, and opportunities for both local and migrant workers alike who are employed in the sector (Li 2015). In the context where the vast majority of landscapes are dominated by oil palm and there are few employment prospects outside the sector, taking up casual employment in oil palm is a manifestation of an absence of choice rather than an exercise of choice. This is especially the case for women whose movements are constrained by domestic responsibilities (Li 2015; Elmhirst et al. 2017).

FIGURE 19.6   Indonesian couple working together in oil palm. (Courtesy of Icaro Cooke Vieira, Center for International Forestry Research, Bogor, Indonesia.)

19.3.2.2   Homeworkers[14]

A common argument is that women are more heavily involved in homework than men, because they prefer to work at home (the "housewife theory")—either for social reasons (as in cultures where female mobility is constrained) or to combine work with domestic and care responsibilities. However, as Carr et al. (2000) note, they are often constrained to undertake homework as employers prefer to subcontract/ outsource as a cost-cutting strategy. Men may also be affected by this strategy: for example, Rani and Unni (2009) found that men's homework rose 14% in India between 1994 and 2000, especially in import-competing industries that were obliged to seek cost-cutting measures.

The global restructuring of food product chains is leading to greater concentration in corporate power and intensifying competition among the few large buyers (mainly northern supermarkets), resulting in purchasing practices that squeeze production and processing costs. This is, in turn, fueling the informalization and casualization of the upstream workforce, particularly with the growth of private factories and even less regulated home processing, as in the case of cashew nuts (Box 19.4). At the same time, stricter quality, hygiene, safety, sanitary and phytosanitary regulations are further squeezing margins with costs being passed to upstream processors (Harilal et al. 2006).

BOX 19.4   INDIA'S CASHEW INDUSTRY

While India was the sole exporter of cashew kernels in the 1920s, the growing demand for cashew nuts in the north, as well as in emerging markets such as Russia and Japan and more recently the Middle East and China, led to rapid growth in production in India, with new entrants including several African countries, Brazil, and Vietnam. India has maintained its leadership position in cashew processing and exports through large imports of raw cashews from Africa. The workforce engaged in processing (shelling and peeling) and grading is almost entirely female with women largely drawn from scheduled castes and poor communities while men do most of the roasting and packing. Almost half of the women processors surveyed by Harilal et al. (2006) were the sole earners in their families, and another 30% paid about half their families' expenses. Their income was insufficient to meet expenses, and almost 80% said they were in debt. Concerned to increase and stabilize their earnings in such a casual industry, many women rotated between government and private factories and home processing, depending on the availability of raw nuts.

Working conditions were recognized as deplorable in Kerala (the historical center of the industry) as early as the 1940s–1960s, inspiring several pieces of legislation, including the banning of cottage processing in 1967. However, factory owners circumvented these laws by moving cottage processing to Tamil Nadu, operating on a clandestine basis or subcontracting through the commission agent. Home processing has been expanding since the 1990s as companies cut costs (Harilal et al. 2006). While some hygiene and safety standards are enforced (largely for the consumers' benefit), particularly in government and some private factories, northern supermarkets are not obliged to maintain labor or environmental standards unless they have agreed to voluntary codes of conduct. The poor enforcement of labor legislation means that workers in private factories and especially in home processing often do not receive the minimum wage or the "dearness allowance" and other benefits that are paid in government factories (Harilal et al. 2006; Kuzhiparambil 2016).

Unions are relatively powerless in such global chains, and the informal work environment reduces women's bargaining strength. Societal attitudes of what is acceptable work for men and women, and women's more limited choices, mean that women tend to accept lower wages and more insecure work than men. Influenced by the stereotype that women are secondary earners, employers have exploited this social and economic discrimination against women to decrease processing costs and increase profits. As Harilal et al. (2006) note, companies that seek to provide decent wages and working conditions may be unable to compete in this increasingly global industry unless policies, including competition policy, force international firms to address the impact of their purchasing practices on labor standards throughout the supply chain.

19.4 INTERNATIONALIZATION OF NONECONOMIC CONCERNS: GENDERED IMPACTS ON EMPLOYMENT AND LIVELIHOODS

Klienshmit et al. (2015) argue that there are two major manifestations of noneconomic concerns: first, emerging global socioecological narratives such as those on planetary boundaries [see also the study by Leach et al. (2013) on "social and planetary boundaries"), action to combat climate change, indigenous peoples' rights, and gender equality; second and relatedly, the rise in ecosystem services thinking, which conceives of nature as a provider of services to humans. These have fostered new concepts and methods for valuating rural landscapes, described in the following section, and led to the economic valuation of forest hydrological or carbon storage "services."

Klienshmit et al. (2015) argue that these valuations provide new legitimacy to particular forms of governance. These include state regulation to protect downstream and global citizens and/or public–private partnerships to market forest carbon offsets and to balance trade-offs between profit and the environment. Emergent actors in this governance system include international and national NGOs advocating for the rights of women, indigenous peoples, and other marginalized actors as well as corporations pursuing voluntary standards to enhance the sustainability of their operations. These changes have meant that forest governance is no longer defined by states alone: the influence and objectives of global and nonstate actors have progressively widened and deepened due to expanding and new frontiers of forest and environmental governance (Chandhoke 2003).

Such valuation and resultant governance modalities pose certain risks, especially to poor or minority actors including communities whose livelihoods have been particularly tied to forest habitats. For instance, they may result in shifting responsibility for mitigating climate change and restoring forestlands onto marginalized populations and/or in framing technoscientific solutions that obscure the experience and challenges faced by these actors (Arora-Jonsson 2014; Bee et al. 2015; Bee and Sijapati Basnett 2016). Women within these communities tend to suffer greater disadvantages due to pre-existing inequalities, the gendered nature of community institutions, and "gender blindness" of discourses justifying these interventions. Interventions in forestry, such as REDD+; growing efforts to restore forests and meet national pledges made under global and regional initiatives such as the Bonn Challenge, AFR 100, Initiative 20×20; and a wide range of certification standards developed to protect the environment and people, serve as illustrations of the gendered outcomes resulting from this internationalization of noneconomic concerns.

19.4.1   REDD+

REDD+ stands out as a particularly relevant illustration of such risks from a gender perspective. Originally developed as a global approach to reducing deforestation and forest degradation in tropical and subtropical countries, REDD+ is increasingly concerned with reconciling multiple objectives including "gender considerations" (as indicated in the 2010 Cancun Agreements) and the rights of indigenous peoples, local communities and women. This is in response to growing criticisms and advocacy by women's rights and indigenous peoples' organizations arguing that the infusion of financial capital without adequate safeguards and opportunities for marginalized groups would exacerbate their vulnerability (Bee and Sijapati Basnett 2016).

However, Westholm and Arora-Jonsson (2015) argue that while decision-making on REDD+ is largely confined to the global and national scales, the responsibility for carrying out those agreements is being shifted towards marginalized groups and women. This is problematic in light of endemic information asymmetries between the genders and limited options for including women on equal terms with men. For instance, global comparative research carried out by the Center for International Forestry Research (CIFOR) in 69 villages in 18 REDD+ sites across five countries (Brazil, Cameroon, Indonesia, Tanzania, and Vietnam) found that women as a group were less knowledgeable about REDD+ projects and less involved in REDD+ piloting than their male counterparts (Larson et al. 2015). In their follow-up study in 2013/2014 in the same REDD+ research sites using a before–after control intervention research design, which permitted clearer attribution of results to REDD+ (Bos et al. 2017), Larson et al. (2018) found that women in REDD+ sites were more likely to perceive a decline in their subjective well-being than in sites with comparable situations but no REDD+ projects. The authors concluded that these results were likely to be explained by the fact that REDD+ had not been gender responsive and that women perceived that they had either not been included in REDD+ and/or not benefitted from it in the way that they had anticipated (Larson et al. 2018). Furthermore, national and global discourses over who should benefit from REDD+ exclude women's views and perspectives altogether, leading to inequalities in policies and programs. For instance, proposals that REDD+ benefits should flow to recognized owners of forests and trees risk excluding women since few have formal titles to land, only enjoying micro-rights and informal rights within existing unequal tenure systems, with their access to land and trees often mediated by their male relatives (Arwida et al. 2017).

Bee and Sijapati Basnett (2015) find that the vast majority of options for embedding "gender considerations" in REDD+ focus on increasing women's participation in community organizations and actions. Having specific provisions and safeguards for women's participation is important. Yet simply increasing their presence by promoting women-only groups or having a few token women in mixed groups does not result in any meaningful influence by women unless accompanied by simultaneous attention to capacity building and the underlying power dynamics that serve to perpetuate gender inequalities (Agarwal 2010; UN Women 2013).

19.4.2 CERTIFICATION

Certification labels, from "organic" to "sustainable," play an increasingly prominent role in nonstate, market-based environmental governance. Proponents argue that certification rectifies market failures to absorb social and environmental costs in the production of agroforestry commodities by appealing to consumer demand for ethical products. Certification systems involve the development of standards for commodity production and trade and third-party monitoring apparatuses to ensure compliance with those standards. Certified products often command a premium price, although in some cases, their continued or enhanced market access serves as compensation for compliance with standards (Elgert 2012).

Despite their potential for improving the socioeconomic and environmental conditions of production and trade, certification systems have been critiqued for the asymmetries they embed in terms of agenda setting: deciding what is "fair," to whom it is fair, who decides the alternatives, and who is empowered to certify whom (Jaffee et al. 2004; Moore 2004; Moberg 2005). Certain criteria or the measures required to achieve them are disjointed from local realities (Elias and Saussey 2013). Fair trade and/or ethical guidelines, for example, can represent a considerable nuisance for producers and their associations (Freidberg 2003). For instance, to achieve the transparency required for fair trade certification, producer associations must maintain extensive bookkeeping. This transparency seeks to favor equitability among producers, but the need to maintain elaborate written records in an international language often excludes all but the few schooled producers from participation in the process (Off 2006). Gendered exclusions result from rural women's often more limited access to formal education than men's. Yet even literate members of producer associations experience difficulties fulfilling certification documentation requirements without the help of an NGO, as "no one unfamiliar with the workings of a European bureaucracy could make any sense of the documents" required to maintain FLO's fair trade certification (Off 2006).[15] The transparency, justice, and proper business ethics these norms are meant to embody thus frequently vanish in their practical translation.

There is very little gender-based discussion in the burgeoning literature on the social dimensions of certification, even as "women's empowerment and gender equality" are increasingly seen as requirements to be included in certification standards (Smaller et al. 2016). The certification of oil palm (Box 19.5) serves as a

BOX 19.5   GENDER IN CERTIFIED PALM OIL STANDARDS

The roundtable on sustainable palm oil (RSPO) and other certification standards such as the Indonesian sustainable palm oil, have developed standards for sustainable palm oil focused on deforestation, lawfulness, transparency, and social impacts of production and trade. The overarching framework of standards, such as the RSPO's "principles, criteria, indicators and guidance" (P&C and Guidance), falls short of integrating gender meaningfully. Standards linked to "free prior and informed consent" that are important for producers and resource owners are couched in gender-neutral language. The only gender issues covered relate to the treatment of workers, specifically "no discrimination" against workers on the basis of gender, and nonexposure of toxic chemicals for pregnant and breastfeeding women so as to safeguard women's reproductive health. In the absence of any mandatory provisions to absorb breastfeeding and pregnant women in other activities, the P&C and Guidance designed to safeguard women's reproductive rights risks becoming de facto discrimination against them, as they are likely to lose their jobs. Moreover, the P&C and Guidance is very safeguards oriented and does not consider how gender-sensitive support can be extended to independent smallholders, contractors, and cooperatives. Finally, there are considerable discrepancies in the ways in which they are interpreted and implemented by 'social auditors', who are tasked with monitoring and validating producers' compliance with these standards. Hence, both the design and implementation of the RSPO principles and criteria are in need of reform to ensure that sustainable palm oil adequately reflects both human and environmental considerations (Sijapati Basnett et al. 2016).

striking illustration of gender-related bottlenecks in certification processes. It also raises wider questions related to the adequacy of certification and the extent to which gender issues will be prioritized in the future, in light of "other pressing" concerns occupying the power-laden process of certification (Elgert 2012).

19.5   CONCLUSIONS

The impact of globalization on employment in forests and tree value chains, including forest-related services, varies considerably across and within countries depending on a host of factors, including the type of product and scale of production; the power of transnational corporations (TNCs), producer, and civil society organizations; and government ideologies. These can result in public policy conflicts between promoting/protecting smallholders and supporting international/national agroindustrials. Governments often prefer to attract agroindustrial investors for large-scale production systems (for example, palm oil) or for managing landscapes for global markets for carbon, biofuels, and biodiversity rather than for promoting the livelihoods of poor/small producers (Vira et al. 2015). This can seriously undermine smallholder employment and livelihoods. In other cases, globalizing markets can create new remunerative employment for both men and women, although the extent to which these new jobs are "decent" varies hugely. Growing consumer pressure, companies' ethical commitments such as to corporate social responsibility, the UN's international standards and regulations, and improved labor legislation in some countries, are causes for hope. However, the gender sensitivity of these standards often requires significant improvements. Moreover, unless companies are operating in an enabling policy environment with propoor and proworker labor laws and good practices, companies that unilaterally follow voluntary codes of conduct could be priced out of the market unless they are operating in a niche market.

While the challenge of "decent" work concerns both men and women, women are often more disadvantaged than men. Many of the reasons stem from the interaction of globalizing value chains with traditional institutions at multiple scales that are intrinsically gendered and discriminate against women. These are reflected in the sort of work that is regarded as acceptable for women; women's more limited access to and control of land and trees, labor, finance, technology, and information; and their rights to control/use income earned by their labor. Occupational segregation by gender within processing and packing plants is often used to legitimize lower wages and less skilled and secure work for women. Yet women's earnings are often crucial to allow them and their families to live decent lives, in dignity and, in many cases, for their families' very survival. Social norms are changing in some cases in response to new income-earning opportunities, formal education, and male outmigration. Nonetheless, government, global companies, and civil society can do much more through more gender-sensitive policies, labor laws, regulations and practices, and advocacy to change these norms and public opinion.

REFERENCES

Agarwal, B. 1994. A *Field of One's Own: Gender and Land Rights in South Asia*. Cambridge, MA: Cambridge University Press.

Agarwal, B. 2002. The hidden side of group behaviour: A gender analysis of community forestry in South Asia. In *Group Behaviour and Development: Is the Market Destroying Cooperation?* eds. Heyer, J., Stewart, F. and R. Thorp, 185–208. Oxford: Oxford University Press.

Agarwal, B. 2003. Gender and land rights revisited: Exploring new prospects via the state, family and market. *Journal of Agrarian Change* 3(1–2):184–224.

Agarwal, B. 2010. *Gender and Green Governance: The Political Economy of Women's Presence within and beyond Community Forestry*. Oxford: Oxford University Press.

Agrawal, A. 2001. Common property institutions and sustainable governance of resources. *World Development* 29(10):1649–1672.

Agrawal, A., Cashore, R., Gill, H., Benson, C., and C. Miller. 2013. *Economic Contributions of Forests*. Background paper 1, Tenth session of the United Nations Forum on Forests. Istanbul: United Nations Forum on Forests.

Arora-Jonsson, S. 2013. *Gender, Development and Environmental Governance: Theorizing Connections*. London: Routledge.

Arora-Jonsson, S. 2014. Forty years of gender research and environmental policy: Where do we stand? *Women's Studies International Forum* 47:295–308.

Arwida, S. D., Maharami, C. D., Sijapati Basnett, B., and A. L. Yang. 2017. *Gender-Relevant Considerations for Developing REDD+ Indicators: Lessons Learned for Indonesia.* CIFOR Info Brief 168. Bogor, Indonesia: Center for International Forestry Research.

Barrientos, S. 2001. Gender, flexibility and global value chains. *IDS Bulletin* 32(1):83–93.

Barrientos, S. 2007. *Female Employment in Agriculture: Global Challenges and Global Responses.* London: Commonwealth Secretariat.

Bee, B. and B. Sijapati Basnett. 2016. Engendering social and environmental safeguards in REDD+: Lessons from feminist research. *Third World Quarterly* 1–18.

Bee, B., Rice, J., and A. Trauger. 2015. A feminist approach to climate change governance: Everyday and intimate politics. *Geography Compass* 9(6):339–350.

Benería, L., Floro, M., Grown, C., and M. MacDonald. 2000. Introduction: Globalization and gender. *Feminist Economics* 6:3, vii–xviii.

Biermayr-Jenzano, P., Kassam, S. N., and A. Aw-Hassan. 2014. *Understanding Gender and Poverty Dimensions of High Value Agricultural Commodity Chains in the Souss-Masaa-Draa Region of South-Western Morocco.* ICARDA working paper, mimeo. Amman: International Center for Agriculture Research in the Dry Areas.

Bolwig, S. and M. Odeke. 2007. *Food Security Effects of Certified Organic Export Production in Tropical Africa: Case Studies from Uganda.* SAFE Policy Brief No. 7. Copenhagen: Danish Institute for International Studies.

Bos, A. B., Duchelle, A. E., Angelsen, A. et al. 2017. Comparing methods for assessing the effectiveness of subnational REDD plus initiatives. *Environmental Research Letters* 12(7).

Brondizio, E., Siqueira, A., and N. Vogt. 2014. Forest resources, city services: Globalization, household networks, and urbanization in the Amazon estuary. In *The Social Lives of Forests*, eds. Hecht, S., Morrison, K., and C. Padoch, 348–361. Chicago, IL, and London: University of Chicago Press.

Bruton, G. D., Ahlstrom, D., and H.-L. Li. 2010. Institutional theory and entrepreneurship: where are we now and where do we need to move in the future. *Entrepreneurship Theory and Practice* 34:421–40.

Carr, M. and M. A. Chen. 2004. Globalization, social exclusion and gender. *International Labor Review* 143(1–2):129–160.

Carr, M., Chen, M. A., and J. Tate. 2000. Globalization and home-based workers. *Feminist Economics* 6(3):123–142.

Chalfin, B. 2004. *Shea Butter Republic.* New York: Routledge.

Chandhoke, N. 2003. Governance and the pluralisation of the state: Implications for democratic citizenship. *Economic and Political Weekly* 38(28):2957–2968.

Colfer, C., Sijapat Basnett, B., and M. Elias. 2016. *Gender and Forests. Climate Change, Tenure, Value Chains and Emerging Issues.* London; New York: Routledge.

Dancer, H. and D. Tsikata. 2015. *Researching Land and Commercial Agriculture in Sub-Saharan Africa with a Gender Perspective: Concepts, Issues and Methods.* Future Agricultures Consortium (FAC) Working Paper 132. Land and Agricultural Commercialization in Africa project. Brighton: Future Agricultures Consortium.

Deere, C. D. 2005. *The Feminization of Agriculture? Economic Restructuring in Rural Latin America.* Occasional paper 1. Geneva: United Nations Research Institute for Social Development (UNRISD).

Dey, J. 1981. Gambian women: Unequal partners in rice development projects? *Journal of Development Studies* 17(3):109–122.

Dey de Pryck, J. and P. Termine. 2014. Gender inequalities in rural labor markets. In *Gender in Agriculture. Closing the Knowledge Gap*, eds. Quisumbing, A., Meinzen-Dick, R., Raney, T. et al., 343–370. Rome: FAO and Springer.

Dolan, C. 2001. The "good wife": Struggles over resources in the Kenyan horticultural sector. *Journal of Development Studies* 37(3):39–70.

Dolan, C. 2005. *Stewards of African Labor: The Evolutions of Trusteeship in Kenya's Horticulture Industry.* Working paper in African Studies 252. Boston, MA: Boston University.

Dolan, C. and K. Sorby. 2003. *Gender and Employment in High-Value Agriculture Industries.* Agriculture and Rural Development working paper series 7. Washington, DC: World Bank.

Dunaway, W. A. 2014. *Gendered Commodity Chains: Seeing Women's Work and Households in Global Production.* Stanford, CA: Stanford University Press.

Eaton, C. and A. Shepherd. 2001. *Contract Farming: Partnerships for Growth.* FAO Agricultural Services Bulletin 145. Rome: FAO.

Elgert, L. 2012. Certified discourse? The politics of developing soy certification. *Geoforum* 43(2):295–304.

Elias, M. 2010. Transforming nature's subsidy: Global markets, Burkinabè women and African shea butter. PhD Dissertation, Quebec: McGill University.

Elias, M. 2016. Gendered knowledge sharing and management of shea. In *Gender and Forests. Climate Change, Tenure, Value Chains and Emerging Issues*, eds. Colfer, C., Sijapat Basnett, B., and M. Elias, 263–282. London; New York: Routledge.

Elias, M. and J. Carney. 2007. African shea butter: A feminized subsidy from nature. *Africa* 77(1):37–62.

Elias, M. and M. Saussey. 2013. "The gift that keeps on giving": Unveiling the paradoxes of fair trade shea butter: Paradoxical narratives of fair trade shea butter. *Sociologia Ruralis* 53:158–179.

Elias, M. and S. Arora-Jonsson. 2016. Negotiating across difference: Gendered exclusions and cooperation in the shea value chain. *Environment and Planning D: Society and Space* 35(1):107–125.

Elmhirst, R., Siscawati, M., and C. Colfer. 2016. Revisiting gender and forestry in Long Segar, East Kalimantan, Indonesia. In *Gender and Forests. Climate Change, Tenure, Value Chains and Emerging Issues*, eds. Colfer, C., Sijapat Basnett, B., and M. Elias, 263–282. London; New York: Routledge.

Elmhirst, R., Siscawati, M., Sijapati Basnett, B., and D. Ekowati. 2017. Gender and generation in engagements with oil palm in East Kalimantan, Indonesia: Insights from feminist political ecology. *The Journal of Peasant Studies* 44(6):1135–1157.

FAO (Food and Agriculture Organization of the United Nations). 2009. *From Land Grab to Win-Win: Seizing the Opportunities of International Investments in Agriculture.* Policy Brief 4. Rome: FAO.

FAO. 2011. *State of Food and Agriculture. Women in agriculture: Closing the Gender Gap for Development.* Rome: FAO.

FAO. 2014. *State of the World's Forests 2014: Enhancing the Socioeconomic Benefits from Forests.* Rome: FAO.

FAO. 2016. *Global Forest Resources Assessment 2015: How Are the World's Forests Changing?* Rome: FAO.

FAO. 2017. *State of Food and Agriculture: Leveraging Food Systems for Inclusive Rural Transformation.* Rome: FAO.

FAO, International Fund for Agricultural Development (IFAD) and ILO (International Labour Organization). 2010. *Agricultural Value Chain Development: Threat or Opportunity for Women's Employment?* Gender and Rural Employment Policy Brief 4. Rome: FAO.

Farcy, C., de Camino, R., Martinez de Arano, I., and E. Rojas-Briales. 2015. External drivers of changes challenging forestry: Political and social issues at stake. In *Ecological Forest Management Handbook*, ed. Larocque, G. R., 87–105. Boca Raton: CRC Press/Taylor & Francis.

Feenstra, R. C. 1998. Integration of trade and disintegration of production in the global economy. *Journal of Economic Perspectives* 12(4):31–50.

Fontana, M. and C. Paciello. 2010. Gender dimensions of rural and agricultural employment: Differentiated pathways out of poverty. In *Gender Dimensions of Rural and Agricultural Employment: Differentiated Pathways out of Poverty: Status, Trends and Gaps*, eds. FAO, IFAD and ILO, 1–71. Rome: FAO.

Fortmann, L. and J. W. Bruce. 1988. *Whose Trees? Proprietary Dimensions of Forestry.* Boulder: Westview Press.

Freidberg, S. 2003. *The Contradictions of Clean: Supermarket Ethical Rrade and African Horticulture.* Gatekeeper Series 109. London: International Institute for Environment and Development (IIED).

Ghosh, J. 2009. Informalization and women's workforce participation: A consideration of recent trends and in Asia. In *The Gendered Impacts of Liberalization: Towards "Embedded Liberalism"?* ed. Razavi, S., 163–190. London: Routledge.

Haight, C. 2011. The problem with fair trade coffee. *Standford Social Innovation Review* 74–79.

Hakizimana, C., Goldsmith, P., Nunow, A. A., Roba, A. W., and J. K. Biashara. 2017. Land and agricultural commercialisation in Meru County, Kenya: Evidence from three models. *The Journal of Peasant Studies* 44(3):555–573.

Hall, R., Scoones, I., and D. Tsikata. 2017. Plantations, outgrowers and commercial farming in Africa: Agricultural commercialisation and implications for agrarian change. *The Journal of Peasant Studies* 44(3):515–537.

Harilal, K. N., Kanji, N., Jeyaranjan, J., Eapen M., and P. Swaminathan. 2006. Power in Global Value Chains: Implications for Employment and Livelihoods in the Cashew Nut Industry in India. Summary Report. London: IIED.

Hecht, S. 2007. Factories, forests, fields and families: Gender and neoliberalism in extractive reserves. *Journal of Agrarian Change* 7(3):316–347.

Hill, R. and M. Vigneri. 2014. Mainstreaming Gender Sensitivity in Cash Crop Market Supply Chains. In *Gender in Agriculture: Closing the Knowledge Gap*, eds. Quisumbing, A., Meinzen-Dick, R., Croppenstedt, A. et al., 315–341. Rome: FAO and Springer.

HLPE (High Level Panel of Experts). 2017. *Sustainable Forestry for Food Security and Nutrition.* A report by the High Level Panel of Experts on Food Security and Nutrition of the Committee on World Food Security. Rome: FAO.

Howard, P. and G. Nabanoga. 2007. Are there customary rights to plants? An inquiry among the Baganda (Uganda), with special attention to gender. *World Development* 35(9):1542–1563.

IFC (International Finance Corporation). 2016. I*nvesting in Women along Agribusiness Value Chains.* Washington, DC: International Finance Corporation, World Bank Group.

ILO (International Labour Organization). 2016. *World Employment and Social Outlook 2016—Transforming Jobs to End Poverty.* Geneva: ILO.

ILO. 2017. *Decent Work.* Geneva: ILO.http://www.ilo.org/global/topics/decent-work/lang —en/index.htm (Accessed February 26, 2017).

Jaffee, D., Kloppenburg Jr., J. R., and M. B. Monroy. 2004. Bringing the "moral charge" home: Fair trade within the North and within the South. *Rural Sociology* 69(2):169–196.

Jarvis, L. and E. Vera-Toscano. 2004. *The Impact of Chilean Fruit Sector Development on Female Employment and Household Income.* World Bank Policy Research Working Paper 3263. Washington, DC: World Bank.

Kabeer, N. 2004. Globalization, labor standards, and women's rights: Dilemmas of collective (in)action in an interdependent world. *Feminist Economics* 10(1):3–35.

Kabeer, N. 2012. *Women's Economic Empowerment and Inclusive Growth: Labor Markets and Enterprise Development.* Discussion Paper 29/12. London: Centre for Development Policy and Research, School of Oriental and African Studies.

Kavoi, M. M., Oluoch Kosura, W., Owuor, P. O., and D. K. Siele. 2003. Gender analysis of economic efficiency in smallholder tea production in Kenya. *Eastern African Journal of Rural Development* 19(1):33–40.

King, R. and J. Bugri. 2013. *The Gender and Equity Implications of Land-Related Investments on Land Access, Labor and Income-Generating Opportunities in Northern Ghana: The Case Study of Integrated Tamale Fruit Company*. Rome: FAO.

Klienshmit, D., Sijapati Basnett, B., Martin, A., Rai, N. D., and C. Smith-Hall. 2015. Drivers of forests and tree-based systems for food security'. In *Forests, Trees and Landscapes for Food Security and Nutrition: A Global Assessment Report*, eds. Vira, B., Wildburger C., and S. Mansourian, 87–110. IUFRO World Series Volume 33. Vienna: International Union of Forest Research Organizations.

Koczberski, G. 2007. Loose fruit mamas: Creating incentives for smallholder women in oil palm production in Papua New Guinea. *World Development* 35(7):1172–1185.

Kuzhiparambil, A. 2016. A secondary informal circuit of globalisation of production: Home-based cashew workers in Kerala, India. *Global Labor Journal* 7(3):279–296.

Larson, A. M., Dokken, T., Duchelle, A. E. et al. 2015. The role of women in early REDD+ implementation: Lessons for future engagement. *International Forestry Review* 17(1):43–65.

Larson, A. M., Solis, D., Duchelle, A. E. et al. 2018. Gender lessons for climate intiatives: A comparative study of REDD+ impacts and subjective wellbeing. *World Development* 108:86–102.

Lastarria-Cornhiel, S., Behrman, J., Meinzen-Dick, R. and A. Quisumbing. 2014. Gender equity and land: Toward secure and effective access for rural women. In *Gender in Agriculture. Closing the Knowledge Gap*, eds. Quisumbing, A., Meinzen-Dick, R., Raney, T., Croppenstedt, A. et al., 117–144. Rome: FAO and Springer.

Leach, M., Raworth, K., and J. Rockström. 2013. Between social and planetary boundaries: Navigating pathways in the safe and just space for humanity. In *World Social Science Report*, 84–89. Paris: International Social Science Council (ISSC), United Nations Educational, Scientific and Cultural Organization (UNESCO).

Le Mare, A. 2008. The impact of fair trade on social and economic development: A review of the literature. *Geography Compass* 2(6):1922–1942.

Li, T. M. 2015. *Social Impacts of Oil Palm in Indonesia: A Gendered Perspective from West Kalimantan*. CIFOR Occasional Paper 124. Bogor, Indonesia: Center for International Forestry Research.

Maertens, M. and J. F. M. Swinnen. 2009. Are African high-value horticulture supply chains bearers of gender inequality? Paper presented at the FAO-IFAD-ILO Workshop on Gaps, Trends and Current Research in Gender Dimensions of Agricultural and Rural Employment: Differentiated Pathways out of Poverty.

Martin, S. 1984. Gender and innovation: Farming, cooking and oil palm processing in the Ngwa Region, South-Eastern Nigeria, 1900–1930. *Journal of African History* 25(4):411–427.

Moberg, M. 2005. Fair trade and Eastern Caribbean banana farmers: Rhetoric and reality in the anti-globalization movement. *Human Organization* 64(1):4–15.

Moore, G. 2004. The fair trade movement: Parameters, issues and future research. *Journal of Business Ethics* 52:73–86.

Mwangi, E., Meinzen-Dick, R., and Y. Sun. 2011. Gender and sustainable forest management in East Africa and Latin America. *Ecology and Society* 16(1):17.

Njuki, J. S., Kaaria, S., Chamunorwa, A., and W. Chiuri. 2011. Linking smallholder farmers to markets, gender and intra-household dynamics: Does the choice of commodity matter? *European Journal of Development Research* 23:426–443.

North, D. C. 1991. Institutions. *Journal of Economic Perspectives* 5(2):97–112.

Off, C. 2006. *Bitter Chocolate: Investigating the Dark Side of the World's Most Seductive Sweet*. Toronto: Random House Canada.

Oya, C. 2012. Contract farming in Sub-Saharan Africa: A survey of approaches, debates and issues. *Journal of Agrarian Change* 12(1):1–33.

Quisumbing, A., Payongayong, E., Aidoo, J. B., and K. Otsuka. 2003. Women's land rights in the transition to individualized ownership: Implications for the management of tree resources in Western Ghana. In *Household Decisions, Gender and Development: A Synthesis of Recent Research*, ed. A. Quisumbing, 169–175. Washington, DC: International Food Policy Research Institute (IFPRI).

Rani, U. and J. Unni. 2009. Do economic reforms influence home-based work? Evidence from India. *Feminist Economics* 15(3):191–225.

Razavi, S., Arza, C., Braunstein, E., Cook, S., and K. Goulding. 2012. *Gendered Impacts of Globalization: Employment and Social Protection*. UNRISD Research Paper. Geneva: UNRISD.

Reardon, T. and C. B. Barrett. 2000. Agroindustrialization, globalization, and international development: An overview of issues, patterns, and determinants. *Agricultural Economics* 23:195–205.

Robinson, G. M. and D. A. Carson. 2015. *Handbook on the Globalisation of Agriculture*. Cheltenham; Northampton, MA: Edward Elgar.

Rocca, V. 2016. *Gender and Livelihoods in Commercial Sugarcane Production: A Case Study of Contract Farming in Magobbo, Zambia*. FAC Working Paper 136. Land and Agricultural Commercialisation in Africa project. Brighton: Future Agricultures Consortium.

Rousseau, K., Gautier, D., and D. A. Wardell. 2016. Renegotiating access to shea trees in Burkina Faso: Challenging power relationships associated with demographic shifts and globalized trade. *Journal of Agrarian Change* 17(3):497–517.

Rubin, D. and C. Manfre. 2014. Promoting gender-equitable agricultural value chains. In *Gender in Agriculture. Closing the Knowledge Gap*, eds. Quisumbing, A., Meinzen-Dick, R., Raney, T. et al., 287–313. Rome: FAO and Springer.

Selwyn, B. 2010. Gender wage work and development in North East Brazil. B*ulletin of Latin American Research* 29(1):51–70.

Sijapati Basnett, B. 2011. Linkages between gender, migration and forest governance: Re-thinking community forestry policies in Nepal. *European Bulletin of Himalayan Research* 38:7–32.

Sijapati Basnett, B., Gnych, S., and A. M. Anandi. 2016. *Transforming the Roundtable on Sustainable Palm Oil for Greater Gender Equality and Women's Empowerment*. CIFOR Info brief 166. Bogor, Indonesia: Center for International Forestry Research.

Sippel, S. 2015. All you need is export? Moroccan farmers juggling global and local markets. In *Handbook on the Globalisation of Agriculture*, eds. Robinson, G. M. and D. A. Carson, 328–349. Cheltenham; Northampton, MA: Edward Elgar.

Smaller, C., Sexsmith, K., Potts, J. and G. Huppé. 2016. Promoting gender equality in transnational agricultural investments: Lessons from voluntary sustainability standards. Paper presented at the 2016 World Bank Conference on Land and Poverty.

Smalley, R. 2013. *Plantations, Contract Farming and Commercial Farming Areas in Africa: A Comparative Review*. FAC Working Paper 55. Land and Agricultural Commercialisation in Africa project. Brighton: Future Agricultures Consortium.

Sunderland, T., Achdiawan, R., Angelsen, A. et al. 2014. Challenging perceptions about men, women, and forest product use. A global comparative study. *World Development* 64:S56–S66.

The Guardian. 2014. From rainforest to your cupboard: The real story of palm oil—interactive. November 10, 2014.

Tsikata, D. and J. A. Yaro. 2014. When a good business model is not enough: Land transactions and gendered livelihood prospects in rural Ghana. *Feminist Economics* 20(1):202–226.

UNCTAD (United Nations Conference on Trade and Development). 2009. *World Investment Report 2009: Transnational Corporations, Agricultural Production and Development*. New York; Geneva: United Nations.

UNDAW (United Nations Division for the Advancement of Women). 2009. *World Survey on the Role of Women in Development: Women's Control over Economic Resources and Access to Financial Resources, including Microfinance.* New York: UN Division for the Advancement of Women, Department of Economic and Social Affairs.

UN Women. 2013. The *World Survey on the Role of Women in Development 2014: Gender Equality and Sustainable Development.* New York: UN Women.

UN Women. 2017. *Columbian Women Growing Coffee, Brewing Peace.* New York: UN Women. http://www.unwomen.org/en/news/stories/2017/10/feature-colombian-women -growing-coffee-brewing-peace (Accessed February 22, 2018).

Vermeulen, S. and L. Cotula. 2010. *Making the Most of Agricultural Investment: A Survey of Business Models That Provide Opportunities For Smallholders.* London/Rome/Bern: IIED/FAO/IFAD/SDC.

Vira, B., Wildburger, C., and S. Mansourian. 2015. *Forests, Trees and Landscapes for Food Security and Nutrition: A Global Assessment Report.* IUFRO World Series Volume 33. Vienna: IUFRO.

von Braun, J. and R. Meinzen-Dick. 2009. *"Land Grabbing" by Foreign Investors in Developing Countries: Risks and Opportunities.* Policy Brief 13. Washington, DC: IFPRI.

von Bülow, D. and A. Sørensen. 1993. Gender and contract farming: Tea outgrower schemes in Kenya. *Review of African Political Economy* 56:38–52.

Westholm, L. and S. Arora-Jonsson. 2015. Defining solutions, finding problems: Deforestation, gender and REDD+ in Burkina Faso. *Conservation and Society* 13(2):198–199.

Whiteman, A., Wickramasinghe, A., and L. Piña. 2015. Global trends in forest ownership, public income and expenditure on forestry and forestry employment. *Forest Ecology and Management* 352:99–108.

Wonani, C., Mbuta, W., and M. Mkandawire. 2013. *The Gender and Equity Implications of Land-Related Investments on Land Access, Labour and Income-Generating Opportunities: A Case Study of Selected Agricultural Investments in Zambia.* Rome: FAO.

World Bank. 2007. *Agriculture for Development World Development Report 2008.* Washington, DC: World Bank.

Yabiku, S., Agadjanian, V., and A. Sevoyan. 2010. Husbands' labor migration and wives' autonomy. *Population Studies (Cambridge)* 64(3):293–306.

Yaro, J. A., Teye, J. K., and G. D. Torvikey. 2017. Agricultural commercialisation models, agrarian dynamics and local development in Ghana. *The Journal of Peasant Studies* 44(3):538–554.

ENDNOTES

1. We refer to "tree product" value chains to recognize that some of the products marketed derive from trees growing in forests, but others outside of forests, such as in agroforestry systems.

2. This contrasts with the concept of "internationalization" in the late nineteenth century, when goods were mainly manufactured in one country (often from imported raw materials) and subsequently exported (Feenstra 1998). As Robinson and Carson (2015) point out, in contrast to "internationalization," the term *globalization* implies a degree of purposive functional integration among geographically dispersed activities.

3. The outsourcing of such jobs to developing countries has an impact on labor in developed countries, where blue-collar workers are increasingly suffering downward pressure on wages or job losses as their factories are closed, resulting in rising political discontent.

4. REDD+ stands for countries' efforts to reduce emissions from deforestation and forest degradation and foster conservation, sustainable management of forests, and enhancement of forest carbon stocks (http://www.forestcarbonpartnership.org/what-redd).

5. Institutions refer to the formal and informal rules, codes of conduct, and norms that define what is an appropriate action and structure human interactions (North 1991; Bruton et al. 2010).

6. See the studies by FAO (2014) and HLPE (2017) for discussions of other estimates.

7. Informal employment, which generates income, is distinct from the time spent by people collecting forest products for subsistence use (FAO 2014).

8. The Montréal Process indicators (covering 12 temperate and boreal countries including the Russian Federation) include a section on wage rates and average income, but many countries do not collect this information (FAO 2014).

9. In many developing countries, small producers/wage laborers might also undertake communal labor or work in labor exchange groups.

10. A nucleus estate model relies on fostering economic interactions between an estate plantation or "modern" farm, usually located near a processing plant and smallholders or landless laborers who supply the same plant and can receive managerial and technological assistance from the estate.

11. Dolan (2005) found in Meru, Kenya, that women provided three-quarters of the labor for French beans but only received one-third of the income. The amount they received was contingent on the leverage they could exert within their households or, more commonly, on the good will of their husbands.

12. Decent work is a UN goal and is defined as "productive work for women and men in conditions of freedom, equity, security and human dignity." Decent work has four interconnected pillars: employment creation, social protection, rights at work, and social dialogue, which are now integral elements of Goal 8 of the 2030 Sustainable Development Agenda (http://www.ilo.org/global/topics/decent-work/lang--en/index.htm).

13. Piece work refers to employment in which a worker is remunerated by a fixed piece rate for a given unit produced.

14. The term *home-based workers* refers to two types of workers who carry out remunerative work within their homes—independent own-account producers and dependent subcontract workers. The latter are also referred to as "homeworkers" (Carr et al. 2000).

15. Fairtrade International (FLO) (also known as Fairtrade Labelling Organizations International) was established in 1997, and is an association of 3 producer networks, 19 national labelling initiatives, and 3 marketing organizations that promote and market the Fairtrade Certification Mark in their countries. Source: Wikipedia.

20 Fragmented Forest Policy
Asset or Concern?

Pauline Pirlot

CONTENTS

20.1 INTRODUCTION

This chapter disentangles the implications of the fragmented nature of international forest policies. The past few decades showed ever-increasing globalization and deepening interdependences between sectoral policies and levels of action. In this context, forest-focused and forest-related sectors were given growing attention, forests being at the crossroads of multiple issues. For instance, forests provide tradable resources, while they also welcome biodiversity. Accordingly, the growing number of international regimes with an interest on forests created policies regulating the way forests are managed and used. For instance, forest management is a cornerstone of agricultural, trade, and biodiversity preservation policies. Forests are also highly relevant to energy sustainability and climate change mitigation. Therefore, the forest regime is now characterized by a mosaic of international venues[1] (Pons 2004) that are different not only in their substantial objective (addressing forests specifically or using forests to reach a nonforest goal) but also in their character (regimes, international organizations, norms etc.), scope (from local to global), and constituencies (public or private) (Biermann et al. 2009). Not only international organizations such as the Food and Agriculture Organization (FAO) and intergovernmental bodies such as the United Nations Forum on Forests (UNFF) shape the international forest regime, but nation states, regional organizations (e.g., Forest Europe or the Commission des Forêts d'Afrique Centrale), local stakeholders (e.g., timber industry), private actors (e.g., the Forest Stewardship Council or the Program for the Endorsement of Forest Certification), networks and nongovernmental organizations (NGOs) are also involved in making, implementing, and influencing international rules for forest governance. They do so at all levels (global, multilateral, regional, bilateral, and even internal policies with external relevance) (Pirlot et al. 2017), using different instruments (market power, persuasion, and coercion). In practice, many of these venues adopt nonforest policies but integrated provisions that target forests directly or indirectly (Bernstein and Cashore 2012). It is the case of climate forest-related policies, aiming at reducing carbon stocks from the atmosphere relying on forest management. Needless to say, the many forest-focused and forest-related interests quarter forests between different approaches. The forest issue is consequently cogoverned by a large array of venues, resulting in a highly fragmented landscape.

The fragmentation of international law (Fischer-Lescano and Teubner 2004), international economics (Bernstein and Ivanova 2007; Biermann 2007), international sociocultural interactions (Menzel 1998), and international regimes (Humphreys 2006; Alter and Meunier 2009; Biermann et al. 2009; Keohane and Victor 2011; Bernstein and Cashore 2012; Colgan et al. 2012; Jinnah 2012; Oberthür and Pożarowska 2013; Orsini et al. 2013; Zelli and van Asselt 2013; Gehring and Faude 2014; van Asselt and McDermott 2016) are widespread issues among scholars. All schools share the same assumption, which constitutes the starting point of this chapter: complexity, hence fragmentation, is a structural characteristic of international relations today (Zelli and van Asselt 2013). While some policy fields are highly centralized, such as the trade regime, regime fragmentation is a widespread phenomenon. The forest, fisheries, and access to genetic resources regimes are a few examples. Ensuing fragmentation provides a specific perspective for understanding and structural frame for carrying out forest policies. This structural frame not only limits the scope of action but also provides enabling features to international actors. On the one hand, the fragmentation of international forest policies is often seen as detrimental to international forest governance, reducing the effectiveness of forest policies. It is why international forest governance is called "failed" by some (Dimitrov 2005; Humphreys 2006). From this perspective, fragmentation should be curtailed and the international forest regime be made more centralized and unified to increase its consistency and its effectiveness. On the other hand, fragmentation provides specific opportunities to the very same actors. For instance, the multiplicity of venues and instruments creates opportunities for tailor-made solutions to problems related to forests. From that perspective, fragmentation should be embraced and made the most out of.

The assets and concerns provided by fragmentation for individual actors and the forest regime are intertwined but not akin. For instance, some features of fragmentation allow individual actors to reach their objectives, but impedes the forest regime to be effective. Likewise, an effective regime does not imply that individual actors benefit the regulatory framework provided. The chapter therefore provides keys to understand specific features of fragmentation that are matters of not only concern but also asset. It also provides a bigger picture of whether fragmentation is an asset or a concern for forest governance. The assessment of policy fragmentation is not only consequently conducted in a discrete manner, focusing on actors, institutions, and policies, but also integrates the larger context of global forest governance. The chapter reveals that fragmentation is a double-edged sword. A given fragmentation characteristic can very well be an asset and a concern, depending on the analytical characteristic (level of action, means of implementation, actor, sector, etc.) we look at.

The chapter is structured as follows: Section 20.2 clarifies what is meant by policy and regime fragmentation. Applied to the forest regime, it discloses the existing debate concerning the nature and the origins of the international forest regime and reveals three main features of regime and policy fragmentation (venue multiplicity, venue smallness, and lack of hierarchy). Building on this, Section 20.3 starts by discussing the concerns ensuing from policy fragmentation. It follows by developing on the assets policy fragmentation might offer. Section 20.4 puts these findings into perspective and explains that concern or asset, fragmentation reduction, as often claimed by practitioners, may not be the point to focus on. Instead, fragmentation management and spontaneous accommodation adjust fragmented regimes to actors' practices. A new—flat—ontology to global governance is also suggested, questioning the relevance of fragmentation as an approach. Finally, Section 20.5 concludes, providing a transversal summary of the key findings of the chapter.

20.2 TAKING STOCK OF FRAGMENTATION

An international regime is a set of "principles, norms, rules and decision-making procedures around which actors' expectations converge in a given area of international relations" (Krasner 1982). The international forest regime is constituted of all institutions, actors, measures, and instruments activated to carry out policies on the forest issue at large. The extensive system of international policies that has developed comprises diverse modes of governance, such as national and international law, nonlegally binding instruments (Chapter 11 of Agenda 21 of the United Nations Conference of Environment and Development, 1992), legally binding treaties specifically on forests (International Poplar Convention) and on forest related topics (United Nations Framework Convention on Climate Change or United Nations Convention to Combat Desertification), nongovernment measures carried out by NGOs, club governance institutions ("Gx"), financial market rules, bilateral and unilateral initiatives [respectively, the Forest Law Enforcement, Governance and Trade Voluntary Partnership Agreements (FLEGT VPAs), and the European Union (EU) Timber Regulation], public–private partnerships (Collaborative Partnership on Forests), agendas and programs of international organizations (UNFF and the World Bank), private instruments (such as certification from the Forest Stewardship Council and the Program for the Endorsement of Forest Certification), and network-like arrangements, emanating from all sectors that have an interest in forests.

Even if "regimes [theoretically] exist in all areas of international relations" (Krasner 1983), the existence of a forest regime does not have unanimous approval. Some scholars believe in the existence of a no regime (yet) (Humphreys 2006; Rayner et al. 2010) or a nonregime (Dimitrov 2005; Giessen 2013). These approaches are built upon the assumption that an international regime is necessarily framed by a multilateral legal framework. Indeed, the "regime" label is not without consequences. It implicitly suggests a hierarchical conception of world politics. No multilateral agreement on forests, which would orchestrate forest policies, has ever been agreed upon, despite several trials. From this perspective, forest governance would not take the shape of a regime—not yet.

When the norms and principles over the same issue/area are produced by an array of different cogoverning actors through a multitude of political venues that are not hierarchically organized, loosely coupled, partially overlapping, and thus potentially conflicting, the international regime is called fragmented (Alter and Meunier 2009; Keohane and Victor 2011; Bernstein and Cashore 2012; Colgan et al. 2012; Gehring and Faude 2014; Zelli and van Asselt 2013) or complex (Jinnah 2012; Oberthür and Pożarowska 2013; Orsini et al. 2013). Fragmentation is characterized by three main features. First, fragmentation emphasizes the multiple interconnected elements comprising the governance architecture of the regime (Biermann et al. 2009). Multiplicity entails diversity of rules, principles, measures, levels of action, and actors in the governance system (Alter and Meunier 2009). Second, fragmentation entails a mosaic of small venues. In small venues, a limited number of case-specific related actors gather to deal with the given issue in a confined environment. Finally, fragmentation suggests uncertainty. Fragmented regimes do not assure principles and norms convergence, neither the consistency of implemented rules.

Building on this, the low degree of interconnectedness and coordination between forest venues, actors, and policies is enough to consider that international forest regime does exist (Humphreys 2005, 2006; McDermott et al. 2010; Rayner et al. 2010; Keohane and Victor 2011; Giessen 2013). There is no focal institution steering and orchestrating international forest policies. However, the manifold of venues, measures, and actors emanating from different regimes and affecting forest-focused and forest-related issues constitute the forest regime.

Fragmentation finds its origins in international and in domestic dynamics. Contrarily to what its name suggests, regime fragmentation is the consequence of an integrative process at the international level (Biermann et al. 2009). The demand for addressing different yet interrelated problems that cannot be unilaterally solved creates a rapprochement between erratic blocs, also called elemental regimes, creating de facto a fragmented regime (Overdevest and Zeitlin 2013). These blocs share (a part of) an issue. In this regard, regime fragmentation is linked to the multifaceted nature of an issue (Humrich 2013; Orsini et al. 2013) over which a comprehensive multilateral treaty would not capture the multiple dimensions (Bernstein and Cashore 2012). Forests are multifunctional, and forest policies address intertwined issues, such as trade, environment, climate change, and indigenous peoples' rights. They also address the management of common goods (such as air quality) and of private goods (such as timber production), the two being inextricable. The integration of the multiple forest-focused and forest-related issues could not lead to a governing top layer that would supposedly take the form of a forest convention or a global forest institution adopting legally binding measures. The borders of the forest issue are blurred, large in number, and different in nature, making a comprehensive agreement on forests hardly achievable. Unifying the forest regime is consequently jeopardized. Moreover, some elemental regimes are not willing to attempt to reach any agreement on forests (such as the agriculture sector), rendering fragmentation structural.

International dynamics alone do not fully explain policy fragmentation. Domestic factors also provide insight into understanding the path to fragmentation (Giessen 2013). At the national level, different national administrations are responsible for forest issues. Forests are handled by the Ministry of Agriculture in France and Brazil, the Ministry of Industry in the United Kingdom, the Ministry of Economic Affairs in the Netherlands, and the Ministry of Environment and Forestry in Indonesia. Each administration approach forests from its own specialized perspective, support different goals, and benefit differently from forest policies. Internally, administration branches with an interest on forests can develop rivalry, paralyzing ambitious internal policy making. At the international level, each of them negotiates on its own terms. Accommodating all domestic approaches (not to mention interests) on forests is a challenge. Therefore, competitions between different branches of administration are reflected in international politics, leading to and crystallizing policy fragmentation (Giessen 2013).

20.3   FRAGMENTED FOREST POLICY: ASSET OR CONCERN?

This section depicts the concerns and assets related to policy fragmentation. It does not aim at providing an exhaustive list, but rather specifies core structural features that are a subject of concern (in Section 20.3.1) and assets (in Section 20.3.2). Three core features of fragmentation are identified: (1) venue multiplicity, (2) venue smallness, and (3) the lack of hierarchy between venues. In reality, their interconnectedness nonetheless appears. They co-define and reinforce one another, all having specific consequences (that also are interconnected). In the following sections, the three features are distinguished for matter of clarity.

Both individual actors and the forest regime as a whole are covered. Building on these findings, the analysis will zoom out on the bigger picture of the forest regime, providing indicative conclusions on policy fragmentation as an asset or a concern. At the end of the chapter, Table 20.1 provides a summary of the main findings of this section, revealing that fragmentation features are both assets and concerns, depending on characteristics the analysis examines.

20.3.1   Concerns

The fragmentation of international regimes and assorted policies is regularly depicted as a matter of concern by scholars and practitioners (Biermann et al. 2009). This section develops on the main impeding features emanating from regime fragmentation, ensuing from the abundance of venues (Section 20.3.1.1), their smallness (Section 20.3.1.2), and the lack of hierarchy among them and their outcomes (Section 20.3.1.3).

TABLE 20.1

Concerns and Assets of Policy Fragmentation

Feature	Concern		Asset	
	Actors	**Regime**	**Actors**	**Regime**
Numerous venues	• Unclarity • Empowers already powerful actors	• Reduces accountability • Reduces ambition • Reduces effectiveness	• Empowers nonstate actors • Increased legitimacy • Increased individual goal achievement	• Increased legitimacy and transparency • Increased innovation
Small venues		• Reduces accountability • Reduces effectiveness	• Develops trust between actors, thus ambitious policies	• Tailor-made solutions • Reactivity
Legal inconsistency		• Reduces effectiveness • Unclarity • Reduces ambition	• Selective implementation increases individual effectiveness	

20.3.1.1 Venue Multiplicity

The forest regime is characterized by a large number of venues. Each venue has different objectives and assorted perspectives on forests. Venue multiplicity generates competition between venues and their outcomes, empowering the already powerful, triggering structural inequity and inefficiency. Forest venues have different interests and perspectives on forests. Each adopts policies accordingly. Therefore, venues compete over forested land and their resources. The larger the number of venues in attendance, the more the competition for access to forests and the fiercer forest resources will be. The competition in a fragmented environment advantages already powerful actors. Actors described as powerful have comparative advantage in their access to resources. Powerful actors can rely on expertise, knowledge, financial, material, and human resources to weigh on the forest regime. Mobilizing these resources, they are able to shape international institutional environments to suit their interests. For instance, the EU is known for pooling member states' expertise to draw its negotiation arguments and have a strong external action (Delreux and Pirlot 2017). Added to important financial capacities and a very entrenched contribution to international development policies, the EU capitalizes strengths to be a powerful actor. In this regard, the EU plays the card of conditionality to achieve its goals. The FLEGT VPAs are a good example. To put it simply, the EU shares expertise and provides technical and material help with the condition that partner countries follow commonly agreed good forest governance practices (Pirlot et al. 2017). The EU consequently mobilizes its material, financial, normative, and knowledge power to achieve its goals and influence other actors. On the international scene, less powerful actors mobilize the same resources, but are able to do so only to a lesser extent. Their financial, material, and technical dependence on more powerful actors further impedes their abilities to carry out self-designed policies and weigh on international negotiations. In this regard, Orsini (2017) points to the fact that private businesses are favored compared to NGOs precisely because the former are more able to mobilize resources. In this context, less powerful actors gain power by forming negotiation blocs (such as the G77 + China) and obstructing or leading negotiations. Due to the multiplicity of venues present, this power is even more important, as powerful actors can activate it in a greater number of political loci, or favoring one venue over another. This matter of equity is particularly concerned in the international forest regime, as a large proportion of forest policies is geographically implemented in less powerful countries and sometimes relies on marginalized communities (van Asselt and McDermott 2016), who clearly have limited access to decision-making. In this sense, fragmentation crystallizes inequity.

20.3.1.2 Small Venues

So-called small venues proliferate in fragmented regimes. Small venues are composed of a limited number of actors. With regard to forest governance, regional organizations such as Forest Europe, clubs of like-minded countries such as the "Gx," bilateral agreements, and small NGOs are illustrative of small venues. On the long run, the patchwork of uncoordinated small venues is concerning. They cannot address and resolve structural forest problems (Biermann et al. 2009), hindering the effectiveness of the forest regime. Hence, forest solutions in small venues address specific problems in isolation from each other, rendering ex post cooperation difficult (Oberthür and Gehring 2006), especially in combination with the large number of venues to coordinate. Aggregating unconnected policies does not provide a key to solve structural forest problems. To put it differently, the effectiveness of the forest regime is not equal to the sum of the effectiveness of its parts.

Small venue proliferation curtails the possibility of reaching large-scale ambitious policies (Biermann et al. 2009). The regime is consequently made of small-scale narrowly focused policies and large-scale shallow statements that generally happen to be a lowest common denominator (Bernstein and Cashore 2012). In this sense, the fragmentation of negotiation loci entails forest regime ineffectiveness.

Moreover, considering the importance of small venues in a fragmented regime, potential blocs of less powerful actors are dispersed. Powerful actors comparatively weigh more in small venues, increasing their capabilities in fragmented regimes.

One could argue that policy making concerns linked to small venues is in practice counteracted on the ground while policy implementation starts. The legal principles being mere words (indeed sometimes assorted with sanctions), the impact of policies depend on the way they are implemented on the ground. However, once fragmented, policies are entrenched in implementation acts, it seems unlikely that they will spread and adjust to one another, for two reasons. First, small venues develop individually, pulling the forest issue in their desired direction, minimizing the acceptance of their policy to the eyes of other groups. Second, if a small group tackles a forest issue from a determined angle, another small group is disincentivized to adhere and embrace the policies of the former small venue precisely because the former small venue is dealing with it. Designing a policy among a small group of actors does not mean it will be upscaled and linked to other regimes and policies.

20.3.1.3   Lack of Hierarchy

Fragmentation and legal inconsistency are the two sides of the same coin (Morin and Orsini 2014; van Asselt 2014). In the absence of hierarchy and coordination procedures, political venues are most likely to adopt policies that are not necessarily consistent with one another (Alter and Meunier 2009), creating conflicts between legal outcomes. This phenomenon is accentuated by the number of venues active in fragmented regimes. Legal inconsistencies (van Asselt 2014), also called substantive conflicts (Morin and Orsini 2014) reveal the legal weakness of fragmented regimes. Legal inconsistencies translate in practice in the impossibility to comply at the same time with two (or more) (conflicting) legal norms. For example, one treaty imposes biodiversity preservation, thus preserving forests the way they are or managing them in a specific way, while another treaty aims at increasing forests carbon sequestration capacities, thus striving for fast-growing monocultural forests. Joint compliance is not possible because the two legal requirements pull forests in opposite directions. Legal inconsistencies also refer to the use of different principles (precautionary approach vs. cost-effective approach) and the use of opposing economic incentives (Clean Development Mechanism vs. Kyoto Protocol). They reveal the dysfunctionality of policy fragmentation and undermine the effectiveness of the international forest regime.

The competition in the context of a lack of coordination creates so-called unintentional reverberation (Alter and Meunier 2009). To put it simply, a change in one venue or sector unpredictably affects other venues or sectors. The competing venues need not to focus on the same field, as seemingly unrelated regimes do overlap. Hence, one regime concluding an agreement on the issue will automatically constrain the other regime. For instance, allowing the trade of timber species will de facto have an impact on the environment, due to the possible pressure of timber logging on biodiversity or simply the environmental costs of timber logging and shipping worldwide. Another unintentional reverberation is highlighted by the tensions between regulatory schemes against illegal logging that may conflict with international trade law (Fishman and Obidzinski 2014) as defined by the World Trade Organization (WTO). Unintended reverberation so contributes to unpredictability and systemic inefficiency.

Therefore, the strategic choice between measures must be operated, first during the policy making phase and second during implementation phase. More generally, the absence of hierarchy and order triggered by fragmentation entails that actors can use regime fragmentation favoring one type of policies over others and pursue their objectives by forum shopping (Zürn and Faude 2013). Forum shopping in this environment may have two negative consequences. First, it may contribute to legal ambiguity, hence deepening fragmentation, with actors individually pursuing their objectives (Alter and Meunier 2009). While legal inconsistencies prevail, actors select which rule to implement (Alter and Meunier 2009). The least implemented rules end up lacking credibility, regardless of its political origin. The legitimacy of the forest regime finds itself reduced. In this regard, actors can adopt a more entrepreneurial strategy voluntarily creating inconsistencies between rules (Raustiala and Victor 2004). In this case, an actor pushes for inconsistent rule adoption in parallel or overlapping venues with the intention to undermine one of this rule, hence weakening the associated institution and regime. Overall, inconsistent legal norms pressure each other (Raustiala and Victor 2004; Raustiala 2006), undermining the clarity of which rule or institution has authority over an issue (Karlsson-Vinkhuyzen and McGee 2013). In turn, accountability is ever less clear.

Second, legal inconsistency may create a race to the bottom. Strategic actors would hypothetically shop for the rule that is most suitable to their interests. One criterion is to decide upon and select to implement only the softest or least costly measures (Biermann et al. 2009). The ambition of the regime as a whole would decrease in the long run (Biermann et al. 2009). In turn, to ensure the implementation of the legal rules adopted, each of these negotiated would strategically end up being vague or ambiguous on how it should be interpreted (Alter and Meunier 2009). It is a way to limit legal inconsistency and consequently afford no possibility for selective implementation compliance, lowering the ambition of the regime.

Altogether, conflicts may arise between large political orientations (that may be crystallized while becoming legal norms) (Sturett et al. 2013), hindering the effectiveness of the forest regime. Different elemental regimes approach forestry from different angles, which can be conflictive, generating contradictory rationales and approaches to forests. For instance, the trade international regime and the climate international regime both have an interest in forest management. Their interests are in practice not necessarily compatible. However, they are core components of the forest regime, which ends up being unclear and lacking accountability.

20.3.2 ASSETS

International regime and policy fragmentation are recognized to be an asset for forest actors and the whole forest issue. Fragmented regimes have specific elements that provide actors with an enabling environment.

20.3.2.1 Venue Multiplicity

A fragmented regime is composed of a multiplicity of unhierarchically organized venues (Alter and Meunier 2009; Biermann et al. 2009; Drezner 2009), providing as many occasions for negotiators to meet. This section discloses that venue multiplicity empowers weaker actors, in particular nonstate actors, and enhances the creativity and ambition of fragmented policies. First, fragmentation provides opportunities for weaker actors (states and nonstate actors such as NGOs, businesses, and stakeholders) to participate in and influence international forest policy making (Arts 2000; Dimitrov 2005; Bass and Guéneau 2007; Alter and Meunier 2009; Biermann 2009; Helfer 2009; Orsini et al. 2013; Van de Graaf 2013; Zürn and Faude 2013). Regime fragmentation is characterized by multiplicity of venues, understood both as abundance (there exists many forest venues) and as variety (the many forest venues are diverse in their subject matter and constituency and provide different types of outcomes). Each venue is an access point for actors to international forest governance. Fragmentation so provides actors with a number of choices concerning where to exercise their action, potentially increasing their international presence (Helfer 2009). When putting in practice such a selective choice, also called venue shopping, nonstate actors chose the venue most suitable for achieving their objectives. Weaker actors are consequently not limited to the well-established obedience to global multilateralism. In global venues, weaker actors are challenged to exert countervailing power. Indeed, the presence of many voices makes every actors' voices proportionally less vocal. With fragmented regime being structured by a large number of venues, limited in their issue area and constituency, weaker actors are provided with contextual and structural advantages that they do not have in larger venues. They may very well exercise their power and voice their claim in a more visible manner in alternative venues. Regime shifting is also used by weaker actors to "relocate rule-making initiatives to international venues concerned with other issue areas" (Helfer 2009) to venues that are better aligned with their interests. In venue shopping and venue shifting, weaker actors strategically relocate or put their negotiations efforts to appropriate venues, increasing their individual goal achievement.

The case of nonstate actors is highly relevant in this regard. Nonstate actors are as important as state actors of international policy-making. Nonstate actors are not only less powerful than states, but are also known to be the most important source of input legitimacy, authority (Arts 2000), credibility, and transparency (Bäckstrand 2008) in international politics. They voice societal and private interests and are generally highly knowledgeable on specific domains. Recent developments in global governance literature also reveal that the presence of nonstate actors positively contribute to the accountability of fragmented regime (Keohane and Victor 2011). In this regard, Alter and Meunier (2009) explain that regime fragmentation makes international venues more permeable and cooperative (Gómez-Mera 2015), generating a more important role for nonstate actors and experts. Their participation in international governance is therefore a strong asset to global governance, increasing the input legitimacy of the forest regime.

Nonstate actors' repeated activity in multiple venues makes them grow over time (Orsini et al. 2013). In turn, global governance finds itself even more legitimate and accountable. The recurrent activation of multiple venues increases NGOs' organizational and ideational powers. They gather expertise across areas, boosting their credibility as an international negotiation partner (Gullberg 2008). The ideational power and expertise is particularly valuable for technical and multifunctional issues (Alter and Meunier 2009) such as forest-related issues.

Second, venue multiplicity stimulates innovative problem-solving. In a context characterized by regulatory diversity, regulatory competition is most likely to be high. Facing regulatory competition, actors and venues create diverse solutions in different contexts. For instance, two venues face the same problem. They use their resources and expertise and rely on different regulation bodies to adopt solutions (Biermann et al. 2009). Policy making in fragmented regimes is consequently innovative. In this sense, regime fragmentation contributes in functionally differentiating between the components of the fragmented regime (Gehring and Faude 2014). On the whole, a fragmented regime foresees the possibility to build an innovative type of comprehensive regime structure that does not rely on unity and hierarchy (Overdevest and Zeitlin 2014).

20.3.2.2 Small Venues

Small venues are characterized by a small number of closely related actors. The smallness of fragmented regime's venues renders deep connections between actors. Limited in number, actors in small venues can engage in profound negotiations on issues for which actors have clearly common interests but discordant perspectives (Hafner 2004). Closely related actors develop group dynamics within which expectations, norms, and objectives tend to converge.

This is reflected by the easiness to reach common ground. Centralized regimes are made of large international venues (as opposed to small venues). In large venues, many interests and preferences must be pooled. The policies adopted therefore take the form of the less costly, broad, and shallow lowest common denominator, hindering ambitious objectives to be reflected in policies (Oberthür and Rabitz 2014). For instance, the United Nations system is based on consensus-based decision-making procedures. As a consequence, its inertia is sometimes great. Conservative preferences are more easily adopted than reformist ones. On the contrary, small venues gather a smaller number of interested actors. Reaching an agreement is consequently less challenging than that in global venues, allowing for more ambitious policies to be adopted, deeply focusing on specific issues (Biermann et al. 2009).

The policies adopted in small venues thus reflect better the interests and objectives of the negotiators (Biermann et al. 2009). For instance, the UNFF counts 193 members. Finding common ground and converging interests between every state of the world appears to be a challenge that is impossible to resolve. This is one of the reasons why the UNFF adopts only nonbinding general statements on forests (Delreux and Pirlot 2017). On the contrary, bilateral negotiations, even between actors with seemingly incompatible approaches on forests, appear to end up on more specific and deeper agreements. For instance, the FLEGT VPAs between the EU and non-EU timber-producing countries focus on agreements targeting specific aspects of forest governance and timber trade. Negotiators found common ground to engage in trade and governance policies. At the actor level, small venues are consequently beneficial.

The mutual trust and loyalty developed between actor increases their willingness to find collective solutions to problems, enhancing creative and tailored problem-solving (Alter and Meunier 2009; Kuyper 2014). Creative and specifically adapted policies contribute to the output effectiveness of the policies. Every measure is consequently adjusted to given situations.

The presence of small venues is also beneficial to the forest regime as a whole. Contrary to highly centralized and hierarchized regimes, fragmented regimes are composed of a large number of small venues. Such small venues encompass the most important actors for a given issue and accelerate the speed of negotiation and entry into force, limiting hindrance and defection from unlike-minded countries (Biermann et al. 2009). Speed and reactivity in decision-making clearly is an asset for global forest governance. It creates a flexible environment.

20.3.2.3 Lack of Hierarchy

Section 20.3 explained that the lack of hierarchy among fragmented policies creates political competition and legal inconsistencies. It turns out to be a structural opportunity, as legal inconsistencies can be used or even created to one's profit (Alter and Meunier 2009), during both negotiation and implementation phases. Legal and political inconsistencies actually enable cross-institutional strategies (Alter and Meunier 2009), increasing flexibility for the actors (Keohane and Victor 2011; Morin and Orsini 2013; Orsini et al. 2013; Kuyper 2014). During the negotiation phase, fragmentation provides actors with the opportunity to create legal inconsistencies. While negotiating simultaneously in two (or more) venues, an actor can strategically shape the outcome of one institution to put the emphasis on (Alter and Meunier 2009), to complement, or not to obstruct another venue (Raustiala and Victor 2004; Raustiala 2006). Such cross-institutional strategies induce hierarchies between international negotiations arenas, spotlighting an institution (and obviously its constituency, political outcome, and instruments) over another. Hence, fragmentation allows individual actors to strategically select the most suitable rule, ranking legal rules according to their own relevance.

During the rule implementation phase, inconsistent rules can be exploited by actors to achieve their goals. Implementing authorities venue shop, prioritizing some rules over others. Legal inconsistency provides actors with leverage on existing policies by shopping for the most suitable one to their interest and capabilities. This feature is particularly interesting for local communities that do not necessarily have the resources to implement all international legal requirements. Some legal rules will emerge as most used compared to others, providing some structure to the fragmented regime, encouraging renegotiation, amendment, or interpretation of the least used rule (Raustiala 2006). The forest regime consequently ends up adapting to and mirroring what is done in reality.

This practice is especially relevant in the forest domain. International forest policies are adopted by a large range of state and nonstate actors, each of them having very specific interests and preferences. The strategic use and creation of legal inconsistencies allow actors to make the most of forest policies, increasing their individual effectiveness.

20.3.3 WRAPPING UP

It appears that regime fragmentation not only unevenly distributes negotiation power and resources between actors, empowering the already powerful. But it also triggers strong political competition over public and private goods and creates legal inconsistencies that further blur an already complex environment. On the whole, these arguments suggest that policy fragmentation impedes the whole international regime from being effective.

Venue multiplicity renders policy making reactive and policies more specific, adapting to the local environment. Moreover, small venues equilibrate the debates. Weaker actors find themselves embedded in negotiation environments made of trust and loyalty. Forest policies are thus possibly more ambitious. The lack of hierarchy triggers cross-institutional strategies from individual actors, enhancing their goal achievement. Finally, it renders policies ambitious, creative, and adaptable. On the whole, this section reveals that fragmentation brings about opportunities, which are concerns for some (as depicted in Section 20.3.1) and may be assets for others. Table 20.1 provides a summary of the key features of fragmentation and related assets and concern.

20.4 BEYOND ASSET AND CONCERN: PRACTICAL EQUILIBRIUM AND THEORETICAL ALTERNATIVE TO FRAGMENTATION

This section attempts to go beyond policy fragmentation as understood in this chapter so far. It focuses on two aspects. The first one has a practical reach. It advocates that fragmented policies spontaneously organize and adjust to create a viable equilibrium, which is different than what we experience in centralized and hierarchical regimes. This phenomenon is labeled by some "experimentalist governance." The second one is theoretical. It provides keys to assess and understand fragmentation adopting a horizontal—"flat"—ontology (compared to a vertical, hierarchical, one) of world politics (Arts et al. 2016). In this regard, grasping world politics from a non-hierarchical perspective sheds a new light on policy fragmentation.

Section 20.3 revealed that fragmentation features are the two sides of the same coin, both asset and concern, depending on where the analytical focus stands. Regularly in practitioner and academic spheres, fragmentation is depicted as an irregularity that ought to be flattened, an obstacle that should be overcome. Policy fragmentation enthusiasts turn out to be less visible. Practitioners seem to be timorous with regard to the fragmentation of the forest policies. They dread fragmentation for being the main cause of the slow progresses made in the forestry sector. Moreover, they fear that fragmentation leads to losing control over institutional identity, thus paradoxically holding on the existing venues and resisting any venue change.

It is against this backdrop that a calling for harmonizing and centralizing the environmental regime at large was launched. The report claims to tackle the drawbacks of regime and policy fragmentation by creating a world environmental organization based on the model of the WTO (Runge 2001). A global centralized institution would orchestrate international environmental initiatives. For their part, scholars often refer to the necessity to reduce fragmentation in order to curtail its drawbacks (Biermann et al. 2009; Oberthür 2009; Young 2011; Oberthür and Pożarowska 2013; Zürn and Faude 2013). Fragmentation would consequently take the path to centralization and hierarchization. However, centralizing and hierarchizing fragmented regimes are assumed not to enhance regime effectiveness. It is in this context that Oberthür and Gehring (2004) respond to the claim that creating a centralized and hierarchically structured environmental regime (a trade-like regime) would not increase decision-making, environmental governance, and coordination. On the whole, they conclude that regime hierarchization and centralization does not ameliorate its effectiveness. Taking this argument, fragmentation should not be reduced, but embraced as another way of making politics.

Literature often refers to managing fragmentation. To put it simply, fragmentation management suggests that the creation of a coordination umbrella to fragmented institutions and policies would neutralize the drawbacks from fragmentation while maintaining its advantages, rendering fragmented regimes stable over time. In practical terms, fragmentation management requires the coordination of regimes and policies that are at odds, within and across issue areas (Zürn and Faude 2013). In the forest domain, coordinations within or across the forest issue are synonymous, as the forest regime is very much entangled in a mosaic of issue areas. Coordination mechanisms would be horizontal, to avoid the weaknesses of hierarchy (Zürn and Faude 2013) and preserve the assets of fragmentation, such as flexibility and adaptiveness.

More generally, fragmentation seems to spontaneously coordinate and stabilize (Gehring and Faude 2014), for two reasons. First, actors themselves, in their cross-institutional strategies, seem to bring about coordination and policy accommodation (Gehrnig and Faude 2014). Multiple members (actors being members of several overlapping institutions simultaneously) exploit institutions by forum shopping to promote their mixed motives. It is assumed that multiple members also have an interest in resolving conflicts between inconsistent institutions and rules. Rule adoption, rule (re)interpretation, and rule implementation induce incremental adjustments (Jupille et al. 2013) of one regime or policy to the overlapping one (Gehring and Faude 2013). Such accommodation of overlapping rules and regimes provide the regime with a flexible decentralized mode of coordination, bringing equilibrium in fragmentation.

Second, to survive competition, international venues are expected to specify their issue area and work to specific *niches* (Aldrich 1999). As a consequence, the spontaneous division of labor would organize the governance activities of the venues (Gehring and Faude 2013; Oberthür and Pożarowska 2013). Regime and policy fragmentation generates problems of coordination and spontaneously provides the opportunity to solve them. This would supposedly end up in a governance system that is coordinated enough to ensure the consistency of rules, norms, and institutions, while enabling the polycentricity and flexibility necessary to govern a complex issue, such as the forest, and deal with global interdependences.

In academic literature, issue and level interdependences have been pictured by experimentalist governance scholars. They contend a more inductive approach to studying fragmentation that is based on positive and productive interactions between regulatory schemes that operate in the same domain of forest governance. Experimentalist governance is a "recursive process of provisional goal setting and revision based on learning from comparison of alternative approaches to advancing these goals in different contexts" (Overdevest and Zeitlin 2014). The scope condition for experimentalism is regime fragmentation. It is driven by a broad overarching goal (e.g., sustainable forest management). Local stakeholders seek the board goals according to their own understanding of it and using their own ways. Local stakeholders must, in turn, report on their goal achievements and feed and participate in the redefinition of global open-ended overarching goals (Sabel and Zeitlin 2012; Kuyper 2014). It is through reflexive learning that innovation diffuses across actors.

Altogether, investigating fragmentation through alternative theoretical lenses could be a way to apprehend fragmentation-related phenomena from a less binary stance. This chapter relies on a hierarchical conception of world politics. Besides the current impossibility to erase deforestation practices, there is no legally binding agreement on forests and no institutionalized top layer that would oversee and orchestrate forest policies. It is according to this hierarchical ontology that the forest regime is called failed. However, the use of a hierarchical ontology is not neutral. Ontologies are performative, meaning that it produces certain images of certain realities (Butler 1988). Global politics are not necessarily hierarchical, but are analytically made to be represented as such (Wendt 1992). Against this backdrop, adopting a flat ontology to world politics (Arts et al. 2016), and forest politics in particular, opens up the door for understanding global forest governance. This ontology goes beyond the hierarchical view of world politics and hence does not apprehend world politics as a political regime (and all it entails). On the contrary, it sees the world flat. Elements of world politics (policies, actors, sectors, norms, etc.) are connected through networks operating at every scale, from global to local. In practice, forest policies range from global to local, all having an impact on forest practices on the ground (Overdevest and Zeitlin 2013). In turn, practices on the ground nourish world politics (Arts 2004). This *glocal* (from the contraction of "global" and "local") nexus suggests, from an inductive perspective, that a hierarchical model may not be the most suitable one to understand the implications of fragmentation.

20.5   CONCLUSION

The picture emanating from the international forest regime is one of a mosaic of international venues, actors, and instruments, at all levels of governance, assembling in a fragmented regime. On the one hand, a fragmented regime is composed of a large number of small venues empowering the already powerful international actors, thus lowering the credibility and legitimacy of international forest policies, in fine undermining the predictability of international politics. Legal inconsistencies reveal the dysfunctionality of a fragmented regime, enabling destructive cross-institutional strategies during negotiation and implementation phases, rendering the resolution of structural problems limited. The fragmented forest regime is consequently seen as inherently ineffective and detrimental to weaker actors.

On the other hand, regime fragmentation has considerable advantages, particularly for individual actors' goal achievement. Fragmented regimes are composed of a multiplicity of venues, all of which constitute opportunities for actors to participate to the forest issue, specifically nonstate actors, increasing the legitimacy of forest governance. Then, strategies across small venues renders policies more specialized, ambitious, and adaptable to specific contexts. Finally, fragmentation enables experimentation, incremental learning, and flexibility. Hard law is thus no precondition to global governance.

Hence, this chapter reveals contrasted outcomes when it comes to regime fragmentation and assorted fragmented policies. Indeed, in most cases, fragmentation features are both assets and concern. For instance, while venue shopping is beneficial for individual actors' goal achievement, it is detrimental to the overall forest regime effectiveness.

At the level of the forest regime, one cannot conclude whether fragmentation is an asset or a matter of concern. The chapter rather tilts the balance toward fragmentation as a concern for forest governance. Fragmentation generates structural inefficiencies that do not allow effectively tackling forest-related problems. However, despite the apparent ambition of some practitioners and scholars, reducing fragmentation seems unlikely. First of all, fragmentation is inherent to forests as a political issue. It is multifaceted as forests are central to several other issues (such as trade, environment protection, climate change mitigation, and energy). The international forest regime is at the crossroads of overlapping regimes. Centralizing and hierarchizing relatively independent international regimes is implausible. In particular, it was shown that a centralized and hierarchized regime is not always more effective than a fragmented yet coordinated regime. Second, fragmentation spontaneously finds balance and stability over time, mostly through mutual adaptation.

In addition, the chapter attempted to go beyond fragmentation as an asset or a concern. From Section 20.4, it is concluded that fragmentation is a reflection of a world of globalization and interdependences. Fragmentation is consequently not avoidable. Coordinated fragmentation, like experimentalist governance, could very well not only reflect the current state of international relations, but also provide a novel model for international governance. In the end, and taking this reflection one step further, it is assumed that fragmentation and its implications are dependent on a hierarchical approach to global politics. Theoretically approaching fragmentation through other lenses, such as relying on a flat ontology, provides another outlook on whether fragmentation is a matter of concern or not.

REFERENCES

Aldrich, H. 1999. *Organizations Evolving.* London: Sage.
Alter, K. J. and S. Meunier. 2009. The politics of international regime complexity. *Perspectives on Politics* 7(1):13–24.
Arts, B. 2000. Regimes, Non-state actors and the state system: A "structurational" regime model. *European Journal of International Relations* 6(4):513–542.
Arts, B. 2004. The Global-local nexus: NGOs and the articulation of scale. *Journal of Economics and Social Geography* 95(5):498–511.
Arts, B., Kleinschmit, D., and H. Pülzl. 2016. Forest governance: Connecting global to local practices. In *Practice Theory and Research: Exploring the Dynamics of Social Life*, eds. Spaargaren, G., Weenink, D., and M. Lamers, 202–228. Abingdon: Routledge.
Bäckstrand, K. 2008. Accountability of networked climate governance: The rise of transnational climate partnership. *Global Environmental Politics* 8(3):74–102.
Bass, S. and S. Guéneau. 2007. Global forest governance: Effectiveness, and legitimacy of market-driven approaches. In *Participation for Sustainability in Trade*, eds. Thoyer, S. and B. Martimort-Asso, 161–183. Aldershot: Ashgate.
Bernstein, S. and M. Ivanova. 2007. Institutional fragmentation and normative compromise in global environmental governance: What prospects for re-embedding. In *Global Liberalism and Political Order: Towards a New Grand Compromise*, eds. Bernstein, S. F. and L. W. Pauly, 161–185. Albany, NY: SUNY Press.
Bernstein, S. and B. Cashore. 2012. Complex global governance and domestic policies: Four pathways of influence. *International Affairs* 88(3):585–604.
Biermann, F. 2007. Reforming global environmental governance: From UNEP toward a world environment organization. In *Global Environmental Governance: Perspectives on the Current Debate*, eds. Swart, S. and E. Perry, 103–123. New York: Center for UN Reform Education.
Biermann, F., Pattberg, P., and H. Van Hasselt. 2009. The fragmentation of global governance architectures: A framework for analysis. *Global Governance* 16(1):81–101.
Butler, J. 1988. Performative acts and gender constitution: An essay in phenomenology and feminist theory. *Theatre Journal* 40(4):519–531.
Colgan, J. D., Keohane, R. O., and T. Van de Graaf. 2012. Punctuated equilibrium in the energy regime complex. *Review of International Organisation* 7:117–143.
Delreux, T. and P. Pirlot. 2017. EU performance in the United Nations Forum on Forests: An analysis of the EU at UNFF11. In *The EU in UN Politics. Actors, Processes and Performance*, eds. Bourantonis, D. and S. Blavoukos, 187–207. London: Palgrave Macmillan.
Dimitrov, R. S. 2005. Hostage to norms: States, Institutions and global forest politics. *Global Environmental Politics* 5(4):1–22.

Drezner, D. 2009. The power and peril of international regime complexity. *Perspective on Politics* 7(1):65–70.

Fischer-Lescano, A. and G. Teubner. 2004. Regime collision: The vain search for legal unity in the fragmentation of global law. *Michigan Journal of International Law* 25:999–1046.

Fishman, A. and K. Obidzinski. 2014. European Union timber regulation: Is it legal? *Review of European, Comparative and International Environmental Law* 23(2):258–274.

Gehring, T. and B. Faude. 2014. A theory of emerging order within institutional complexes: How competition among regulatory international institutions leads to institutional adaptation and division of labor. *Review of International Organiz*ation 9(4):471–498.

Giessen, L. 2013. Reviewing the main characteristics of the international forest regime complex and partial explanations for its fragmentation. *International Forestry Review* 15(1):60–70.

Gómez-Mera, L. 2016. Regime complexity and global governance: The case of trafficking in persons. *European Journal of International Relations* 22(3):566–595.

Gullberg, A. T. 2008. Rational lobbying and EU Climate Policy. *International Environmental Agreements: Politics, Law and Economics* 8(2):161–178.

Hafner, G. 2004. Pros and cons from fragmentation of international law. *Michigan Journal of International Law* 25(4):849–863.

Helfer, L. 2009. Regime Shifting in the international intellectual property system. *Perspectives on Politics* 7(1):39–44.

Humphreys, D. 2005. The elusive quest for a global forest convention. *Review of European Community and International Environmental Law* 14(1):1–10.

Humphreys, D. 2006. *Logjam—Deforestation and the Crisis of Global Governance.* London: Earthscan.

Humrich, C. 2013. Fragmented international governance of artic offshore oil: Governance challenges and institutional improvement. *Global Environmental Politics* 13(3):79–99.

Jinnah, S. 2012. *Managing Institutional Complexity: Regime Interplay and Global Environmental Change.* Cambridge, MA: MIT Press paper.

Jupille, J., Mattli, W., and D. Snidal. 2013. *Institutional Choice and Global Commerce.* Cambridge, MA: Cambridge University Press.

Karlsson-Vinkhuyzen, S. I. and J. McGee. 2013. Legitimacy in an era of fragmentation: The case of global climate governance. *Global Environmental Politics* 3(3):56–78.

Keohane, R. O. and D. G. Victor. 2011. The regime complex for climate change. *Perspectives on Politics* 9(1):7–23.

Krasner, S. D. 1982. Structural causes and regime consequences: Regimes as intervening variables. *International Organizations* 36(2):185–205.

Krasner, S. D. 1983. *International Regimes.* Ithaca, NY: Cornell University Press.

Kuyper, J. W. 2014. Global democratization and international regime complexity. *European Journal of International Relations* 20(3):620–646.

McDermott, C. L., Humphreys, D., Wildburger, C. et al. 2010. Mapping the core actors and issues defining international forest governance. In *Embracing Complexity: Meeting the Challenges of International Forest Governance*, eds. Rayner, J., Buck, A., and P. Katila, 19–36. Vienna: International Union of Forest Research Organizations (IUFRO), World Series Volume 28.

Menzel, U. 1998. *Globalisierung versus Fragmentierung.* Frankfurt: Suhrkamp.

Morin, J.-F. and A. Orsini. 2013. Regime complexity and policy coherency: Introducing a co-adjustments model. *Global Governance: A Review of Multilateralism and International Organizations* 19(1):41–51.

Morin, J.-F. and A. Orsini. 2014. Policy coherency and regime complexes: The case of genetic resources. *Review of International Studies* 40(2):303–324.

Oberthür, S. 2009. Interplay management: Enhancing environmental policy integration among international institutions. *International Environmental Agreements: Politics, Law and Economics* 9(4):371–391.

Oberthür, S. and T. Gehring. 2004. Reforming international environmental governance: An institutionalist critique of the proposal for a world environmental. *Politics, Law and Economics* 4:359–381.

Oberthür, S. and T. Gehring. 2006. International interaction in global environmental governance: The case of the Cartagena Protocol and the World Trade Organization. *Global Environmental Politics* 6(2):1–31.

Oberthür, S. and J. Pożarowska. 2013. Managing institutional complexity and fragmentation: The Nagoya Protocol and the global governance of genetic resources. *Global Environmental Politics* 13(3):19.

Oberthür, S. and F. Rabitz. 2014. On the performance and leadership of the European Union in global environmental governance: The case of the Nagoya Protocol. *Journal of European Public Policy* 21:39–57.

Orsini, A., Morin, J.-F., and O. Young. 2013. Regime complexes: A boom, bust or buzz for global governance? *Global Governance* 19(2):27–39.

Orsini, A. 2017. The negotiation burden of institutional interactions: Non-state organizations and the international negotiations on forests. *Cambridge Review of International Affairs*.

Overdevest, C. and J. Zeitlin. 2013. Constructing a transnational timber legality assurance regime: Architecture, accomplishments, challenges. *Forest Policy and Economics* 48:6–15.

Overdevest, C. and J. Zeitlin. 2014. Assembling an experimentalist regime: Transnational governance interactions in the forest sector. *Regulation and Governance* 8:22–48.

Pirlot, P., Delreux, T., and C. Farcy. 2017. A multi-sectoral and multi-level approach to sustainable forest management. In *EU External Environmental Policy. Rules, Regulation and Governance Beyond Borders*, eds. Adelle, C., Biedenkopf, K., and D. Torney, 167–187. London: Palgrave Macmillan.

Pons, X. 2004. El régimen forestal internacional: La cooperación internacional para la ordenación, conservación y desarrollo sostenible de los bosques. Madrid: Instituto Nacional de Investigaciones Agronómicas.

Raustiala, K. 2006. *Density and Conflict in International Intellectual Property Law.* University of California, Los Angeles, School of Law Research Paper No. 06-31. Los Angeles, CA: University of California, Los Angeles, School of Law.

Raustiala, K. and D. G. Victor. 2004. The regime complex for plant genetic resources. *International Organization* 58(2):277–309.

Rayner, J., Buck, A., and P. Katila. 2010. *Embracing Complexity: Meeting the Challenges of International Forest Governance.* Vienna: IUFRO, World Series Volume 28.

Runge, C. F. 2001. A global environment organization (GEO) and the world trading system. *Journal of World Trade* 35(4):399–426.

Sabel, C. F. and J. Zeitlin. 2012. Experimentalist Governance. In *The Oxford Handbook on Governance*, ed. Levi-Faur, D., 169–183. Oxford, UK: Oxford University Press.

Sturett, M. J., Nance, M. T. and D. Armstrong. 2013. Navigating the maritime piracy regime complex. *Global Governance: A Review of Multilateralism and International Organizations* 19(1):93–104.

van Asselt, H. 2014. *The Fragmentation of Global Climate Governance: Consequences and Management of Regime Interactions.* Camberley: Edward Elgar.

van Asselt, H. and C. L. McDermott. 2016. The institutional complex for REDD+: A benevolent jogsaw? In *Research Handbook on REDD+ and International Law*, ed. Voigt, C., 63–88. Research Handbooks in Climate Law, Cheltenham: Edward Elgar Publishing.

Van de Graaf, T. 2013. *The Politics and Institutions of Global Energy Governance.* Basingstoke: Palgrave Macmillan.

Wendt, A. 1992. Anarchy is what states make of it: The social construction of power politics. *International Organization* 46(2):391–425.

Young, O. 2011. *Overcoming Fragmented Governance: The Case of Climate Change and the MDGs*. Governance and Sustainability Issue Brief 2. Boston, MA: University of Massachusetts.

Zelli, F. and H. van Asselt. 2013. The institutional fragmentation of global environmental governance: Causes, consequences, and responses. *Global Environmental Politics* 13(3):1–13.

Zürn, M. and B. Faude. 2013. On fragmentation, differentiation, and coordination. *Global Environmental Politics* 13(3):119–130.

ENDNOTE

1. In this chapter, a venue is a political locus that can be activated to create rules and norms. It can take the shape not only of an international organization, but also of an international agreement, a network, a program, a local authority, etc.

Section V

Lessons Learned

21 Conclusions

*Christine Farcy, Inazio Martinez de Arano,
and Eduardo Rojas-Briales*

In the current period that humanity is going through, paradoxes and contradictions are more prevalent than agreed truths. From one hand, the speed of today's societal changes is higher than the ability of human institutions to react and adapt. From another one, the diversity of their impacts and their deepness shakes and destabilizes certainties and identities and accentuates defensive positioning.

Forestry, which embraces both theory and practice of all that constitutes the management and use of forests, including their preservation and restoration, does not escape this phenomenon as shown through the various contributions shared in this book which is focusing on societal changes affecting the relationship not only between the society and the forests but also between humans themselves when being related to the forests.

Since the enlightenment, forestry has relied on some basic elements or building blocks. First is science, including academic education. Forestry students soon learn the multiple ramifications of forestry that stretch from botany and ecology to silviculture, infrastructure planning, and logistics to hydrology, game management, and applied conservation, all in the area of natural sciences. In comparison and though increasing (e.g., *Journal of Forest Policy and Economics*), social sciences in forestry are minority. Another is the reliance in a *bequest effect* stating that securing the supply of wood secures all other functions. And a final one is a rationalist top-down approach to governance, framed by strong command and control structures. Those elements have often faced contradictions and resistance by different social groups but have somehow persisted until present times, even if, as more or less subtle underpinnings of forestry institutions and deontology.

In the meantime, the world has become a different place. This book invites the forestry community and all those interested in forests and forestry to leave their comfort zone and take stock of the intensity and deepness of changes. Some contributions present a disciplinary state of play whose limits appears sometimes implicitly or are revealed by comparison with others; others dare interdisciplinary prospect or offer useful while uncomfortable transgressions toward new horizons. The whole encourages the review of some pillars of forestry deontology and promotes the adoption of renew principles of dialogue and partnership.

Since the industrial revolution, and the generalized used of fossil fuels, the production of goods, trade, capital, technology, and information flows has rapidly expanded, leading to a globalized, urban, and tertiarized system that is becoming big enough to rival nature in shaping Earth's functioning due to an exponential growth in the use and consumption of natural resources, land use change, and the emission of greenhouse gases, pollution, and waste. The irruption of emerging economies (China, India, Brazil, etc.), which are rapidly adopting western production and consumption patterns, brings in a change of gear, a new acceleration of global change and its global challenges.

The recent adoption of the sustainable development goals and the Paris agreement on climate can be understood as symptoms of some sort of global awareness on the need to rethink development paths and the use and misuse of natural resources and, more in general, the relationships between people and the natural environment.

In this context, which is placing the world's forests at the crossroad, forestry should offer parts of required solutions. Today, the world's forests represent a green infrastructure that helps in mitigating climate change and adapting to it. They provide raw materials for a low carbon economy, invaluable services including in or near cities (e.g., human health and well-being and regulation of water cycles); support local and global economic development; and host a big share of Earth's biodiversity. These various goods and services that forest can provide have different values and implications for different groups of local and distant stakeholders that look at forests from different perspectives. They are also differently affected, or perceived to be affected by management decisions, as not all goods and services can be simultaneously provided in optimal quantities and qualities. As *social-ecological systems*, forests are then deeply inserted in global change megatrends, which are affecting the core of societal attitudes, needs, and demands toward them and the modalities of production and exchange of forest-related goods and services.

While environmental issues have already received a lot of attention (e.g., potential conflict between climate change mitigation and adaptation, biodiversity, and the bioeconomy), this book has placed the focus on three societal megatrends and how they are impacting the current pillars of forestry. It argues that globalization, urbanization, and tertiarization represent a qualitative change in the social component of forest systems. This is because they modify the fundamentals of our relationships with forests, the ethical stands, the way we understand and appreciate their value, the functioning of markets, and, as a consequence, the way decision are taken and the design and performance of forest operations in the field. Nevertheless, all components of global change are intimately interrelated. While climate change will increasingly affect forest ecology, *climate change policies* are already affecting the economics and governance of forestry. This is, in fact, a good example of the increased relevance of services (e.g., carbon sequestration) that tertiarization brings along. Some of the main findings of the book are summarized in the following paragraphs.

A critical element of global urbanization and tertiarization is a change in lifestyles. Nature, forests, and agroecosystems are becoming immaterial, virtual, distant, and typically a pale backdrop where life takes place, if at all. Urban dwellers do not live from the land and are deprived from daily contact with nature, its cycles, and its rules. The distance and lack of constant contact and personal experience in forests change people's understanding of what forests are, what they have to offer, and how they function and react to natural disturbances and human action, including forestry. Except from nature-based tourism and recreation that in many countries are practiced by a small share of the population, information and knowledge about forests and forestry are massively received through media channels, including not only news but also movies, video games, or advertising, where messages are fragmentary, emotional, and not always geographically or socially contextualized. The fragmented, diverse channels of media communication favor and reinforce audience fragmentation and the emergence of contradictory, often appearing irreconcilable, social representations of forests and forestry, held by different urban communities. These social representations are shared values, knowledge, and perceptions, exchanged mostly by education, communication, and practices and are rooted in ethical and ideological substrates. As, for example, for large portions of the urban population, forests and natural ecosystems are to be preserved and left alone, unmanaged. For others, maybe more directly related to forests industries, forest are mainly the natural source of sustainable raw material that must be actively managed in order to be preserved. This divergence can be seen as a symptom of the lack of integration of the three pillars of sustainability as defined by the United Nations Conference on Environment and Development (1992) and recalled in the Rio+20 Summit.

The coexistence of "irreconcilable" social representations of forests linked to different interests, perceptions, and understandings on forestry, deserve very close attention. There is an urgent need to further investigate the origins and resilience of those representations. They are frequently based on a mismatch between reality and perception in all aspects related to forest, from the ownership structure to the history of vegetation, current trends in forests cover, whether forests are currently managed or not, the environmental impacts of forest management, or the economic dimension of forest industries. However, it would be an error to think that different social representations of forest emerge solely from a lack of formal knowledge or a set of vested interests. In fact, deep social representation divides are found within the very well-trained and highly acknowledged professionals in policy and within the academic community. They rely on deeply rooted symbols, and can materialize when attention, personal experience, and the conversation around forests focus only on certain aspects or areas enclosed within a given social group bound by profound ethical and ideological stands.

The coexistence of sometimes radically different understandings on what forests are makes it very difficult to arrive to a social consensus or even at a compromise on the very meaning of sustainable managed forests or even on whether forest management is desirable. Obtaining a *social license* to manage forests, specially for wood supply, is becoming a critical challenge for forest managers (public or private) as society perceives tree harvesting as a synonym of destruction and where there is a new mystification of untouched nature as the most desirable element of the landscape.

Forest authorities, industries, and managers are well aware of it, and significant efforts have been put in place to try to revert this situation and to influence people's perceptions of forest and forestry. Most often these campaigns are based on one-way strategic communication approaches developed in marketing, based on targeted message or narratives, in order to promote certain perceptions, values, and attitudes. They are typically based on short, emotional messages that fail to grasp the complexity of forest-related issues. They are frequently ineffective as those messages are filtered and decoded by different population groups that reinforce their own social representations through their own conversation and give credibility to specific channels and messages. Similarly, messages promoted by environmental activists, for example, against game, fail to reach opposing societal groups closer to hunters and hunting.

Three main approaches should be followed to try to overcome this paradox or fail communication in the era of Internet and social media. On the one hand, there is a need to move away from slogans and to embrace a more pedagogic, educational approach to communication and, especially, there is a need for the development of media literacy, so the audiences can acquire the knowledge, competences, and mental frames required to participate in the coconstruction of the meaning of media messages and their critical evaluation. There is no other choice than taking the hard road: betting on the intelligence of the audience and supporting them in not only appropriating these complex forest issues, but also in paying much bigger attention to their own media literacy. It also means moving away from a unidirectional communication targeting individuals to engage in a conversation with different societal groups based on a much better understanding of their different social representations on forests. This, in turn, requires cultivating empathy and understanding the views, values, and needs of others. In order to move forward, along with global resource assessment and sectoral surveys, there is an urgent need for quantitative and qualitative surveys on social representations at different levels (regional, national, and global) and, possibly, a media observatory. These are basic tools necessary to research and better understand social representation on forests, their resilience, the available leeway, and the effectiveness of actions and initiatives undertaken. Secondly, it is necessary to reconnect people with nature, recovering and actualizing dendroculture or forest culture. This means improving the understanding of historical and current human–forest interactions and restabilizing emotional links. Arts can play a key role in revising and actualizing forest culture and as mediators with urban and tertiary society. A sustained effort is needed, also involving the educational institutions in which outdoor education must play a much bigger role. Urban and periurban forests are increasingly recognized for their role in human health and well-being. They should also become a strategic educational infrastructure. Despite some encouraging examples, mainly for primary education, this is an area quite underdeveloped. Specific attention must be placed to the risk of oversacralizing nature, as what happens when only *virgin* or pristine forests are considered true forests or when, as often happens, "killing tree" is symbolically associated with "killing the forest," which reinforces a generalized reaction against managing forests. A third important element to overcome the dysfunctional communication is to give citizens a bigger role and responsibility on forests and forestry and greater opportunities to create value from their goods and services. While traditional forestry is based on top-down technical prescriptions, is product-oriented, and relied on strong command and control policies, it becomes increasingly important to reverse the scheme by developing consumption-oriented and participatory management approaches and to embrace social innovation as an important element of forest governance. Social innovation refers to the reconfiguration of social practices, in response to societal challenges, which seeks to enhance outcomes on societal well-being and which is especially well suited to develop business models and approaches for the provision of forest goods and services in areas with fragmented ownerships.

An additional difficulty to raise interest, awareness, and stakes about forestry is its generally low contribution to the gross domestic product (GDP). However, a low contribution to the GDP does not indicate low relevance. Agriculture is typically just one or a few percent points in the GDP, yet it produces what we eat, nothing less than one of our most basic needs: food! Similarly, only a small portion of the contribution of forests to the economy is captured in current accounting approaches. This is not only because forest services, both social and environmentally-related, are generally non-marketable public goods but also because many tangible forest services are currently accounted for, and their value is captured in other sectors of the economy. As those forest services become increasingly relevant, it is important that they also become increasingly visible. Evidence nevertheless recalls that here also, apparent paradoxes exist and nuances are needed; economic levers are indeed not always those activating the motives and motivations of policy makers or forest service producers, and can, on the contrary, produce opposite effects to those pursued or expected.

Tertiarization, the predominance of the services in the economy, has, in fact, multiple consequences for forest and forestry and needs further careful conceptual and methodological efforts. On the one hand, increasingly the largest part of the value produced by forests managed for wood, cork, resin, or other goods is produced in the service economy that encompasses not only planning, research, monitoring, but also engineering, product development, marketing, etc. This intensifies the forest-based bioeconomy development, and a more sophisticated product portfolio is produced. While the cost of wood represents most of the value of a wooden pallet, it is a minor fraction of a prefabricated multistorey wood building. This means that value creation across alternative, maybe complementary, maybe competing, value chains, needs to be well understood. It also means that, increasingly, the value of forests activities is captured in the downstream application of forest goods with the risk of creating unfair value chain arrangements and reducing the social relevance of forestry itself (this refers, for example, to the links of log exporting countries vs. importing countries with sophisticated wood industries and access to final markets).

One of the consequences of tertiarization is that forests services are becoming increasingly more important than forest goods, not only in terms of mentalities, social preferences, and perceived values by people, but also in mainstream economic and policy terms. One good example of this trend is climate change policy and specifically reduction in emissions from deforestation and degradation, in which carbon services of tropical and subtropical forests have dominated the debate on international forest policy for years. A second example is the emergence of tourism that already represents 10% of the global GDP with a very important development of nature-based tourism activities. In some regions, cultural, heritage, and recreational services of forests are the basis of relevant value chains that are contributing to employment and livelihoods.

As a consequence, there is a mounting demand to manage forests for their services, both social as environmentally-related, including those than can be derived from the natural and cultural heritage. However, since its origins, forestry science and forestry institutions have been primarily focused on production of goods (mainly wood). It was assumed that optimizing wood production would also optimize these other generally subordinated goals (e.g., soil protection, water services, and recreation). Consequently, education, institution, regulation, sectoral associations, etc., have been shaped, organized, and developed to serve this goal and are ill prepared to respond to the increasing need to place forest services in the foreground. This demand for *service-first* approaches clashes with the traditional focus on *wood first* (the bequest effect) that is deeply embedded in the deontology of forest professionals.

It is important to realize that an increased relevance of services does not mean that wood (and other goods) become irrelevant. On the contrary, the demand for renewable raw materials is very likely to increase in a postoil era. In this context, managing multiple goods and services becomes a renewed challenge, especially where synergies and trade-offs across services of different forestry activities can be understood and interpreted in radically different ways, by different societal groups and persons.

The emergence of a global service economy has another far-reaching consequence for forests and forestry, which is the financialization of forest governance. It refers to the increasing role of financial motives, financial markets, financial actors, and financial institutions in the operation of the domestic and international economies. In forestry, main impacts have to do with changes in ownerships and changes in behavior of forest corporations and the increased relevance of sophisticated financial instruments as future markers. An increased portion of forests, at least in the *new world* (e.g., the United States and New Zealand), is owned by institutional and financial investors (e.g., pension funds) and managed through timber investment management organizations and real estate investment trusts. Their interest is to create value for the shareholder in typically short cycles of few years. These companies are likely to stay in the forest commodity market rather than developing more sophisticated value chains. They are contributing to implementing commodity futures markets in the forest sector, adding increasing market volatility and complexity. Financialization is also a cultural change in the management of forest corporations, focused on creating value for a frequently absent or remote shareholder, and this can increase the contradictions between short vs. long term objectives, social vs. economic objectives, and local vs. global benefits. On the other hand, large multinational corporations may be more sensitive to environmental and social scrutiny and more likely to develop stronger corporate social responsibility programs. They can also speed up the transition to a service-oriented forestry as they are also more willing to engage in service forestry as, for example, in carbon forestry or biodiversity forestry projects.

Globalization and tertiarization also open new, and very interesting, political challenges. Having remained during centuries under the domain and responsibility of forest authorities, forest policy today is fragmented in multiple components and is addressed by various subnational, national, and international settings, which deal with different portions of the policy cake (e.g., carbon related, biodiversity related, energy related, poverty related, etc.). This phenomenon is consequently diluting and dispersing power levers while incurring contradictions between strategic directions. Moreover, traditional forest authorities are still centered on wood and wood supply and risk losing centrality vis-à-vis new policy actors better addressing the global demand for services. Overcoming this dispersion of policies and policy loci, securing coherence and retaining centrality are becoming a policy priority for forest institutions (e.g., as reflected in long-lasting efforts to create a legally binding instrument on forests at the global or regional level). Achieving this through a new hierarchical institutionalization of the international and even national forest policy domains seems illusory and vain. However, other methods well known in political sciences but to be explored by the forest community, are more promising. Notably more brains and resources should be invested in the strategic management of the forest-related policy fragmentation. This would constitute a deep cultural change and imply a complete change of posture, relying on more structured coordination of forest-related dossiers and on more voluntarist transversal vision and action at all levels. Implications on the organization, structure, and role of forest-related institutions is a corollary that should be also further investigated.

Globalization frequently refers to the consequences of strong economic integration through increased trade and capital flows across borders. The liberalization of trade coupled with the emergence of new global players in the forestry sector is producing major impacts in the markets of forest products and their related value chains. Forested regions and countries lacking technological capacities or sophisticated forest based industries are (or risk becoming) suppliers of roundwood or low added-value commodities and, thus, unable to generate significant value from traditional forest value chains. In developed countries, the low prices of international commodities, a fortiori when combined with difficult ecological conditions, can lead to forest abandonment as it happened with resin and cork production in large parts of Europe. In less developed countries, it favors the corporate control of local forest resources and impacts employment. The intensity and direction of these impacts depend on multiple factors, such as the scale of production, the power of transnational corporations, producer and civil society organizations, and government ideologies. Frequently, governments often prefer to attract large industrial investors for large-scale production systems (for example, palm oil and large biorefineries) or for managing landscapes for global markets for carbon, biofuels, and biodiversity rather than for promoting the livelihoods of local peoples. Those investments create employment with various degrees of decency, which is generally lower for women.

On the other hand, large companies can be subject of greater consumer attention and be more likely to develop ethical commitments and promote fair salaries and better working conditions. Research shows that even when those ethical commitments exist, the gender dimension is not well addressed. This is definitely an area that requires increased attention and decisive action. A gender approach is also necessary in urban forestry and in the design and delivery of recreational services as women can be involuntarily secluded as forests can be seen as risk areas for them. In comparison to large-scale forest industrial investments, place-based development strategies and social innovation approaches may be better suited to generate equitable development, putting into value the full range of forest goods and services. Unfortunately, the social dimension of forests is receiving far less consideration than the economic, and the environmental dimensions that have captured and still capture most of the attention from both providers and consumers of forest goods and services, and also from the forest policy and research communities.

Forests are located overwhelmingly in deep rural areas. Urbanization induces important rural decline processes that can risk the loss of rural social capital and in the most extreme cases the collapse of rural population. This brings about new challenges and opportunities. In many developed countries, rural abandonment leads to an important recovery of forests and generates multiple environmental benefits, such as improving climate change mitigation through increased carbon sequestration. On the other hand, increased continuity of forests and high biomass build-ups across large landscape can increase the frequency of catastrophic forest fires in Mediterranean and other high risk areas. Rural landscapes are increasingly determined by urban lifestyles, consumption patterns, and, of course, policy decisions and is important to uncover and make those urban-rural linkages more visible and explicit. Both urban growth and rural decline need adequate planning in order to take advantage of the new opportunities and tackle the emerging risks. In this sense, it is crucial to understand and address divergent social representations within rural societies, and along the rural-urban gradient.

A clear consequence of the preceding discussion is a need to rethink and reequip forest professionals and forestry institutions. The tools, approaches, and deontology that was crystalized at the eve of the industrial revolution need to be adapted to global change. Just panel and paintwork will not be enough.

Scientific forestry was born around the principles of sustained yield of wood and the protection of the resource. This was achieved through strong command and control structures (strong forest regulation, policing or militarized bodies, etc.) and the delegation of all management decisions and prescriptions in a well-trained elite of forest professionals imbued with public authority and acting as symbolic owners of the forest. Frequently, the needs of the forests were more important than the needs of local inhabitants as in a forest derivative of an enlightened absolutism. The success of this approach, especially in countries where its principles where socialized beyond the forest elites, has been remarkable (e.g., in central Europe). How those ideas and institutional models expanded through the world is a clear indicator. In the past centuries and especially in the past decades, society has changed in many significant ways. While the firm convictions on secure sustainable yields are as valid as ever, there are many other aspects of forest deontology that need a deep revision. As society demands greater participation in all aspects of public life, command and control governance structures need to evolve, be complemented, and sometimes even partially replaced with social innovation and participatory approaches.

This is also important to be able to capture the value of multiple ecosystem services, to produce more equitable livelihoods, and to overcome ownership fragmentation. It is of special relevance in regions with strong rural abandonment and expansion of nonmanaged forests. In addition, mind-sets and institutional frameworks need to better reflect societal demands for forest goods and services. This will frequently mean a *service-first* approach, far away from a traditionally primary focus on wood supply and the bequest approach. This is not only self-evident in the case for urban and periurban forests, but is also relevant elsewhere. Ultimately, it is necessary to recognize that sustainability, sustainable forest management, is technically framed but socially defined and endorsed. Breaking this umbilical cord with the enlightenment and rationalism is necessary to develop a new generation of forest institutions. Part of the way is already done, as in the participatory certification standards.

An additional important element is to better solve the contradiction, between the slow growth of trees and rapidly changing societies. While developing and maintaining the forest resource requires long-term planning, this should not be an excuse for immobilism. The supposed inertia and permanence of forests does not preclude a dynamic update on the arrangements, valuation, commercialization, and use of the goods and services generated.

Moreover, forestry today is one of the domains that link the natural sciences with the social sciences and humanities. Forestry is increasingly more about people than about trees. The forest professional needs to understand the social–ecological nature of forests and master both subsystems. It needs to draw upon the perspectives and expertise from a wide range of disciplines. This requires new ways of training and education, a new deontology, and new basis for dialogues and decision-making processes where ethical, cultural, and symbolical dimensions would be made explicit instead of remaining largely encrypted for the benefit of discourses only based on science and economics.

It is also maybe time to hand over more voluntarily to the youth who grew up in this changing world and knows their codes and practices. They could actively contribute to continue what was outlined in this book by, for example, conceiving, implementing, and coordinating creative societal foresight exercises that could inspire ways for tomorrow's forestry.

"It is hard for many to remember a slower time" are the words opening this book. We could close it after having explored some paradoxes of our time, by suggesting to further develop the idea that forests would be in some way the mirror of human society, as their attractiveness in today's speedy society would rely on the ability of trees to remain at the same place during centuries, admired and relevant for all.

Index

Page numbers followed by f and t indicate figures and tables, respectively.

413

T - #0099 - 111024 - C446 - 234/156/21 - PB - 9780367570736 - Gloss Lamination